# CROP REACTIONS TO WATER AND TEMPERATURE STRESSES IN HUMID, TEMPERATE CLIMATES

## Westview Special Studies in Agriculture/Aquaculture Science and Policy

*Crop Reactions to Water and Temperature Stresses in Humid, Temperate Climates*

edited by C. David Raper, Jr., and Paul. J. Kramer

Aimed at improving agricultural production by providing a better understanding of the interaction between crops and the environment, this book presents the latest research findings on the effects of water and temperature stresses on plants in humid temperate regions. It also covers management practices and breeding programs that may reduce crop sensitivity to the vagaries of weather.

**Dr. Raper** is associate professor of soil science at North Carolina State University. **Dr. Kramer** is James B. Duke Professor of Botany, Emeritus, Duke University.

# CROP REACTIONS TO WATER AND TEMPERATURE STRESSES IN HUMID, TEMPERATE CLIMATES

EDITED BY C. DAVID RAPER, JR., AND PAUL J. KRAMER

WESTVIEW PRESS / BOULDER, COLORADO

*Westview Special Studies in Agriculture/Aquaculture Science and Policy*

**Proceedings of a Workshop on Crop Reactions to Water and Temperature Stresses in Humid, Temperate Climates, held 13 to 15 October 1980 at Duke University, Durham, North Carolina, co-sponsored by NSF Grant PFR 80-12152 and Duke University.**



Published in 1983 in the United States of America by
Westview Press, Inc.
5500 Central Avenue
Boulder, Colorado 80301
Frederick A. Praeger, President and Publisher

Library of Congress Catalog Card Number: 82-51118
ISBN: 0-86531-176-5

Printed and bound in the United States of America.

Composition for this book was provided by the editors through the facilities of
Paragraphics, Inc.
5205 Oak Park Road
Raleigh, North Carolina 27612

# CONTENTS

# PREFACE

The papers published in this volume were presented at a workshop organized to discuss the importance of water and temperature stress on plant growth and crop yield in humid, temperate climates. They should be of interest to all scientists involved in crop production because they not only deal with the effects of water and temperature stresses on plants but also with management practices and breeding programs intended to reduce the damage caused by unfavorable weather conditions.

Several recent symposia and workshops on water and temperature stress have dealt with the stresses encountered in severe environments, but little attention has been given to the effect of these stresses in mild climates. However, most of the world's food is produced in relatively mild climates where short periods of water and temperature stress occur almost at random during the growing season and produce significant, but often unrecognized, decreases in yield.

Potential yields of crops have been greatly enhanced in recent decades by development of improved cultivars and pest management programs. However, farmers know that even when the highest yielding cultivars are planted in good soils and the best cultural practices are used, actual yields are dependent on "weather," chiefly the amount and distribution of rainfall and the temperature. Thus, crop plants seldom attain their full potential for yield because of limitations on their physiological processes imposed by unfavorable weather.

Insects, diseases, and weeds sometimes reduce yields in spite of good control methods, but in the long run such reductions are small compared with those caused by unfavorable weather. Boyer, in the introduction, reports that the average yield of major crops in the United States is only 13 to 35% of the record farm yields. Only about 15% of this reduction is caused by pests, the remainder being caused by environmental stresses, including unfavorable soil and atmospheric conditions. According to J. P. Hudson [1977. Plants and the weather. p. 1-20. In J. J. Landsberg and C. V. Cutting (eds). Environmental effects on crop physiology. Academic Press, London], in England, with its more intensive agriculture, the average yield of potatoes is only 34% of the record yield, that of wheat 40%. He concluded that, regardless of the method used to analyze crop yields, the results indicate that yields are determined largely by weather, even in the mild climate of England. It therefore seems well established that the perturbations in rainfall and temperature encountered in mild climates probably are reducing crop yields much more than is generally realized.

In view of these facts it appears that an important area in crop production is being neglected. Agricultural scientists have been urged to do more basic research on the process of photosynthesis in the hope of increasing the photosynthetic efficiency of crop plants. There also is a lively interest in the direct and indirect effects of the increasing concentration of atmospheric $CO_2$ on plant growth. It seems probable that the full photosynthetic potential of crop plants is seldom utilized because of environmental constraints. It therefore is likely that the greatest increases in yield can be obtained by finding ways to decrease the inhibitory effects of "weather" on crop plants. As this area is receiving relatively little attention, it was decided to bring together agricultural engineers, meteorologists, crop and soil scientists, plant physiologists, plant breeders, and geneticists to evaluate the amount and kind of injury caused by moderate environmental stresses and to discuss ways of minimizing the injury.

The program was divided into seven sections and the papers grouped in the same manner in this book. The groups deal with crop climatology in the humid, temperate regions of the United States, the yield reductions caused by stress, the physiological basis of stress injury and stress tolerance, reduction of injury by management, reduction of injury by plant breeding, stress research in controlled environments, and finally, a discussion of research needs and priorities.

Two important features of this workshop were its concentration on stress in mild climates and its interdisciplinary nature. Most of the participants were from regions with relatively mild climates and possess first-hand knowledge of the problems. Also, there were representatives from a half-dozen fields talking with one another about problems of common interest. We believe the participants left the workshop with a better appreciation of the importance of the stress problem and of the research needed to solve it. We hope that some of them were stimulated to start research in that area.

We are indebted to many people for advice and support in connection with this workshop. The Program Analysis Section of the National Science Foundation provided the major financial support. We also appreciate the cooperation of the U.S. Department of Agriculture in facilitating the participation of several of their staff. Duke University provided space and hospitality for the meetings and the Department of Botany provided clerical assistance. We wish to acknowledge the assistance of Mrs. Patricia James and Doctors A. H. Markhart and N. Sionit in handling problems during the meeting.

We particularly thank the advisory committee consisting of John S. Boyer, Wayne L. Decker, Oliver E. Nelson, and Joe T. Ritchie for their advice concerning the program and George E. Brosseau, Jr., of the National Science Foundation, for advice concerning the general planning of the workshop.

Most importantly, we sincerely thank the participants for the time and effort expended in preparing their papers.

C. David Raper, Jr.
Paul J. Kramer

# INTRODUCTION

*Crops seldom are grown in ideal environments. Farmers know too well that even if they plant potentially high yielding varieties in good soil and use the best cultural methods, their actual yields depend chiefly on the amount and distribution of rainfall and the occurrence of favorable temperatures. Crop plants seldom attain their full yield potential because of limitations on physiological processes imposed by environmental stresses. It is true that insects and diseases sometimes drastically reduce yields, but in the long run such reductions are small compared with those caused by unfavorable weather.*

*Boyer points out that the average losses from unfavorable environmental factors, especially water stress, are much greater than the average losses from insects and diseases. He reports that certain varieties of soybeans consistently have midday leaf water potentials lower than other varieties and these potentials are low enough to materially reduce the rate of photosynthesis. Boyer suggests that we should be breeding for soybean types which show the least midday reduction in leaf water potential.*

1

# ENVIRONMENTAL STRESS AND CROP YIELDS

**J. S. Boyer**

Record yields of major U.S. crops are 2.8 to 7 times the average agricultural yields of the same crops (Wittwer, 1975). These record yields illustrate that the genetic potential for high productivity is already present in the crops of today. Consequently, even a modest increase in the ability to realize this potential would boost agricultural productivity in the U.S. significantly.

The discrepancy between record yields and average yields must be attributed to factors that prevent the expression of the genetic potential of the plants. In general, these can be classified into biological factors and physical factors. Adverse biological factors consist of plant disease and depredation by insects and other animals, primarily, and adverse physical factors include competition by weeds, unfavorable soil conditions, and unsuitable climate. Losses attributable to biological factors amount to about 15% of the discrepancy between average yields and record yields (USDA, 1965). The remainder must be attributable to physical factors and, consequently, the removal of all physical limitations to plant growth should result in about a four-fold increase in the yields of major crops in the United States. This potentiality for increased yield is so dramatic that much research attention should be aimed at the effects of physical limitations on plant growth.

The U.S. soybean crop illustrates these points. The average U.S. yield is approximately 21% of the record yield, and disease and insect damage exact only small losses on the crop. Because soybean is a legume, it frequently does not respond significantly to large inputs of inorganic nitrogen, and its mycorrhizal roots generally assure that adequate phosphorus is available for crop growth. Where soils are deep and well supplied with water, as in the Midwest, soybean often does not respond to irrigation. Nevertheless, improvement in the yield of soybean has occurred during the last 40 years, which implies an improvement in the ability of newer genotypes to tolerate the midwestern environment. Apparently, plant breeders, in selecting for yield, have unwittingly selected for physiological traits favoring production.

It is the purpose of this chapter to point out some of the physiological characteristics of soybean that have been altered by plant breeding. The results show that an important alteration is the improved ability of modern genotypes to utilize water efficiently (Boyer et al., 1980).

## METHODS

It is possible to determine the genetic improvement in yield by growing a range of cultivars in a common field environment. This approach was used in the present work, and plants were grown on the Agronomy South Farm, Urbana, Illinois, for three years (1975 to 1977). Soil types were typical of the midwest, and a range of soybean cultivars from maturity group II and maturity group III were tested. The four soybean [*Glycine max* (L.) Merr.] cultivars in maturity group II were Beeson, Harosoy, Hawkeye, and Richland, and in maturity group III were Wayne, Ross, Shelby, and Manchu. These cultivars were released from 1911 to 1968. Plots were arranged in a 2x3 factorial split plot design and planted at 350,000 seeds/hectare. Measurements of grain yield, plant height, lodging, leaf water potentials, vascular disease incidence, and root density were made. Leaf water potentials were measured in midday (between solar noon and 1600 h) because it was expected that water deficits would be most severe during this time and photosynthetic activity of the crop would be closest to maximum and therefore most sensitive to inhibitory water potentials.

## RESULTS AND DISCUSSION

Rainfall during the three years of the experiment ranged over the conditions expected for central Illinois. Soil water potentials were always higher than -0.6 bar in the lower two-thirds of the root profile and a water table was present at 2 m. Therefore, soil water should always have been available to the crop and high yields were recorded by farmers in central Illinois in all three seasons. Table 1 shows that highest yields were obtained in the newest cultivars in the field experiment. Richland, the oldest cultivar in group II, yielded only 3391 kg/ha on average whereas Beeson, the newest cultivar in group II, yielded 3950 kg/ha. Similarly in group III, Manchu yielded 2871 kg/ha but Wayne yielded 3731 kg/ha. Thus, genetic progress is apparent. Luedders (1977) also showed significant genetic improvement of soybean

**Table 1. Mean cultivar characteristics measured at maturity for 1975 to 1977 in the field.**

| Cultivar[a] | Plant Height | Lodging[b] | Vascular Disease[c] | Water Deficit[d] | Grain Yield |
|---|---|---|---|---|---|
| | *–cm–* | | | *–bar/day–* | *–kg/ha–* |
| Group II | | | | | |
| Beeson (1968) | 110 | 2.7 | 2 | 1.1 | 3950 |
| Harosoy (1951) | 118 | 3.5 | 2 | 1.7 | 3647 |
| Hawkeye (1947) | 112 | 2.9 | 4 | 1.8 | 3619 |
| Richland (1938) | 92 | 2.4 | 3 | 2.8 | 3391 |
| Group III | | | | | |
| Wayne (1964) | 122 | 3.2 | 4 | 0.7 | 3731 |
| Ross (1960) | 131 | 3.8 | 2 | 1.6 | 3573 |
| Shelby (1958) | 128 | 3.3 | 2 | 1.1 | 3453 |
| Manchu (1911) | 99 | 4.0 | 4 | 2.1 | 2871 |

[a]Number in parenthesis is the year the cultivar was released in the United States.

[b]1 = erect; 5 = prostrate.

[c]1 = trace; 5 = extensive vascular browning. Measured in mid-August.

[d]Water deficit = –[Leaf water potential – (–11 bars)]. Data are means obtained by dividing cumulative water deficit by number of days observed.

yield in a range of cultivars and noted that the improvement was associated with increased height and resistance to lodging. However, in our experiment, there appeared to be little correlation between height, lodging and yield.

Table 1 also shows that the newest cultivars had higher water potentials in midday than the oldest cultivars. The measurements were made with leaves that were fully illuminated and perpendicular to the incoming light, and consequently they represented the driest leaves in the canopy. They also would have represented those leaves that were carrying on rapid photosynthesis. It is known that water potentials below –11 bars inhibit photosynthesis in soybean (Ghorashy et al., 1971; Boyer, 1970; Huang et al., 1975). Therefore, we defined any leaf water potential below – 11 bars as a water deficit according to: water deficit = –[Leaf water potential – (–11 bars)]. Table 1 shows that water deficits were largest in the oldest cultivars. Therefore, average afternoon water deficits were correlated with the average yield of the cultivars. In group II, yield was described by the equation $y = 4237.6 - 315.4x$ where x is the water deficit in bars per day ($r = 0.95$). For group III, yield was described by $y = 4097.1 - 508.2x$ with $r = 0.84$.

We attempted to determine the mechanism of protection against midday water deficits in new cultivars. In Richland and Wayne, which exhibited extremes of water deficit, measurements throughout the plant canopy indicated that the water deficiency was not associated with only a few leaves but rather was distributed throughout the canopy whenever the leaves were exposed directly to light. Thus, the entire shoot was affected. Observations of vascular disease were made in all cultivars and, although vascular diseases could be observed in lower stems and roots toward the end of the growing season, the effects were too late and had too little relationship to yield to be of significance. Therefore, vascular and root disease could not account either for midday water deficits or for the differences between cultivars and yield.

Since the water deficits were found throughout the shoot, we determined whether there was a relationship between midday water deficits and root densities for Wayne and Richland. Both cultivars rooted to the same depth but Richland had fewer roots than Wayne at all soil levels. Consequently, it appeared that high rooting density was associated with low midday water deficits and that dense rooting may have been selected simultaneously with selection for high yield by plant breeders.

It therefore seems clear that plant breeders have, in selecting for high yield, altered the physiological attributes of soybean cultivars in regard to water use. In effect, selection for high yield has selected for those plants with a balance between root and shoot that favors minimization of water deficits in the evaporative environment of the midwest. The minimization of these water deficits in the shoot is important even though the soil contains optimum water for the plants, since water deficits occur in leaves even in moist soil and the shoot must obtain enough water to cope with the local atmospheric demand without inhibitory water deficits in the leaves.

Although the water deficits in midday varied by only a few bars among the cultivars of this study, they have a large potential impact on photosynthesis and other factors with similar sensitivity to low water potentials. The measurements of photosynthesis in soybean both in the field and the laboratory (Ghorashy et al., 1971; Boyer, 1970; Huang et al., 1975) show that there is a steep inhibition of photosynthesis at water potentials below -11 bars. Thus, water deficits of the magnitude encountered in this study would be sufficient to reduce midday photosynthesis in illuminated leaves by 50% in the cultivars Richland and Manchu. On the other hand, Wayne and Beeson, the newest cultivars, would show photosynthetic inhibition of only 8 to 10% in illuminated leaves.

It is important to consider whether faster breeding progress can be made if problems such as these in soybean are recognized and incorporated into breeding programs. I am convinced that it could. Genotypes having otherwise favorable characters should be screened for midday water potentials early in the growing season. Those individuals avoiding low water potentials would be adapted to the local evaporative conditions and could be retained. Thus, the yield inhibiting effects of shoot water deficiency would be identified early in the growing season and the generation time for cultivar development would be shortened.

In soybean, much of the improvement in shoot water status has already been accomplished by traditional methods. In the group II cultivars, for example, complete elimination of midday water deficits would further improve yield by only about 8%. Thus, the future gains from the above approach will likely be in crops that have not had extensive selection for local conditions or in soybean cultivars now being selected for adaptation to extreme evaporative environments.

## NOTES

J. S. Boyer, USDA/SEA/AR, Departments of Botany and Agronomy, University of Illinois, Urbana, Illinois 61801.

## LITERATURE CITED

Boyer, J. S. 1970. Differing sensitivity of photosynthesis to low leaf water potentials in corn and soybean. Plant Physiol. 46:236-239.

Boyer, J. S., and S. R. Ghorashy. 1971. Rapid field measurement of leaf water potential in soybean. Agron. J. 63:344-345.

Boyer, J. S., R. R. Johnson, and S. G. Saupe. 1980. Afternoon water deficits and grain yields in old and new soybean cultivars. Agron. J. 72:981-986.

Ghorashy, S. R., J. W. Pendleton, D. B. Peters, J. S. Boyer, and J. E. Beuerlein. 1971. Internal water stress and apparent photosynthesis with soybeans differing in pubescence. Agron. J. 63:674-676.

Huang, C. Y., J. S. Boyer, and L. N. Vanderhoef. 1975. Acetylene reduction (nitrogen fixation) and metabolic activities of soybean having various leaf and nodule water potentials. Plant Physiol. 56:222-227.

Luedders, V. D. 1977. Genetic improvement in yield of soybeans. Crop Sci. 17:971-972.

USDA. 1965. Losses in agriculture. United States Department of Agriculture. Agriculture Handbook No. 291.

Wittwer, S.H. 1975. Food production: technology and the resource base. Science 188: 579.

## SECTION I

# CROP CLIMATOLOGY IN HUMID, TEMPERATE REGIONS

*This section contains two papers on weather and climate. The first points out that there are three methods for determining the intensity and duration of drought: the probability of rainfall, the impact on crop yields, and the amount of available soil water. The first two are not useful as management tools, but monitoring the available water content of the soil can be useful. This method requires a better knowledge of the soil water storage capacity, the depth of rooting, and the rate of water extraction. With adequate information concerning these points in specific areas, the onset of drought can be predicted accurately enough to time irrigation. The second paper deals with temperature perturbations. During the growing season diurnal variations in temperature are much greater than day to day variations, and knowledge of short-term variations is much more important than monthly averages. The timing with respect to the stage of crop development is very important because optimum temperatures for vegetative growth and for reproductive development are different for some crops. The rate of air temperature change and soil temperature are important factors in crop response. Both of these papers indicate the need for numerous local observations of temperature and soil moisture as an indicator of growing conditions and a guide to management practices. There also is a need for accurate models of crop growth as tools essential for assessing the impact of temperature stress and drought on yield.*

# 2

# PROBABILITY OF DROUGHT FOR HUMID AND SUBHUMID REGIONS

**Wayne L. Decker**

There are no shortages of qualitative and quantitative definitions of drought (Tannehill, 1947; Hounan et al., 1975). Although the number of technical definitions causes confusion among professionals and scientists, there is general agreement concerning the usage of the word drought in both technical and nontechnical literature. In the natural progression of weather events, a few rain free days often lengthen into a "dry spell." When the dry spell extends into a period of weeks with inconsequential precipitation, then a drought is said to be occurring. Obviously, the beginning and ending points and intensity of a drought depend on which of the analytical definitions of the event is being used. How probable a drought is in a given region and season depends upon the specific definition adopted for drought.

Drought may be defined in three separate ways. Each of these methods, which are listed below, can be analytically defined. There are, of course, possibilities of using combinations of these approaches in defining drought.

## APPROACHES FOR THE DEFINITION OF DROUGHT

1. A definition based on climatic expectations. Under these definitions the "weather" must be much drier in arid regions than in humid regions to be classified as a drought.
2. A definition based on impacts of prolonged dry weather, such as the water level in a major reservoir or the depression of yield for a major crop.

3. A definition based on a given water shortage in agricultural soils. By this definition drought occurs when the level of soil moisture reaches a predetermined level, say 50 or 25% of the total available water in the rooting zone.

## DROUGHT DEFINITIONS AND PROBABILITIES FROM CLIMATIC EXPECTANCIES

The simplest climatic statistics defining drought are based on rainfall amounts. Under definitions of this type, whenever the precipitation for a given time period is less than a given quantity, a drought is said to have occurred. The probability of drought is calculated from the frequency distribution of precipitation. Shaw et al. (1960) have presented probabilities of rainfall amounts for different periods for the north central United States.

The departure of precipitation from normal and the percent of normal precipitation have been used as indices of dry weather or drought. Recently the National Weekly Crop and Weather Bulletin published by NOAA and the USDA have adopted this technique for identifying regions in the world with deficient and excess precipitation. This technique has the advantage of providing an index with meaning for both arid and semiarid regions. The departures can also be accumulated through time to show whether the drought condition is getting worse or better.

The best known climatic index for drought was designed by Palmer (1965). Palmer's index recognized five intensities of drought ranging from incipient drought (an index value $\leqslant$-0.5) to extreme drought ($\leqslant$-4.0). It was Palmer's intention for these indices to normalize drought for different climates so that an extreme drought in the humid regions would have the same impact on the economy of that region as an extreme drought in an arid region.

Using Missouri data, Hamidi (A. R. Hamidi. 1979. The probability analysis of the Palmer Drought Index and the Crop Moisture Index. M.S. Thesis, University of Missouri-Columbia) noted that the different categories of drought as defined by the Palmer method had the same probability of occurrence in both northwest and southeast Missouri. An analysis of the frequency of the Palmer index for each month of the year and at each Missouri location is presented in Figures 1 and 2. These probability estimates for the critical values of the index were obtained by drawing a smooth curve through the points on the accumulative frequency diagram.

The percentage probability of drought index values less than -4.0, -3.0, -2.0, -1.0, -0.5 are 5%, 10%, 20%, 31%, and 44% for the month of December for the location in northwest Missouri, and 4%, 10%, 20%, 32%, and 41% for the same month at the southeast Missouri location. These figures show that the probability of a drought index less than a specified value are about the

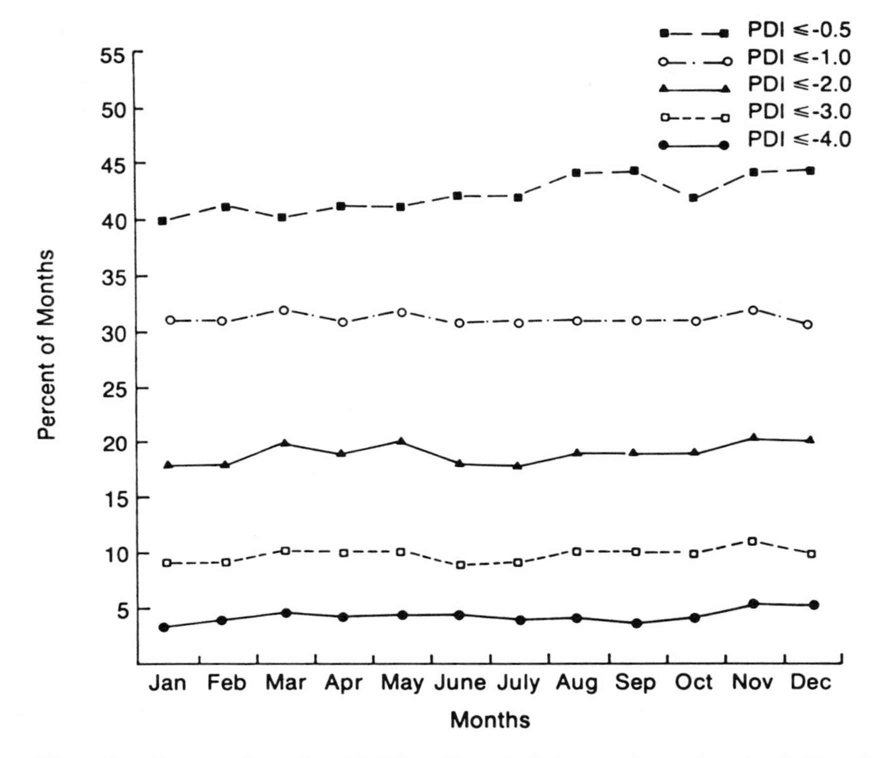


**Figure 1. Percent of months with Palmer Drought Index equal to or less than indicated value for a location (Tarkio) in Northwest Missouri.**

same during all months at both locations. Severe drought occurs on 2 to 4% of the months while 7 to 10% of the months have values of -3.0 or below. The frequencies of -2.0 or below, -1.0 or below, and -0.5 or below are in the range of 16 to 20%, 30 to 32%, and 44%, respectively.

The Palmer indices are shown in Table 1 for various return periods in the humid and subhumid regions of eastern United States (Dickerson and Dethier, 1970; McWhorter, 1974). The values for Missouri were obtained from a plot of accumulative frequency of drought for the two locations used in Figures 1 and 2. Although the variation from region to region is considerable, there appears to be sufficient uniformity to encourage calculating an average drought index for each reoccurrence interval. A linear relationship occurs between these average drought indices on the return periods on a logrithmic scale (Figure 3). From Figure 3 it can be seen that in the eastern United States an extreme drought (-4.0) has a return period of 13 years, while a severe drought (-3.0) will occur every 6.5 years.

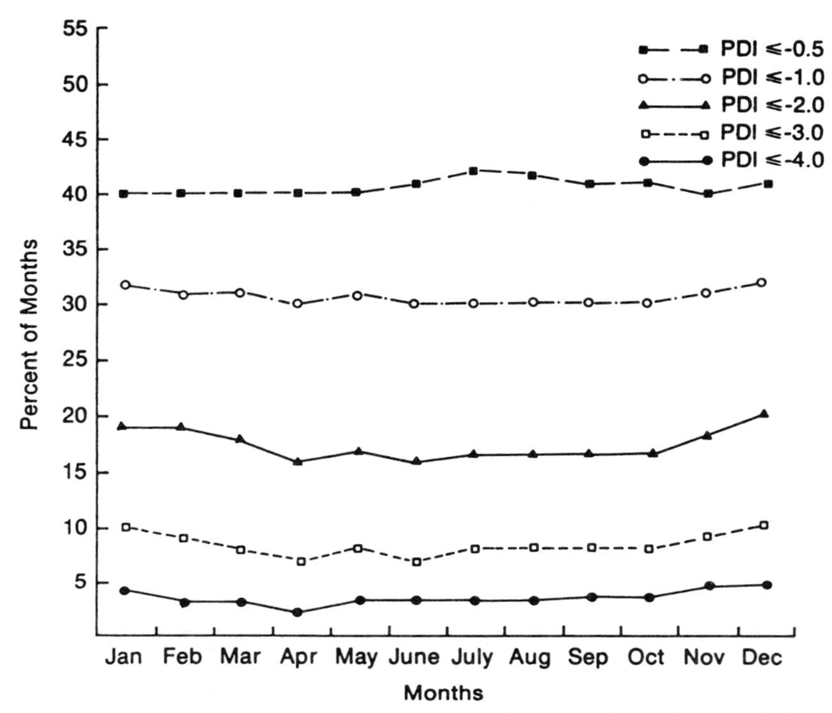


**Figure 2.** **Percent of months with Palmer Drought Index equal to or less than indicated value for a location (Caruthersville) in Southeast Missouri.**

**Table 1.** **Magnitude of the Palmer Drought Index for selected return periods for selected states in the humid and subhumid areas of Eastern U.S.**

| Location | Return Periods (Years) | | | | |
|---|---|---|---|---|---|
| | 2 | 5 | 10 | 25 | 50 |
| New England | - 1.20 | - 2.49 | - 3.44 | - 4.64 | - 5.75 |
| New York | - 1.33 | - 2.65 | - 3.64 | - 4.55 | - 5.95 |
| Pennsylvania and New Jersey | - 1.28 | - 2.64 | - 3.62 | - 4.94 | - 5.98 |
| Delaware, Maryland and Virginia | - 1.12 | - 2.78 | - 3.94 | - 5.54 | - 6.75 |
| Ohio and West Virginia | - 1.18 | - 2.86 | - 4.13 | - 5.88 | - 7.07 |
| North and South Carolina | - 1.13 | - 2.26 | - 3.10 | - 4.22 | - 5.07 |
| Missouri | - 1.26 | - 2.60 | - 3.63 | - 4.94 | - 6.00 |
| Mississippi | - 1.95 | - 2.90 | - 3.60 | - 4.60 | - 5.35 |
| Average | - 1.31 | - 2.64 | - 3.64 | - 4.91 | - 5.99 |

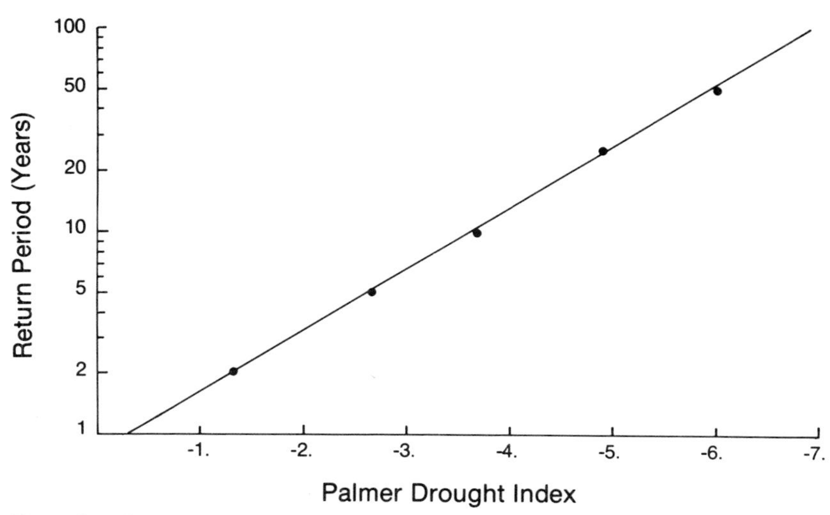


**Figure 3. Return periods and drought intensities for the humid and subhumid regions of the Eastern United States.**

In spite of rather exact appraisals of drought probabilities provided in Figure 3, there is doubt concerning the value of such presentations. It is not apparent how one can utilize this information in developing management strategies, in planning for the best utilization of the available water resource, or in assessing the impact of water deficiencies on various soils within a geographic region.

## DROUGHT DEFINITIONS AND PROBABILITIES BASED ON IMPACTS OF DRY WEATHER

If one knows the effect on seasonal or annual weather on yields, then an evaluation of the variation of climate can be interpreted in terms of the resulting production of agricultural crops. A definition of the percent reduction in yield to define drought can then be made. Such an analysis involves three steps:

1. The definition of the functional relationship between climate and yields. This mathematical expression must include either terms for all the production inputs (fertilizers, pest management treatments, genetics, etc.) or a term or terms for advances in technologies.
2. Holding technologies constant (say at a recent year or for recent levels of production inputs), yields are simulated for the entire climatic record.
3. The number of occurrences of simulated yields below a predetermined level constitutes the frequency of drought.

Such an analysis was prepared by NOAA (1973) for wheat and corn production in the midwestern United States. In this case, one standard deviation from the average of the simulated yields was defined as a drought occurrence. For wheat production 17 out of 80 years produced droughts, while for corn 6 out of 73 years were droughty. This translates into drought probabilities of 0.21 and 0.08, respectively.

Runge and Benci (1975) performed a similar analysis for corn, although this method used planting dates and soil moisture at planting time as added variables. Again drought was defined as years with yields less than one standard deviation from the average of the simulated yields. By this example, for Missouri there were eleven drought years for corn in the 55-year period used in the study. This translates into a probability for drought of 0.20.

If the simulated yields were normally distributed, the use of one standard deviation departure from the mean simulated yield would produce a probability of drought of about 0.17. When the actual frequency deviates from this value by a significant amount, the frequency distribution for simulated yields must be skewed. The simulated yields for corn by the NOAA method are skewed to the right since there are fewer than the expected 17% of years with drought.

Obviously, the estimated probability of drought from the simulated impacts of drought on yields can be only as good as the model for estimating yields. Although both the NOAA and the Runge and Benci models considered climatic impacts for an entire production region, a simulation model designed for estimates of local yield could also be used. Through a simulation analysis, drought probabilities appropriate to particular soils of a region could be designed. This adaptation would be a very appropriate use of the models for crop yields now being developed for estimating yields of agricultural crops for single fields.

## DROUGHT DEFINITION AND PROBABILITIES BASED ON SOIL WATER BALANCE

In a very real sense drought definitions based on arguments using climatic data or on the depression of yield provide an interesting commentary on climatic variabilities, but these estimates do little to provide a guide for more effective use of the water resource. It appears more reasonable to base the definition of drought on a soil moisture indicator. This definition and the associated probabilities will be more directly related to stress in agricultural plants. To quote from van Bavel (1959):

> *"The focal point of the problem (the definition of drought) is the lack of suitable soil moisture conditions for plant growth that occurs from time to time. Occurrence of these undesirable conditions varies with geographic location, soil properties, cropping systems, and, in particular, weather conditions. In*

*the latter category, the distribution of rainfall and consumptive use of water by crops are of principal significance."*

In the middle to late 1950's van Bavel and his associates (van Bavel and Verlinden, 1956; van Bavel and Carreker, 1957; van Bavel, Forest, and Peele, 1957; van Bavel and Lillard, 1957) completed analyses of drought for the coastal states of southeastern United States and the south central states of the Mississippi Valley. (Apparently Florida was not included in this regional analysis.) The analysis was based on a soil water balance constructed from rainfall measurements at cooperative climatic stations, evapotranspiration estimates from a simplified Penman model and varying capacities of the soil to hold water. A drought day by the van Bavel definition was a day with the soil water balance at zero. The probability assessments concerned the likelihood of the number of drought days.

In Table 2 are listed the minimum number of drought days for various probability levels at selected locations in the lower Mississippi Valley. Even in the locations near the Gulf of Mexico (see values for Covington, LA), drought days occur with considerable frequency. As many as 100 drought days a year occur with a frequency of one in ten years in western Arkansas and northwestern Louisiana. On half the years, even with soil water holding capacities of five inches, up to 50 drought days occur in much of this area. Similar values occur in the Atlantic Coast states. For example, in farming regions of North Carolina on soils with five inches of available water, nearly ten drought days occur during August in one out of ten years (van Bavel and Verlinden, 1956).

There are several problems dealing with the details of the drought definition based on soil moisture balance suggested by van Bavel and his associates in the 1950's. These uncertainities concern the establishment of a soil water

**Table 2. Expected number of drought days as defined by van Bavel for selected stations, soil water holding capacities and probability levels.**

| Location | Soil Water Holding Capacities | | | | | |
|---|---|---|---|---|---|---|
| | Three Inches | | | Five Inches | | |
| | No. of Years out of 10 | | | No. of Years out of 10 | | |
| | 1 | 3 | 5 | 1 | 3 | 5 |
| Covington, LA | 54 | 36 | 23 | 21 | 8 | 0 |
| Ozark, AR | 94 | 76 | 63 | 78 | 59 | 46 |
| Sikeston, MO | 99 | 81 | 68 | 88 | 76 | 53 |
| State College, MI | 98 | 78 | 70 | 90 | 59 | 54 |

content of zero as the threshold for drought and the model for evapotranspiration estimates. The problems can be overcome by the adoption of more refined methods, such as those suggested by Shaw (1964).

The importance of an analysis of this type rests in the establishment of drought as a major problem in the humid and subhumid regions of the United States. One can look at these figures and recognize the importance of drought in humid and subhumid regions as limitation to agricultural production. Further, these methods reduce the definition to specific soil types so a farmer can apply the result to his own farm or even fields within his farm. But, most important, this definition for drought can be interpreted in terms of management alternatives to avoid, alter, or ameliorate the impacts of drought.

## SUMMARY

Whatever the definition, drought does occur in humid and subhumid regions. The exact probabilities depend on the definitions employed, but drought is certainly not a "rare event" to the important humid and subhumid regions of North America.

When one focuses on management alternatives and strategies to avoid or alter drought, the definition of drought should be based on the soil water balance. The following recent developments have made possible and advisable a new analysis of the expectancies of drought based on rational soil moisture consideration:

1. There are now more than thirty years (1948 to 1980) of climatic records available on magnetic tape for many climatological stations of every state in the United States. These data, which are available at the National Climatic Center (NOAA) at Asheville, NC, of maximum and minimum temperatures and precipitation can be used in simulating soil moisture conditions.
2. Soil surveys have been expanded and digitized so that the extent of soil areas with varying moisture capacities can be estimated.
3. Much progress has been made since the 1960's in defining methods for estimating evapotranspiration from climatic data and soil moisture conditions. In many cases these estimates can be made specific to particular crops.

It is time to use the advances of the past two decades to redefine the incidence of soil moisture deficiencies in the eastern United States. The analysis should estimate the probability of drought for important agricultural soils. But, most importantly, the project should be completed in a manner allowing the interpretation of the impacts of stress imposed by drought on the adoption of management practices such as irrigation, minimum tillage, etc. It appears that such an analysis could be undertaken by one or more of the Regional Research Projects of the State Agricultural Experiment Stations.

## NOTES

Wayne L. Decker, Department of Atmospheric Science, University of Missouri-Columbia, Columbia, Missouri 65211.

Contribution from the Missouri Agricultural Experiment Station Journal Series Number 8672.

## LITERATURE CITED

Dickerson, W. H. and B. E. Dethier. 1970. Drought frequency in the Northeastern United States. W. Va. Agric. Exp. Stn. Bull. 595.

Hounam, C. E., J. J. Burgos, M. S. Kalik, W. S. Palmer, and T. C. Rodda. 1975. Drought and agriculture. World Meteorological Organization Tech. Note 138, Geneva.

McWhorter, J. C. 1974. Severity and frequency of drought in Mississippi. Water Resources Institute, Miss. State Univ.

National Oceanic and Atmospheric Administration. 1973. The influence of weather and climate on U.S. grain yields, bumper crops or droughts. Report to the Administrator from the Assoc. Administrator for Environmental Monitoring and Prediction.

Palmer, W. C. 1965. Meteorological drought. Research Paper 45, U.S. Weather Bureau, Washington, D.C.

Runge, E. C. A., and J. F. Benci. 1975. Modeling corn production—estimating production under variable soil and climatic conditions. Proceedings of the 30th Annual Corn and Sorghum Res. Conf., pp. 194-214.

Shaw, R. H. 1964. Prediction of soil moisture under meadow. Agron. J. 56:320-324.

Shaw, R. H., G. L. Garger, and R. H. Dale. 1960. Precipitation probabilities in the North Central States. Mo. Agric. Exp. Stn. Bull. 753.

Tannehill, I. R. 1948. Drought, its causes and effects. Princeton University Press.

van Bavel, C. H. M. 1959. Drought and water surplus in agriculture soils of the lower Mississippi Valley area. U.S.D.A. Tech. Bull. 1209.

van Bavel, C. H. M. , and J. R. Carreker. 1957. Agricultural drought in Georgia. Ga. Agric. Expt. Stn. Bull. N.S. 15.

van Bavel, C. H. M., and J. H. Lillard. 1957. Agricultural drought in Virginia. Va. Agric. Expt. Stn. Bull. 128.

van Bavel, C. H. M., and F. T. Verlinden. 1956. Agricultural drought in North Carolina. N.C. Agric. Expt. Stn. Bull. 122.

van Bavel, C. H. M., L. A. Forest, and T. C. Peele. 1957. Agricultural drought in South Carolina. S.C. Agric. Expt. Stn. Bull. 447.

# 3

# TEMPERATURE PERTURBATIONS IN THE MIDWESTERN AND SOUTHEASTERN UNITED STATES IMPORTANT FOR CORN PRODUCTION

**R. F. Dale**

The first two chapters in this book are intended to show the need for the workshop on environmental stress. Even in the humid temperate climates of the United States—the most extensive area of favorable climate and soils for crop production in the world—droughts and unusually low or high temperatures are still sufficiently frequent to reduce crop production below its potential. The climatologists are asked to present the climatic risk of moisture and temperature occurrences detrimental to crop production, the critical thresholds of which are to be discussed later in this volumn. Ideally, if we had the necessary climatological data files on the computer, we should improve these risk assessments to take advantage of ideas or information on the thresholds critical to plant growth revealed in subsequent chapters. While we may narrow our idea of the critical thresholds for some of the more complicated soil-plant-air relations, however, I suspect that we will not have a consensus on exactly what standard levels we would use for climatological summaries. This is the primary reason climatological data are not in greater demand. We do not know how to exploit them properly. Increased efforts in physiological modeling are forcing better quantification of the soil-plant-weather relations and, with increasingly better definition of the more illusive joint temperature-light-moisture thresholds, climatology is becoming increasingly useful in providing the base for making all types of long and short range management decisions.

In this contribution, I have four objectives: first to look at the usual presentations and measures of temporal patterns of temperatures; second, to discuss briefly some fairly well accepted 10 and 32 C temperature thresholds for corn growth and development and to use these to provide examples of spatial patterns of the temperature perturbation risk in the Corn Belt; third, to point out some of the intercorrelations between weather variables which complicate the interpretation of the climatology of any single weather variable; and fourth, which in practice should be considered first, to review some of the cautions or problems in the use of available files of temperature data.

## TEMPORAL PATTERNS OF TEMPERATURE

The climate of a location includes the seasonal march of the daily mean air temperature, the diurnal and interdiurnal temperature variability, and the year-to-year variability from these "normals." The relative magnitudes of the temperature components vary with latitude, elevation, continentality, and time of year. Our estimates of these climatic parameters also vary with the period of record used. Generally, however, daily maximum temperatures in July average between 28 and 33 C in the Corn Belt and the Southeast. Daily minimum temperatures in July average from near 16 C in the northern Corn Belt to about 24 C along the Gulf Coast. The normal diurnal temperature range (daily maximum to daily minimum) averages from 10 to 14 C. The inter-diurnal temperature change is defined as the absolute difference from one day's mean temperature to the next (Landsberg, 1966). For the summer these differences average about 2 C in the Corn Belt and slightly less in the Southeast. Probably less than 5% of the days will have interdiurnal temperature changes greater than 6 C.

To better appreciate the relative magnitudes of the various sources of air temperature variability, the mean interdiurnal temperature range and mean daily maximum and minimum temperatures, with their departures from normal, are shown in Table 1 for the months of May through August in 1979 and 1980 at West Lafayette 6NW, Indiana. Note that the average diurnal temperature range is about 12 C. This is about six times the magnitudes of the average interdiurnal variability and the largest monthly departures from normal in July and August. July and August, 1980, averaged about 2 C above normal, and were the warmest since 1955. The same two months for 1979 averaged about 1 C below normal, and demonstrates one of the most contrasting year-to-year July and August temperature differences observed at West Lafayette.

Our usual measure of year-to-year temperature variability, such as the departures from normal (Table 1) and the standard errors of monthly mean temperatures (Thom, 1968), and even the interdiurnal variability do not describe the within-month temperature pattern. The daily maximum and minimum temperatures for the 1979 growing season are shown in Figure 1, together with their 7-day moving averages and "normals." While we have

**Table 1.** Mean daily maximum (Max.) and minimum (Min.) air temperatures in C, mean interdiurnal (ID) temperature range, and mean daily total incident solar (S) radiation (ly/day) with respective departure from normal for indicated month and year, West Lafayette 6NW, Indiana.

| | | May | | | | June | | | | July | | | | August | | | |
|---|---|---|---|---|---|---|---|---|---|---|---|---|---|---|---|---|---|
| | | Max. | Min. | ID | S | Max. | Min. | ID | S | Max. | Min. | ID | S | Max. | Min. | ID | S |
| 1979 | Mean | 22.0 | 9.1 | 3.8 | 484 | 26.9 | 15.1 | 2.8 | 512 | 26.9 | 16.1 | 1.8 | 469 | 26.4 | 15.9 | 2.2 | 422 |
| | Dep. | +0.2 | -0.7 | | -24 | 0 | +0.2 | | -17 | -2.1 | -0.6 | | -53 | -1.5 | +0.3 | | -33 |
| 1980 | Mean | 22.6 | 10.3 | 2.4 | 443 | 25.8 | 13.7 | 2.9 | 454 | 30.8 | 18.7 | 2.3 | 506 | 29.8 | 18.2 | 2.1 | 409 |
| | Dep. | +0.8 | +0.5 | | -19 | -1.1 | -1.2 | | -66 | +1.8 | +2.0 | | +9 | +1.9 | +2.6 | | -46 |

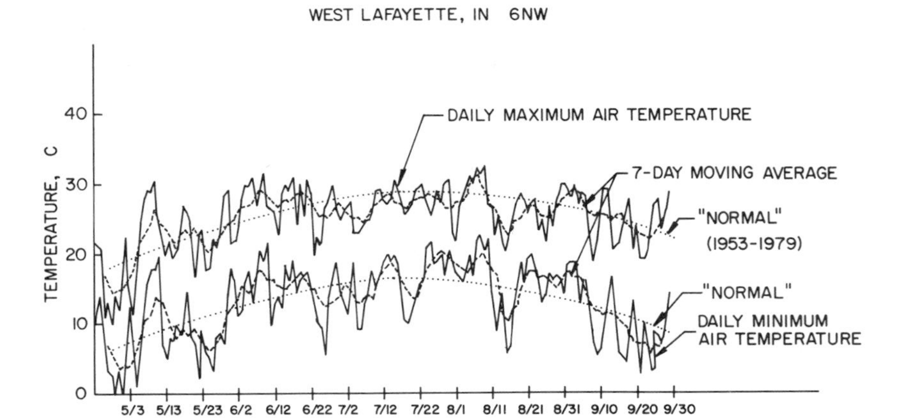


**Figure 1. Daily maximum and minimum air temperatures and "normals" (1953 to 1979 for indicated day, and seven-day moving averages for middle day of period, May through September 1979, for West Lafayette, Indiana.**

stressed the magnitude of the diurnal temperature range relative to those of the interdiurnal and year-to-year variability, note that the combination of all three sources of variability resulted in a temperature increase of almost 30 C in 5 days, from a minimum of 1.1 C on 5 May to a maximum of 30.6 C on 10 May. The moving 7-day averages of the maximum and minimum temperatures for the 1979 growing season were replotted in Figure 2 with those of 1980 to show the year-to-year differences in the timing of the temperature anomalies. This timing, sychronized with crop ontogeny, is important in interpreting the effect of the temperature regime on corn growth and development. For example, in the West Lafayette area in 1980, the weather was relatively warm during corn planting and emergence and up to about 2 June. The temperatures then fell below normal during the establishment period from 2 June to about 22 June. But, during the grand growth period ending about 22 July, temperatures were much above normal which hastened the growth and development of the corn crop. There was even some silking and pollination late in the period, but fortunately most of the pollination occurred during the cooler period in late July. Note that the temperature patterns during the establishment and grand growth periods in 1980 were reversed from those in 1979. This resulted in very rapid crop growth and development, and caused pollination, grain filling, and maturity to be seasonally ahead of 1979. The 1980 temperature anomaly during the last week of June and the first three weeks of July had an observable negative impact on corn yields by significantly shortening the grand growth, pollination and grain filling

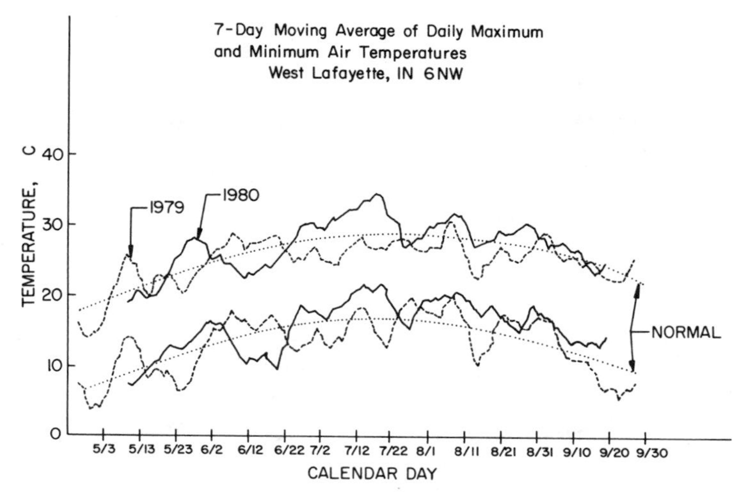


**Figure 2. Seven-day moving averages of daily maximum and minimum air temperatures for middle day of period, May through September, 1979 and 1980, for West Lafayette 6NW, Indiana.**

subperiods in the 1980 corn crop. Probably, only day-by-day crop simulation can properly interpret the joint effects of crop ontogeny and weather anomaly timing.

## SPATIAL PATTERNS OF TEMPERATURE PERTURBATION

There are four commonly recognized temperature perturbations which act to reduce corn production: (1) a late freeze in the spring, (2) the occurrence of anomalously low temperatures in the spring which put the corn crop behind a desirable schedule, (3) the occurrence of unusually high or low temperatures in the summer which decrease the yield potential, and (4) an early freeze in the fall. Since the spring and fall freeze risks are fairly well understood and determined for all areas, only (2) and (3) will be discussed here.

Persistent cool weather in late May and June leads to low accumulation of growing degree days above base 10 C and results in reduced growth or delayed plant development and final maturity. This delayed development increases the risk of freeze damage in the fall. Decker (1967) presented charts for the North Central states showing the probability of having runs of 5-or-more and 15-or-more days with minimum temperatures below 10 C for each week during April, May, and June. The 5-day charts for 24 May to 20 June are reproduced in Figure 3. For example, the chances of having at least 5 consecutive days with minimum temperatures below 10 C beginning during the week of 24 to 30 May increases from less than 10% in Kansas, Missouri, and southern portions of Illinois and Indiana to about 40% in the northern

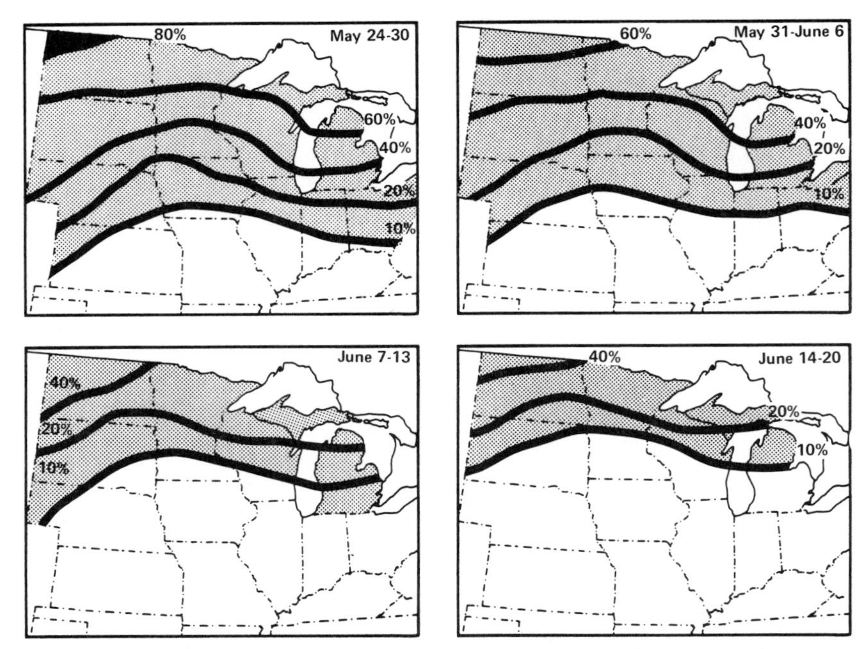


**Figure 3. Probability of 5 or more consecutive days with minimum temperatures below 10 C beginning in the indicated week (Decker, 1967).**

fringe of the Corn Belt. By mid-June, and through July and August, chances are less than 10% over most of the Corn Belt, but the cumulative effects of low temperatures anytime during the spring or summer are observed in delayed corn development almost every year. Periods of persistent warm weather, however, have offsetting effects by speeding crop development. Temperature summation indices, such as the accumulation of growing degree days above a crop growth threshold, are commonly used to predict the cumulative effect of temperature on crop development (Newman, 1971).

Unusually high temperatures during the summer decrease corn yields. Lehenbauer (1914) found that the elongation of corn seedlings kept in the dark at constant temperatures and with water not limiting decreased linearly as the temperature increased above 32 C (Coelho and Dale, 1980, Figure 1). Thompson (1968) found the best single weather variable for estimating corn yields to be an accumulation of the daily maximum temperatures above 32 C in July and August. In general, he found that state average corn yields decreased about 63 kg/ha for each 5.5 C accumulated above 32 C. The probabilities of having at least 5 consecutive days with maximum temperatures above 32 C in each week during July in the North Central states are reproduced in Figure 4 from Decker (1967). This period usually covers the time of

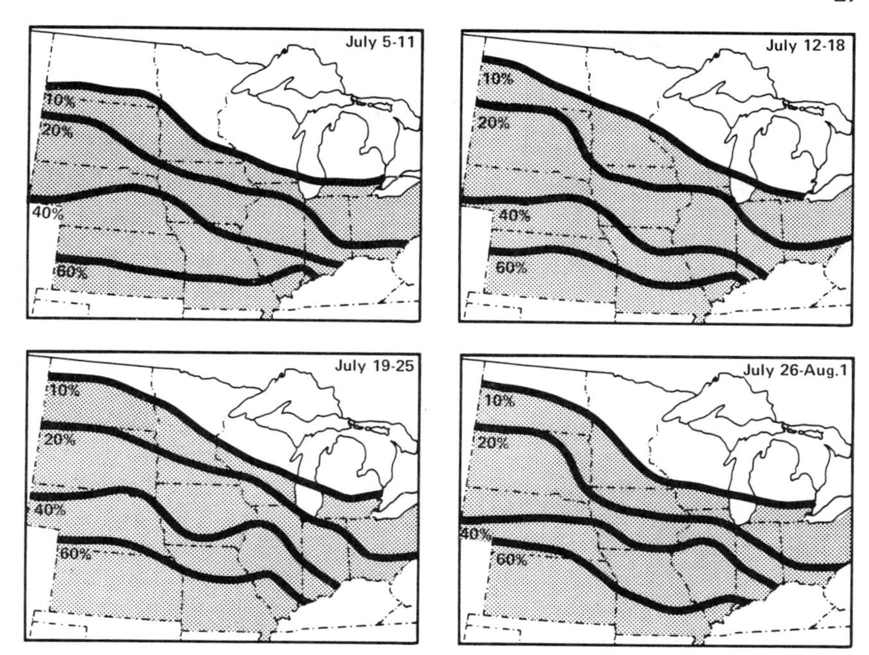


**Figure 4. Probability of 5 or more consecutive days with maximum temperatures above 32 C beginning in the indicated week (Decker, 1967).**

corn silking, an especially critical period for determining the number of kernels on the ear. The chance for having at least 5 consecutive days with maximum temperatures above 32 C beginning in the week of 19 to 25 July ranges from 10% in southern Michigan and the Corn Belt areas of Wisconsin and Minnesota to greater than 60% in Kansas and southern Illinois and Missouri. For this same week in these southern Corn Belt areas, the chances of beginning a run of at least 15 consecutive days above 32 C range from 10 to 20%.

The same "blocking high" synoptic situation of persistent extension of the Bermuda High into the southern United States which causes persistent hot weather during the summer, cuts off the flow of moist air from the Gulf into these areas and causes dry weather. This is a self-feeding system. Soil moisture shortages force less of the net radiation to be expended in latent heat and more in sensible heat. This further increases the temperature and potential evapotranspiration and accelerates the decrease in potential corn yield with combined moisture and temperature stress. Wallace (1980) discusses the state of the art and the needs for research in predicting these seasonal and interannual climate perturbations so important to crop production.

## INTERCORRELATIONS BETWEEN WEATHER VARIABLES

The limiting factor concept and joint effects of several weather variables upon plant growth complicate the interpretation of the climatology of a single variable. For example, mean daily solar radiation and its departures from normal are shown in Table 1. Note the contrasting anomalies of solar radiation and temperature. In August 1980, the mean temperature was 2.3 C above normal while the incident solar radiation was below normal. In July 1980, the temperature averaged 1.9 C above normal, but the radiation was slightly above normal. The mean temperatures for July 1979 averaged 1.4 C below normal and the solar radiation was also below normal. Although the monthly departures might suggest a negative correlation between solar radiation and temperature, the correlation of daily maximum temperature with daily total solar radiation from May through September of 1979 was not significantly different from 0 (-0.02). Hollinger (1980) correlated the daily maximum temperature with daily solar radiation in 7-day moving periods, as plotted in Figure 5. The correlation coefficients ranged from greater than 0.90 for weeks in early June and late August of 1979 to between -0.5 and -0.6 during several periods in 1979. Thus, during the growing season, insolation and temperature sometimes are positively correlated and sometimes negatively correlated. Although these correlations can be explained in changing advective conditions, these interactions pose problems in statistical and simulation models.

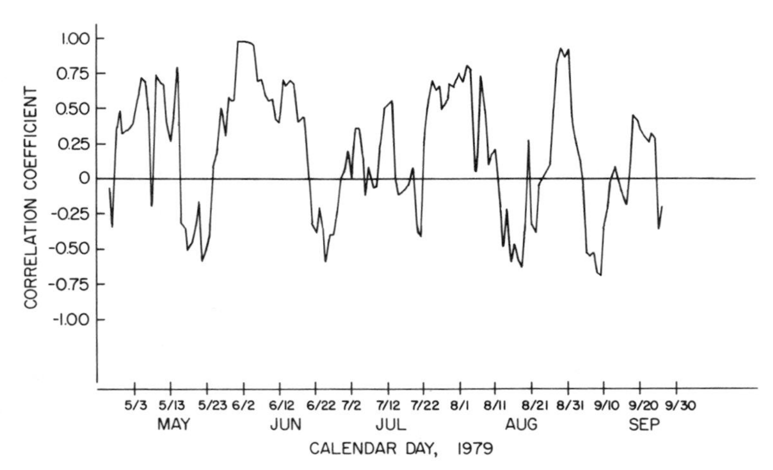


**Figure 5. Correlation of daily maximum air temperature with daily total solar radiation for indicated seven-day period, plotted on middle day, May through September 1979, for West Lafayette, Indiana.**

The interaction of temperature or the closely-related potential evapotranspiration and soil moisture or precipitation on corn yields is most often considered, as shown by the chapter by Shaw (1982). Schaal (L. A. Schaal, unpublished data, State Climatologist, Agronomy Department, Purdue University) has provided an interesting summary of the relation between summer mean temperature and precipitation anomalies for the Indiana Central Crop Reporting District reproduced in Figure 6. Although the means for the three summer months can obscure important within-season perturbations, some interesting observations can be made. Long-term agricultural experiments in Central Indiana—or more of the Corn Belt—may be subject to summer weather typical of almost any area of the United States. The years near the elliptical envelope of the scatter diagram had the most persistent anomalies and are identified with all four numerals, starting from the hottest, driest summer in 1936 and proceeding clockwise to 1934, 1900, 1896, 1958 (coolest, wettest), 1915, 1927, and 1967 (coolest, driest). Field research in Central Indiana in 1967 provided results more representative of those for Montana climate! This shows dramatically the necessity for quantitative consideration of the weather factor in interpreting the results from field experimentation.

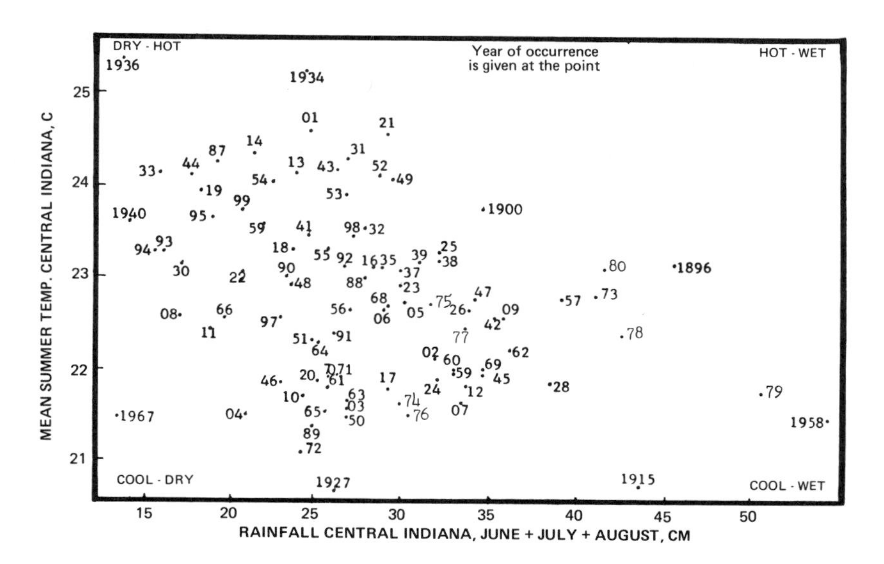


**Figure 6. Scatter diagram of summer (June, July, and August) mean temperature on summer total precipitation for Indiana Central Crop Reporting District, 1887 to 1980 (L. A. Schaal, unpublished, State Climatologist, Purdue University).**

## USE OF AVAILABLE FILES OF TEMPERATURE DATA

Most climatology of air temperature is based upon records from liquid-in-glass thermometers located in a standard Cotton Region Shelter. Although the maximum observed in the shelter may be slightly higher than the true air maximum temperature on a day with clear skies and little wind, the shelter temperatures serve as a base to which experimentally measured temperature should be calibrated for proper comparison. Even in the example given for West Lafayette, where temperature was measured with thermisters and electronically averaged within the hour, the published observations are from the maximum and minimum thermometers in the shelter. The readings are also published on the day of observation. Since observations at West Lafayette are taken at 0800 EST, the daily maximums presented here were set back to the previous day. However, there is a bias in the mean daily maximum, minimum, and the mean, if the observation is not taken at midnight. Schaal and Dale (1977) showed that this bias changed throughout the year. Fortunately, during the summer months it is almost 0 at a station with a 0800 LST observation. If the observations are taken at 1900 LST, the observed maximum and minimum temperatures have the advantage of usually being on the right day, but the means are biased upwards, i.e. about 0.3 C at Indianapolis. If the observation is taken earlier, say at 1700, the bias increases. These biases cause disturbing discrepancies for anyone trying to draw charts of seasonal accumulations of growing degree days or heating degree days. For example, a station with a 0700 observational time accumulates about 1.3% too few growing degree days for the season, while a station with a 1900 observational time accumulates about 3.4% too many, a discrepancy of 4.7% between two adjacent stations with the same "real growing degree days."

Most researchers become aware of temperature bias problems when working with data from a single station. In the late 50's and early 60's the National Weather Service remoted its temperature observations at First Order stations to the middle of the airport runway complex and also changed its instrumentation from liquid-in-glass maximum and minimum thermometers to the hygrothermometer. Although in many cases the temperatures may be more representative for agricultural purposes, these readings were so different that new station temperature "normals" had to be established. Most researchers working with Crop Reporting District or state average temperatures and crop yields assume that the temperature biases at individual stations average out. Nelson et al. (1979) showed that this is not necessarily true. In Indiana, the percentage of "AM" observing stations in the climatological network increased from 11% in 1935 to 55% in 1975. For the Central Crop Reporting District, they showed that -0.8 C of the "climatic cooling" of the summers from 1950 to 1970 was caused by changes in network configuration and time of observation. Thus, the points in Figure 6 for the years before 1950 should be reduced by 0.8 C to make them comparable with those in the 70's, e.g., reduce the 1936 mean temperature from 25.3 to 24.5 C. The published (USDC, 1887 to 1980) Crop Reporting District means were usually based on about 20 stations.

Perhaps, our knowledge of the effects of temperature on plant growth is not sufficiently precise to catch these subtle changes in the published temperature record, but I would be remiss if I did not convey to you some of the problems we have encountered in our use of these data. We should also recognize that soil temperature may be more of a limiting factor for plant growth and development than air temperature. For example, when predicting the time from planting to corn silking, Coelho and Dale (1980) found that the use of soil temperature from planting to the time when the growing point of the corn plant rose above the soil surface and then use of air temperature to silking provided slightly more accurate results than those obtained when only air temperature was used. Although the temperature of the soil surface usually has a much greater diurnal range than that of the air temperature, below a depth of 10 cm soil temperature generally has less diurnal and interdiurnal variability than air temperature. This variability in soil temperature decreases rapidly with depth, and approaches zero below 50 cm. At West Lafayette, Indiana, at the 10-cm depth under a bare soil surface in Toronto silt loam (Udollic Ochraqualf), on the average the daily maximum soil temperature rises above 10 C in the spring about a week later than the daily maximum air temperature and drops below 10 C in the fall about a week earlier than does the air temperature (McGarrahan and Dale, 1980). The daily minimum 10-cm soil temperature rises above 10 C in the spring about two weeks before the daily minimum air temperature reaches this level, and remains above 10 C about three weeks longer in the fall than the minimum air temperature. This is a characteristic air-soil temperature lag pattern, although soil temperature climatology will depend upon the specific soil type, soil moisture, and weather regime.

I have not included in this discussion the climatology of growing degree days or other temperature functions and their use in plant growth and development models, because they all derive from the basic temperature data. I should point out, however, that the use of any nonlinear temperature-plant response functions, or the 10 and 30 C cutoff thresholds in the modified growing degree day approach, may require the preparation of climatology specifically for the pertinent temperature function rather than interpreting it from the climatology of maximum and minimum temperatures.

## NOTES

Robert F. Dale, Department of Agronomy, Purdue University, West Lafayette, Indiana 47907.

Journal paper number 8317 of the Agricultural Experiment Station, Purdue University, West Lafayette, Indiana 47907.

## LITERATURE CITED

Coelho, D. T., and R. F. Dale. 1980. An energy-crop growth variable and temperature function for predicting corn growth and development: Planting to silking. Agron. J. 72:503-510.

Decker, W. L. 1967. Periods with temperatures critical to agriculture. North Central Regional Research Publ. No. 174, Missouri Agric. Exp. Stn. Bull. 864.

Hollinger, S. E. 1980. Environmental effects on corn ear morphology, planting to silking. Ph.D. Thesis, Agronomy Dept., Purdue Univ., West Lafayette, Indiana.

Landsberg, H. 1966. Physical Climatology. Fifth Printing, Revised. Gray Printing Co., Inc., DuBois, Pennsylvania.

Lehenbauer, P. A. 1914. Growth of maize in relation to temperature. Physiol. Researches 1(5):247-288.

McGarrahan, P., and R. F. Dale. 1980. Predicting soil temperatures with air temperature and soil moisture. Proc. Indiana Acad. Sci. 89:386-393.

Nelson, W. L., R. F. Dale, and L. A. Schaal. 1979. Non-climatic trends in divisional and state mean temperatures: A case study in Indiana. J. Appl. Meteor. 18:570-760.

Newman, J. E. 1971. Measuring corn maturity with heat units. Crops Soils, June-July: 11-14.

Schaal, L. A., and R. F. Dale. 1977. Time of observation, temperature bias and "climatic change." J. Appl. Meteor. 16:215-222.

Shaw, R. H. 1982. Estimates of yield reductions in corn caused by water and temperature stress. p. 49-66. In C. D. Raper, Jr., and P. J. Kramer (eds.) Crop reactions to water temperature stresses in humid, temperate climates. Westview Press, Boulder, Colorado.

Thom, H. C. S. 1968. Standard deviation of monthly average temperatures. ESSA Tech. Report. EDS 3. NOAA, EDIS, Silver Spring, Maryland.

Thompson, L. M. 1968. Weather and technology in the production of corn. p. 3-19. In Purdue Top Farmer Workshop, Corn Production Proceedings. Purdue Univ., West Lafayette, Indiana.

U.S. Dept. of Commerce, NOAA, Environmental Data and Information Service. 1887 to 1980. Climatological Data, Indiana. EDIS National Climatic Center, Asheville, North Carolina.

Wallace, J. M. 1980. Seasonal and interannual climate predictions. p. 73-78. In C. E. Leith et al. (eds.) The atmospheric sciences: National objectives for the 1980's. Natl. Acad. Sci., Washington, D.C.

SECTION II

# YIELD REDUCTIONS CAUSED BY STRESS

*This section deals with the effects of water and temperature stress on yield. According to Mederski, both soil temperature and day/night temperature have important effects on soybean yield. There also is need for more information concerning the effects of temperature on specific processes, such as seed germination, vegetative growth, nodulation, and nitrogen fixation. Soil water supply is very important and Mederski concludes that drought tolerance can be determined only under field conditions. Shaw states that soil water supply and July weather are extremely important for corn in Iowa. Water stress delays silking more than pollen formation, resulting in poor yield. High night temperatures probably reduce yield by causing excessive use of carbohydrate in respiration. Shaw also states that although adequate storage of soil water in the spring is important in Iowa, corn yield is reduced more often by excess of soil water in the spring than by drought. Excess water in the spring delays planting and the reduced aeration decreases root growth and the availability of nitrogen. Shaw also concludes from the large differences in corn yields in adjacent fields that there are large genetic differences in stress tolerance among varieties. These genetic differences can be used as a basis for breeding programs to increase drought tolerance. However, we must still learn why some corn hybrids yield well under a wide range of weather conditions while others yield well only in a narrow range.*

*Kozlowski states that yield is the integrated response to the complex of environmental factors operating through a complex of physiological processes. Processes such as cell enlargement and cambial activity respond more promptly to water stress than metabolic processes such as photosynthesis and nitrogen metabolism. He states that practically all trees, even those in rain forests, are subjected to water stress. He suggests that search for morphological modifications such as deep root systems and good control of transpiration, which prevent or postpone dehydration, is the most hopeful approach to increasing drought tolerance of woody plants.*

# 4

# EFFECTS OF WATER AND TEMPERATURE STRESS ON SOYBEAN PLANT GROWTH AND YIELD IN HUMID, TEMPERATE CLIMATES

**Henry J. Mederski**

## GERMINATION

### Temperature Stress

Seed of each plant species appears to have a definite temperature requirement for germination. Soybeans will germinate in the range of 10 to 40 C, but maximum rates of emergence occur at 25 to 30 C (Hatfield and Egli, 1974; Edwards, 1948). There does not appear to be a significant interaction between cultivar seed germination and temperature (Stuckey, 1976). Rate of emergence decreases with planting depth at 5.0, 7.5, and 10 cm, but the differences in rate decrease as temperature increases in the range of 16 to 32 C. Although depth of planting and temperature affected the rate of emergence, final emergence was not markedly affected by treatment (Stuckey, 1976).

Mechanical impedence as well as depth of planting interact with temperature to affect growth rates of hypocotyls and radicals. The negative effect of mechanical impedence on reducing growth rates are greatest at low temperatures and least at optimum temperatures of 25 to 30 C. As temperatures decrease from 30 to 15 C, growth rates of the radical decrease, but the effect appears to be independent of soil moisture (Bowen and Hummel, 1980). Optimum or near optimum soil temperatures appear to lessen the rate limiting effects of deep planting or high mechanical impedence. Grabe and Metzer (1969) in a study with 25 cultivars concluded that poor germination may be a consequence of interactions between planting depth, temperature, and cultivar.

Although some kinds of seeds require alternating temperatures for germination, the rate and percent germination of soybeans appear to be the same under either alternating or constant temperatures (Delouche, 1953).

## MOISTURE STRESS

The first process to occur prior to germination is the absorption of water by the seed. This is followed by increased metabolic activity and finally the emergence of the radical. Because of the importance of water to seed germination, considerable work has been done on soil-seed-water relations.

Changes in soil moisture content affect two components of soil water potential, the osmotic potential (due to solutes) and the matric potential (the binding force of soil solids for water) as well as hydraulic conductivity. A decreasing matric potential, which occurs as the soil loses water, increases the amount of energy and time required by the seed to obtain a unit mass of water. As soil water content decreases the hydraulic conductivity decreases and reduces the rate of movement of water to the surface of the seed. Several studies (Collis-George and Sands, 1959, 1961, 1962) show that the germination of seeds is influenced by matric potential, osmotic potential, and hydraulic conductivity. In later work, Collis-George and Hector (1966) indicate that matric potential is of importance because of its direct effect on the rate of water uptake by the seed and its indirect effect on controlling the wetted area of contact between soil and seed.

Studies in which soybean, corn, sugar beet, and rice were germinated over a wide range of soil moisture conditions indicate that the minimum seed moisture content for germination was 30% for corn, rice, and sugar beets and 50% for soybeans. At an optimum temperature of 25 C the soil should have a water potential of not less than -6.6 bars for soybean germination to occur within 5 to 8 days. At soil moisture contents too low to ensure germination, the seed may imbibe some water making them susceptible to fungal attack and eventual decay (Hunter and Erickson, 1952). Although soybeans will germinate over a relatively wide range of soil moisture contents from -0.3 to -10 bars, high levels of mechanical impedence and unfavorable soil temperatures interact with soil moisture to affect germination.

The osmotic component of soil moisture potential is a negligible part of total moisture potential in non-saline soils but may be of significance when fertilizers are applied in a band very close to the seed. Poor germination usually occurs when fertilizers are placed in contact with the seed, particularly when soil moisture content and temperature are unfavorable. The poor germination is attributed in part to a high osmotic water potential that limits water absorption by imbibition. Toxic effects of high concentrations of fertilizer salts in solution near the seed also may contribute to poor germination.

The problem of poor seed-to-soil contact can be overcome by compacting the soil below the seed or pressing the seed into compacted soil and covering the seed with loose soil. Good contact between seed and firm soil will increase capillary conductivity of the soil and improve water transport to the seed (Hillel, 1972).

## GROWTH AND YIELD

### Water Stress

During the past decade there has been a rapid increase in the number of studies of the effects of soil-plant water stress on physiology, development, and yield of soybean plants. The increased research effort is due in part to improvement in instrumentation, but more importantly, to a growing recognition of the complexity and importance of effects of water stress on plant growth. In this review, stress is defined as a lack of sufficient water to maintain maximum growth rates and yields.

Since water is essential for plant growth, it is axiomatic that water stress, depending on its severity and duration, will affect plant growth and yield. Statistical studies (Runge and Odell, 1960) of the relationship between long-term weather data and soybean yields in Illinois show that precipitation and daily maximum temperature explained about 70 percent of the variation in yield and that the soybean plant is particularly sensitive to water stress occurring during the grain filling period. A later study (Thompson, 1960) of soybean yield in five Corn Belt states shows that highest yields were associated with normal preseason precipitation from September to June. Above normal temperatures in June and lower than normal temperatures in July and August were also associated with the highest yields.

There are numerous studies to show that soybean yields are increased by irrigation during years of below average precipitation. During the past 10 years or so, there has been a growing interest in the effects of water stress occurring at various stages of development, its effect on yield components, and the physiological mechanisms that underlie yield reductions caused by increased soil and plant water stress.

In field studies where water stress was imposed on soybeans at various stages of development, limiting water to early flowering reduced yields about 3%, whereas limiting water from flowering to maturity reduced yields about 50% (Doss et al., 1974). The investigators concluded that the pod fill stage of development was most sensitive to soil moisture stress. Similar conclusions were drawn from a potometer study conducted under field conditions where soybeans were subjected to controlled water stress during flowering, pod initiation, and bean filling stages of development (Shaw and Laing, 1960). Maximum yield reductions occurred when soil water stress occurred during bean filling and late pod development. Stress during flowering reduced yields, but to a lesser extent than at later stages of development. Stress during flowering and pod development increased abortion of flowers and pods, and stress during bean filling reduces yield and seed size (Martin et al., 1979; Constable and Hearn, 1978).

An understanding of how stress affects components of yield has been the subject of several studies. When a range of stress levels was applied during each of four periods of reproductive development, the number of seeds, the number of harvestable pods, and pods per node decreased with increasing degree of stress. Weight per seed was decreased by stress occurring during pod and seed development, but nodes per plant appeared to be relatively insensitive to stress (Monen et al., 1973).

Ample soil moisture supply during the preflower stage of development may produce relatively large amounts of vegetative dry matter without affecting yield. Studies show that full-season irrigation and irrigation commencing at flowering produced the same yield. However, full-season irrigation produced from 50 to 100% greater weight of leaves, petioles, stems, and branches (Ashley and Ethridge, 1978). Clearly, there was no proportionality between vegetative mass and final yield. Beginning irrigation during reproductive development had little effect on vegetative dry weight but increased the number of pods and seed size when compared with the nonirrigated control.

Although experimental evidence indicates that the soybean plant is particularly sensitive to stress during reproductive stages of development, severe stress during vegetative development may limit yield by limiting plant growth below that required for maximum yield. However, the relatively high soil moisture conditions coupled with low atmospheric demand mitigate the development of severe plant water stress during late spring and early summer.

There is a growing interest in determining whether there are genotypic differences among soybean cultivars in their resistance to water stress and whether stress resistance can be achieved through breeding programs.

Several studies have been made to determine differential response of soybean cultivars to soil moisture stress. Mederski and Jeffers (1973) grew 32 soybean cultivars, 8 in each of four maturity groups, under irrigated and nonirrigated conditions. A significant interaction between cultivar and soil stress level was found in all maturity groups. The yields of the most stress-susceptible cultivars grown under nonirrigated conditions were reduced by about 30%, while the yields of the most stress-resistant cultivars were reduced by only 10 to 15%. The difference in yield between stress levels was about 400 kg/ha for the most susceptible cultivars and about 1200 kg/ha for the stress resistant cultivars. Relative leaf water content, determined several times during the growing season, was the same for all cultivars grown under stress conditions indicating that the yield differences among cultivars were not associated with differences in internal plant water stress.

Sammons et al. (1978, 1979) described two methods of screening soybeans for drought resistance. In both they used the same cultivars used by Mederski and Jeffers (1973) with the objective of developing rapid screening techniques that would identify characteristics associated with stress tolerance. In one, a growth chamber study (Sammons et al., 1978), they measured root and shoot water potential, leaf surface area, leaf dry weight, and photosynthetic rate per unit leaf area. Although they found a significant interaction between cultivar and soil moisture treatment for all cultivars measured, no single parameter provided a satisfactory indication of stress resistance exhibited by these cultivars when they were grown in the field. In the second study (Sammons et al., 1979), the same authors grew plants in boxes in a greenhouse and subjected them to gradually increasing soil water stress for 30 days. Leaf expansion and plant growth rates were determined periodically. The data indicated that neither leaf expansion nor plant growth rates can be used alone for predicting stress resistance under field conditions. The authors concluded that drought-stress resistance must be determined under full-season field conditions so that reproductive responses can be measured.

A rapid screening technique confined to the early vegetative stage of development may be inadequate because factors important during one stage of development may be relatively unimportant during another. For example, screening during vegetative development may provide a measure of stress effects on cell enlargement and leaf area development, but would not profide information on stress as it affects photosynthetic rates during seed development or the effects on pod and flower abortion. A definitive assessment of difference in stress resistance among cultivars may require assessment of stress effects during each stage of plant development. Moreover, very little is known about the interacting effects of cultivars with soil water stress on mineral uptake and symbiotic nitrogen fixation.

Research on screening for drought resistance deserves more attention than it has received. One observation from the field study by Mederski and Jeffers (1973) is that the rank in yield of the cultivars grown under high moisture-stress conditions is not, with a couple of exceptions, different from the rank in yield under low stress. The highest yielding cultivars under low stress also produce the highest yields under high stress but the differences in yield among cultivars under high stress are small. One can speculate that since there is no significant change in yield ranking among cultivars under a range of stress conditions, the selection of high yielding cultivars grown under low stress conditions would ensure that these same cultivars would, under stress, produce relatively higher yields than their stress-susceptible counterpart.

Studies with oats indicate that selecting cultivars under low stress conditions is advantageous because the large genotypic variation will afford good genotypic differentiation of lines (Johnson and Frey, 1969). Perhaps yield improvement for soybeans during the past several decades may be due in part to improved water stress resistance. Although the yield of the recently developed cultivars is reduced by limited rainfall during the growing season, they appear to produce higher yields over a range of soil moisture stress than their counterparts developed several decades ago.

**Soil Temperature Stress**

In contrast to water stress research, relatively few studies, particularly field studies, have been made on the effect of soil and air temperature on soybean growth. This may be because of the difficulties and cost of controlling temperature under field conditions. The following brief discussion, drawn largely from greenhouse and growth chamber studies is intended to review some of the more typical studies that have been made.

Earley and Carter (1945) appear to be among the first to study the effects of root temperature on soybean growth. Under greenhouse conditions they found the optimum root temperature for soybean shoot and root growth was about 27 C. Plant height, however, increased when temperature was increased from 2 to 17 C, remained constant to about 27 C, and then declined. Root:shoot ratios were relatively insensitive to root temperatures in the range of 12 to 32 C.

Under field conditions, soil temperature was regulated during the growing season to achieve mean temperatures of 11 and 31 C and compared with an 18 C control treatment (Mack and Ivarson, 1972). A root temperature of 31C increased seed yield 43% while 11 C decreased yield 82%. Although shoot weight increased about 35% between temperature extremes, it was less sensitive to temperature than yield. The study also showed very large increases in phosphorus and potassium uptake with increasing soil temperature.

In greenhouse studies on the effect of controlled root temperatures on growth, there was a four-fold increase in plant dry weight between 15 and 20 C and an additional 50% increase between 20 and 25 C (Trang and Giddens, 1980). Nodule number and size increased as temperature increased from 15 to 20 C but decreased rapidly at 25 C or above. In another study (Duke et al., 1979), a comparison between plants grown at 20 and 13 C root temperatures indicates that all parameters of vegetative and reproductive development increased by a factor of 2 to 20 at the higher root temperature. Poor plant development at 13 C was attributed to poor nodulation and a severe nitrogen deficiency. Interestingly when the plants at 13 C were switched to 20 C after 42 days of growth, they developed nodules and began to fix nitrogen. A similar study also showed that a root temperature of about 25 C produces maximum plant dry weight and nodule weight per plant (Lindeman and Ham, 1979).

In Ohio, diurnal soil temperatures at the 10-cm depth, averaged for several positions extending from the row, are seldom at or above 25 C from mid June to mid July. Temperature extremes ranged from 15 C to about 25 C during this period. If the optimum soybean root temperature is about 25 C, soil temperatures are probably well below the optimum during most of the growing season.

**Air Temperature Stress**

Thermoperiodicity has been defined in several ways but the most recent is an enhanced growth response to a diurnal change in temperature in comparison with the response measured under constant temperature having the same mean as the varying diurnal temperatures (Friend and Helson, 1976). Diurnal variations in air temperature are important as they affect photorespiration and dark respiration rates, enzymatic reactions, and the physical processes of diffusion and transpiration. Plants may respond to heat units accumulated over a period of time, others may be sensitive to day or night temperatures, and some may respond to a day-night differential generally referred to as a thermoperiod irrespective of the mean temperature.

In studies with soybeans grown at the same mean temperature of 23 C, but with thermoperiods of 23/23, 26/20, and 29/17 C (day/night), there were no large effects on vegetative growth rates due to diurnal variations in temperature (Warrington et al., 1977). The authors indicate that sugar beets, sugar cane, tomatoes, and wheat also appear to be insensitive to thermoperiod.

The imposition of different diurnal temperatures during the reproductive stage of development has been shown to affect seed yield and seed characteristics (Sato and Ikeda, 1979). Comparisons of day and night temperatures in the range of 17/12 (day/night) to 30/25 C show that day temperatures of about 25 to 30 C in combination with night temperatures of 15 to 20 C were optimum for total seed production and maximum number of pods and seeds. Temperatures of 30/25 or 17/12 C decreased seed yield about 30%.

The effects of different diurnal temperature cycles are complicated by changing length of photoperiod (Raper and Thomas, 1978). Under short-day conditions during pod fill, seed yield decreased about 30% as day/night temperatures decreased from 30/26 to 22/18 C. Under long-day conditions, created by interrupting the night period with low level light, seed yield increased as day/night temperatures decreased. At either photoperiod, the duration of pod fill increased by 5 days as temperature decreased.

A significant interaction between day and night temperatures from 14/10 to 30/26 C during floral induction on morphology of soybean also has been reported (Thomas and Raper, 1978). Under short-day conditions leaf area and vegetative dry weight were greatest at 26/10 C but pod weights were greatest with higher temperatures of 26/22 C. Lowest carbon dioxide exchange rates occurred at 14/26 C and the highest at 26/14 and 26/18 C. Warm night temperatures at any day temperature reduced the time from the start of inductive photoperiod to anthesis.

Warm days and cool nights of 28/17 C appear to produce leaves with a much higher specific leaf weight than warm days and nights of 30/26 (van Volkenberg and Davis, 1977). The higher specific leaf weights may account for the greater carbon dioxide exchange rates at constant day temperatures. Night temperatures in the range of 14 to 18 C combined with a day temperature of about 25 C appear to be optimum for carbon dioxide exchange rate (Thomas and Raper, 1978).

Day/night temperatures influence the time intervals between developmental events (Hesketh et al., 1973). In the range of 17/11 to 32/29 C, with mean temperatures of 13 to 30 C, the accumulated degree days required for the development of successive trifoliolates remained constant in the range of 13 to 30 C. Trifoliolate emergence per day increased linearly with increasing daily mean temperature. Days between trifoliolate emergence decreased as temperature increased over the entire mean temperature range. Days from flowering to the onset of pod maturity were not affected by mean temperature in the range of 21 to 30 C included in the investigation. Days from the third trifoliolate emergence to flowering decreased linearly with increasing mean temperature from 13 to about 22 C and then remained constant to 30 C. Other studies show that the rate of plant development from planting to flowering attains a maximum at about 26 C (Brown, 1960). The rate of development during the flowering and post flowering period is related to soil moisture as well as temperature. High moisture stress increases the rate of plant development in the post flowering period (Brown and Chapman, 1960). A temperature stress of 40 C occurring from the onset of flowering results in severe flower and pod abortion, and a 40% reduction in number of seeds per plant (Mann and Jawarski, 1970).

Although generalizations on the effect of air temperature on soybean growth are difficult because of interactions with photoperiod and stage of development, day temperatures of 25 C and night temperatures of 15 C appear optimum or near optimum. Virtually all growth chamber studies reporting the effect of air temperature on soybean growth are not independent of soil temperature. Soil and air temperature usually vary together, although soil temperature lags behind changes in air temperature. There is a need for studies in which air and soil temperatures are controlled independently since the root and shoot may have different optima.

## NITROGEN FIXATION

### Water Stress

Sprent (1976) found a nearly linear relationship between leaf dry weight and nodule weight of plants grown under varying degrees of water stress, although the maximum specific activity of the nodules was not affected by stress imposed during their development. She also found that the volume and surface area of the individual nodules decreased with increasing stress. Water stress causes nodule shrinkage and large reductions in nodule surface area, nodule porosity, and rate of oxygen diffusion per unit area of the nodule surface (Parkhurst and Sprent, 1975). The restricted uptake of oxygen was assumed to be an important factor in reducing the rate of $N_2$ fixation. No mention was made of the possibility that the rate of diffusion of $N_2$ into the nodule also may be important in limiting $N_2$ fixation.

Studies (Sprent, 1971) using detached soybean nodules show that acetylene reduction and $N^{15}$ reduction cease when water loss by the nodule was equivalent to 25% of the fully turgid nodules (-8 to -10 bars). Below 80% relative turgidity, irreversible changes occurred in the nodule and the nitrogen fixing ability of the nodule did not recover upon rehydration. Although detached nodules or detached nodulated root systems have been commonly used in experiments, there is evidence that their behavior may be different from those of intact plants, and the results should be considered with caution (Huang et al., 1975). The remaining portion of this discussion will be drawn from work with intact plants.

Greenhouse studies (Kuo and Boersma, 1971) in which soybean plants were grown at several levels of soil moisture stress show that $N_2$ fixation (total N per plant) decreased about 40% as soil moisture stress increased from 0.35 to 2.50 bars. However, percent nitrogen in the tissue remained constant indicating a close coupling between dry matter production and nitrogen fixation.

The effect of soil moisture stress on physiological processes vital to $N_2$ fixation are not fully understood. However, there is very good evidence that as water stress reduces the rate of photosynthesis, the carbohydrate supply to the nodules decreases below the level required for maximum $N_2$ fixation. Studies by Huang et al. (1975) with intact plants show a close relationship between soil, nodule, and leaf water potential. The decrease in leaf and nodule water potential were closely coupled with a decrease in $CO_2$ assimilation and acetylene reduction ($N_2$ fixation). They also showed that in the absence of water stress, acetylene reduction varied with $CO_2$ assimilation rates. They concluded that inhibition of photosynthesis accounted for the inhibition of acetylene reduction at low soil and plant water potentials. Although the authors emphasize a close relationship between current photosynthesis and nitrogen fixation, there is evidence that carbohydrate pools stored within the shoot and/or roots contribute to nitrogen fixation (Mederski and Streeter, 1977; Ahmad, 1978). Possibly, these storage pools could maintain high levels of nitrogen fixation for a short time when severe moisture stress limits photosynthesis. When soil moisture stress is severe, carbohydrate reserves in roots and shoots may not ensure high levels of $N_2$ fixation since nodule activity is negatively affected by stress independent of carbohydrate supply. Our present state of knowledge indicates that nitrogen fixation depends upon concomitant photosynthesis and on carbohydrate from storage pools with the relative importance of each source varying as water stress affects the primary source of carbohydrate, photosynthesis.

**Temperature Stress**

Aprison et al. (1954) were the first to show that nitrogen fixation by excised nodules was temperature sensitive with maximum fixation occurring at 25 C. At 15 and 30 C, $N_2$ fixation decreased 20 and 40%, respectively. Studies with intact soybean plants grown at several root temperatures and constant shoot temperature showed that the total nitrogen per plant was maximum at a soil temperature of 24 C but declined rapidly above or below 24 C. This temperature optimum appeared to be independent of soil moisture stress in the range of 0.3 to 2.5 bars (Kuo and Boersma, 1971). However, studies with detached nodules show that the temperature optimum declined from about 25 C for fully turgid nodules to about 15 C for water stressed nodules (Pankhurst and Sprent, 1976). Further evidence for an interaction between water stress and root temperature was shown by studies indicating a greater decline in acetylene reduction with increasing plant water stress at 30 C than at 20 C. The more rapid decline at the higher temperature is believed to be the result of more rapid utilization of nodule carbohydrates associated with higher respiration rates (Ahmad, 1978).

Day/night temperature regimes during plant development appear to affect the nodule's subsequent response to temperature (Gibson, 1976). A determination of nodule specific activity ($C_2H_2$ reduction/gm·dry weight) when incubated at temperatures in the range of 15 to 35 C show that the specific activity, at all incubation temperatures, was higher for nodules from plants that had a prior history of day/night temperatures of 21/16 C than it was for plants grown at 32/28 C. Also, the incubation temperature optimum was shifted from 25 C for the 21/16 C plants to 30 C for the 38/28 C plants. Irrespective of prior history, nodule specific activity declined very rapidly when incubation temperatures exceeded 30 C (Gibson, 1976).

The optimum temperature for nitrogen fixation may not be optimum for nodule development. Plants grown at 18 C produced twice the nodule weight per unit time as those grown at 25 C, but the nitrogen fixation rate (mg $N_2$/

mg·nodule/day) was doubled at the higher temperature. In this case the total nitrogen fixed per plant was the same at each temperature and indicates a compensatory response to temperature differences during growth (Gibson, 1976). Studies with dry beans show that, although the early development of nodules (21 to 28 days after planting) was not affected by variation in temperature (root and shoot at same temperature) further development between the 28 and 42 days was reduced or arrested at 35/25 C (Graham, 1979). At temperatures of 25/15 C nodule development continued at a constant rate. Cooler growth temperature also produced the greatest nodule specific activity and increased the duration of high $N_2$ fixation rates.

Short term exposure of shoots to various temperatures appears to have an effect on the specific activity of nodules. Shoot/root temperatures were maintained independently at 27/27 and 18/27 during a pretreatment in light, and *in situ* nodule activity was measured in light at the respective root temperature. The specific activity of the 18/27 treatment was about 50% of the 27/27 control indicating a significant effect of shoot temperature on nodule activity. This effect may be due to changes in the level of carbohydrates in the shoot or their translocation from shoot to root in response to temperature (Schwitzer and Harper, 1980).

In general, the root and shoot temperatures that appear to be optimum for plant growth also appear to be optimum for nitrogen fixation. Temperatures exceeding 30 C reduce nodule activity and above 35 C the nodules may be irreversibly damaged.

## RESEARCH NEEDS

Moisture stress and temperature stress pose intractable problems. Control of moisture stress by irrigation is limited by the availability of ground water supplies and cost of application. Soil and air temperature are beyond practical control. In view of this, there is a need for the development of new cultivars with a broad tolerance to deviations from normal temperature and precipitation patterns. Comparisons between cultivars which appear to vary widely in their resistance to moisture stress should be subjected to detailed growth analysis, including net $CO_2$ exchange and photosynthetic partitioning at all stages of plant development. Excepting growth chamber studies, very little is known about plant response to temperature variations under field conditions. Neither do we know whether cultivars respond differently to air and soil temperatures. It seems appropriate to determine whether plant genotypes vary in their response to temperature and termine the reason for the variation.

Although symbiotic nitrogen fixation provides the soybean plant with most of the nitrogen required for optimum growth and yield, the plant does exhibit some dependence on available soil nitrogen. The proportion of the total nitrogen in the plant derived from the soil and from symbiotic fixation may vary with water and temperature stress. The process of symbiotic nitrogen fixation and nitrate absorption may have different temperature optima and water stress responses. There is a need for quantitative studies of the effect of soil moisture and temperature stress on the concomitant uptake of soil nitrogen and symbiotically fixed nitrogen by the soybean plant. That soybeans do not respond to nitrogen fertilizers is due to systems of crop and fertilizer management that ensure sufficient levels of soil nitrogen to supplement symbiotic fixation. Future cropping and fertilizer practices may change in a way that will make the soybean plant increasingly dependent on its symbiotic association, and knowledge of the effect of environment on $N_2$ fixation may become more essential with time.

## NOTES

Henry J. Mederski, Department of Agronomy, Ohio Agricultural Research and Development Center, Wooster, Ohio 44691.

## LITERATURE CITED

Ahmad, R. A. 1978. The effects of water stress, root temperature, and carbohydrate supply to the nodules. Ph.D. Thesis. Ohio State University.

Aprison, T. H., W. E. Magee, and R. H. Burris. 1954. Nitrogen fixation by excised soybean root nodules. J. Biol. Chem. 208: 29-34.

Ashley, D. A., and J. W. Ethridge. 1978. Irrigation effects on vegetative and reproductive development of three soybean cultivars. Agron. J. 70: 467-471.

Bowen, H. D., and J. W. Hummel. 1980. Critical factors in soybean seedling emergence. p. 451-469. In F. T. Corbin (ed.) World soybean research conference II: Proceedings. Westview Press. Boulder, Colorado.

Brown, D. M. 1960. Soybean ecology. I. Development-temperature relationships from controlled environment studies. Agron. J. 52: 493-495.

Brown, D. M., and L. S. Chapman. 1960. Soybean ecology. II. Development-temperature-moisture relationships from field studies. Agron. J. 52: 496-499.

Collis-George, N., and J. B. Hector. 1966. Germination of seeds as influenced by matrix potential and by area of contact between the seed and oil water. Aust. J. Soil. Res. 4: 145-164.

Collis-George, N., and J. E. Sands. 1961. Moisture conditions for testing germination. Nature 190: 367.

Collis-George, N., and J. E. Sands. 1961. Moisture conditions for testing germination. Nature 190: 367.

Collis-George, N., and J. E. Sands. 1962. Comparison of the effects of the physical and chemical components of soil water energy on seed germination. Aust. J. Agr. Res. 13: 575-584.

Constable, G. A., and A. B. Hearn. 1978. Agronomic and physiological responses of soybean and sorghum crops to water deficits. I. Growth, development, and yield. Aust. J. Plant Physiol. 5: 159-167.

Delouche; J. C. 1953. Influence of moisture and temperature levels on the germination of corn, soybeans, and watermelons. Proc. Assoc. Offic. Seed Analysts. 43: 117-126.

Doss, D. B., R. W. Pearson, and H. T. Rogers. 1974. Effect of soil water stress at various growth stages in soybean yield. Agron. J. 66: 297-299.

Duke, S. H., L. E. Schrader, C. A. Henson, J. C. Servaites, R. D. Vogelzang, and J. W. Pendleton. 1979. Root temperature effects on soybean nitrogen metabolism and photosynthesis. Plant Physiol. 63: 956-962.

Earley, E. B., and J. L. Carter. 1945. Effect of temperature of the root environment on growth of soybean plants. Agron. J. 37: 727-735.

Edwards, T. J. 1934. Relations of germinating soybeans to temperature and length of incubation time. Plant Physiol. 9: 1-35.

Friend, D. J. C., and V. A. Helson. 1976. Thermoperiodic effects on the growth and photosynthesis of wheat and other crop plants. Bot. Gaz. 137: 75-84.

Gibson, A. H. 1976. Recovery and compensation by nodulated legumes to environmental stress. p. 385-403. In P. S. Nutman (ed.) Symbiotic nitrogen fixation in plants. Cambridge University Press, Cambridge.

Grabe, D. F., and R. B. Metzer. 1969. Temperature-induced inhibition of soybean hypocotyl elongation and seedling emergence. Crop Sci. 9: 331-333.

Graham, P. N. 1979. Influence of temperature on growth and nitrogen fixation in cultivars of *Phaseolus vulgaris* L., inoculated with *Rhizobium.* J. Agric. Sci., Camb. 93: 365-370.

Hatfield, J. L., and D. B. Egli. 1974. Effect of temperature on the rate of soybean hypocotyl elongation and field emergence. Crop Sci. 14: 423-426.

Hesketh, J. D., D. L. Myhre, and C. R. Willey. 1973. Temperature control of time intervals between vegetative and reproductive events in soybeans. Crop Sci. 13: 250-254.

Hillel, D. 1972. Soil moisture and seed germination. p. 65-84. In T. T. Kozlowski (ed.) Water deficits and plant growth. Vol. III. Plant responses and control of water balance. Academic Press, N.Y.

Huang, C-Y, J.S. Boyer, and L.N. Vanderhoeff. 1975. Limitation of acetylene reduction by photosynthesis in soybean having low water potentials. Plant Physiol. 56: 228-232.

Hunter, J. R., and A. E. Erickson. 1952. Relation of seed germination to soil moisture tension. Agron. J. 44: 107-109.

Johnson, G. R., and K. J. Frey. 1967. Heritibilities of quantitative attributes of oats (*Avena* sp.) at varying levels of environmental stress. Crop Sci. 7: 43-46.

Kuo, T., and L. Boersma. 1971. Soil suction and root temperature effects on nitrogen fixation in soybeans. Agron. J. 63: 901-904.

Lindeman, W. C., and G. E. Ham. 1979. Soybean plant growth, nodulation, and nitrogen fixation as affected by root temperature. Soil Sci. Soc. Am. J. 43: 1134-1137.

Mack, A. R., and K. C. Ivarson. 1972. Yield of soybeans and oil quality in relation to soil temperature and moisture in a field environment. Can. J. Soil Sci. 52: 225-235.

Mann, J. D., and E. G. Jawarski. 1970. Comparison of stresses which may limit soybean yield. Crop Sci. 10: 620-624.

Martin, C. A., D. K. Cassel, and E. J. Kamprath. 1979. Irrigation and tillage effects on soybean yield in a coastal plain soil. Agron. J. 71: 592-594.

Mederski, H. J., and D. L. Jeffers. 1973. Yield response of soybean varieties grown at two soil moisture stress levels. Agron. J. 65: 410-412.

Mederski, H. J. , and J. G. Streeter. 1977. Continuous, automatic acetylene reduction assay using intact plants. Plant Physiol. 59: 1076-1081.

Monen, N. N., R. E. Carlson, R. H. Shaw, and O. Arjimand. 1979. Moisture stress effects on the yield components of two soybean cultivars. Agron. J. 71: 86-90.

Pankhurst, C. E., and J. I. Sprent. 1975. Effects of water stress on the respiratory and nitrogen fixing activity of soybeans roots nodules. J. Expt. Bot. 26: 287-304.

Pankhurst, C. E., and J. I. Sprent. 1976. Effects of temperature and oxygen tension on the nitrogenase and respiratory activities of turgid and water-stressed soybean and french bean root nodules. J. Expt. Bot. 27: 1-9.

Raper, C. J., Jr., and J. F. Thomas. 1978. Photoperiodic alteration of dry matter partitioning and seed yield in soybeans. Crop Sci. 18: 654-656.

Runge, E. C. A., and R. T. Odell. 1960. The relation between precipitation, temperature, and the yield of soybeans on the Agronomy South Farm, Urbana, Illinois, Agron. J. 52: 245-247.

Sammons, D. J., D. B. Peters, and T. Hymowitz. 1978. Screening soybeans for drought resistance. I. Growth chamber procedure. Crop Sci. 18: 1050-1054.

Sammons, D. J., D. B. Peters, and T. Hymowitz. 1979. Screening soybeans for drought resistance. II. Drought box procedure. Crop Sci. 19: 719-722.

Sato, K., and T. Ikeda. 1979. The growth responses of soybean plant to photoperiod and temperature. IV. The effect of temperature during the ripening period on the yield and character of seeds. Japan J. Crop Sci. 48: 283-290.

Schwitzer, L. E., and J. E. Harper. 1980. Effects of light, dark, and temperature on root nodule activity (acetylene reduction) of soybeans. Plant Physiol. 65: 51-56.

Shaw, R. H., and D. R. Laing. 1966. Moisture stress and plant response. p. 73-94. In W. H. Pierre, D. Kirkham, J. Pesek, and R. H. Shaw (ed.) Plant environment and efficient water use. American Society of Agronomy and Soil Science Society of America. Madison, Wisconsin.

Sprent, J. I. 1971. The effect of water stress on nitrogen fixing root nodules. I. Effect on the physiology of detached soybean nodules. New Phytol. 70: 9-17.

Sprent, J. I. 1976. Water stress and nitrogen-fixing root nodules. p. 291-315. In T. T. Kozlowski (ed.) Water deficits and plant growth. IV. Soil water measurement, plant responses, and breeding for drought resistance. Academic Press, New York.

Stuckey, D. J. 1976. Effect of planting depth, temperature, and cultivar on emergence and yield of double cropped soybeans. Agron. J. 68: 291-294.

Thomas, J. F., and C. D. Raper, Jr. 1978. Effect of day and night temperatures during floral induction on morphology of soybeans. Agron. J. 70: 893-898.

Thompson, L. M. 1970. Weather and technology in the production of soybeans in the central United States. Agron. J. 62: 232-236.

Trang, K. M., and J. Giddens. 1980. Shading and root temperature as environmental factors affecting growth, nodulation, and symbiotic $N_2$ fixation. Agron. J. 72: 305-308.

van Volkenburg, E., and W. J. Davies. 1977. Leaf anatomy and water relations of plants grown in controlled environments and in the field. Crop Sci. 17: 353-358.

Warington, I. J., M. Peet, D. T. Patterson, J. Bunce, R. M. Haslemore, and H. Hellmers. 1977. Growth and physiological responses of soybean under various thermoperiods. Aust. J. Plant Physiol. 4: 371-380.

# 5

# ESTIMATES OF YIELD REDUCTIONS IN CORN CAUSED BY WATER AND TEMPERATURE STRESS

**R. H. Shaw**

In reviewing the recent literature on yield reductions in corn related to water and temperature stress, one notes the large number of articles on soybeans and the relatively few on corn. And while considerable number of articles on corn involve some aspect of the environment, relatively few relate the environment to the final yield. This review, therefore, will include few very recent articles.

In examining the effect of environmental factors on yield, it is essential that the stage of development be established, inasmuch as the same environmental factor may have different effects at different stages of development. Hanway (1971) has used 10 stages of development, ranging from 0 when the plant tip emerges from the soil to 10 when the plant is physiologically mature (Figure 1). Although I will not use these 10 stages as such, I will refer to them to provide a common base of reference. During the early period of growth after emergence, the growing point of corn is below the soil surface. Under favorable growing conditions, the entire stem primordia and differentiated tassel are formed underground about two weeks after seedling emergence (Kiesselbach, 1949). At stage 1.5, the growing point is at the soil surface, the corn plant is about 30-cm tall, and an average time of three weeks has elapsed since emergence. The period until stage 3 includes the seedling stage and early leaf growth up to five to six weeks after emergence, or early July in the Corn Belt. By the end of that period, the corn plant has set the maximum number of leaves, vascular bundles, and ovules on the major embryonic ear shoot, and the maximum yield potentialities of the plant have been determined. Although this period is of considerable theoretical importance,

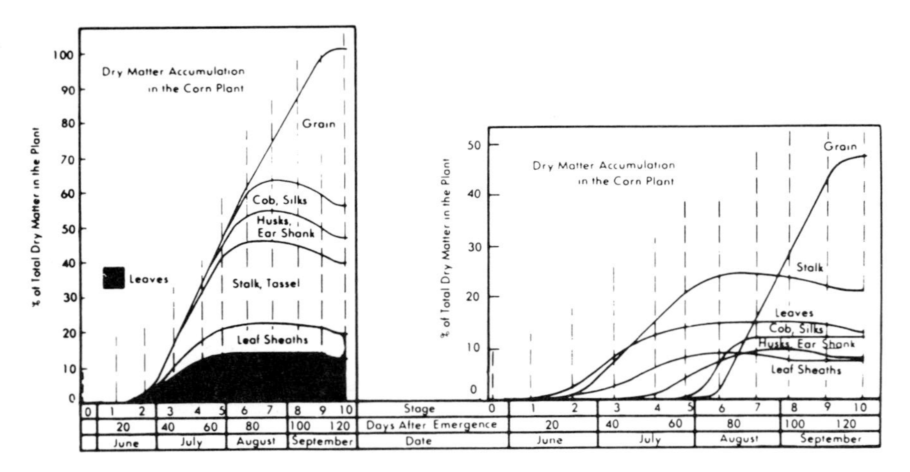


**Figure 1. Dry matter accumulation at various stages of corn development (Hanway, 1971).**

weather in the major corn growing areas seldom is seriously limiting during this period except in affecting stand. Because the corn plant seldom reaches its maximum potentialities under field conditions and has good ability to recover from early-season setbacks, yields cannot be predicted accurately from early season observations.

During stages 3 and 4, leaf area of the plant becomes fully developed and the tip of the tassel emerges at the end of stage 4. The upper internodes of the stalk are elongating rapidly, and the top one or two ears are undergoing rapid enlargement and elongation. Maximum stalk height, stalk diameter and leaf area are reached at the end of stage 4.

Stage 5, which includes tasseling, silking, and pollination, is a very critical stage in the corn plant since the number of fertilized ovules is determined. In the Corn Belt, this stage occurs in mid to late July in an average year.

From stage 5.5 to 8.5, there is a rapid increase in grain weight. In about a 5-week period, almost 85% of the grain dry weight may be produced (Hanway, 1966; Shaw and Loomis, 1950). Near stage 6.5, the maximum unshucked ear size is determined already and the maximum ear size (length x diameter) is reached about 40 days after silking near stage 8. By stage 10, physiological maturity (i.e. the maximum dry weight of the grain) has been attained (Daynard and Duncan, 1969; Rench and Shaw, 1971).

To examine the factors of rainfall and evaporative demand that relate to the adequacy of moisture for the plant, some understanding of the water need is necessary. A number of different experiments are summarized in Figure 2 (Shaw, 1977). Water use is low early in the season and peaks near

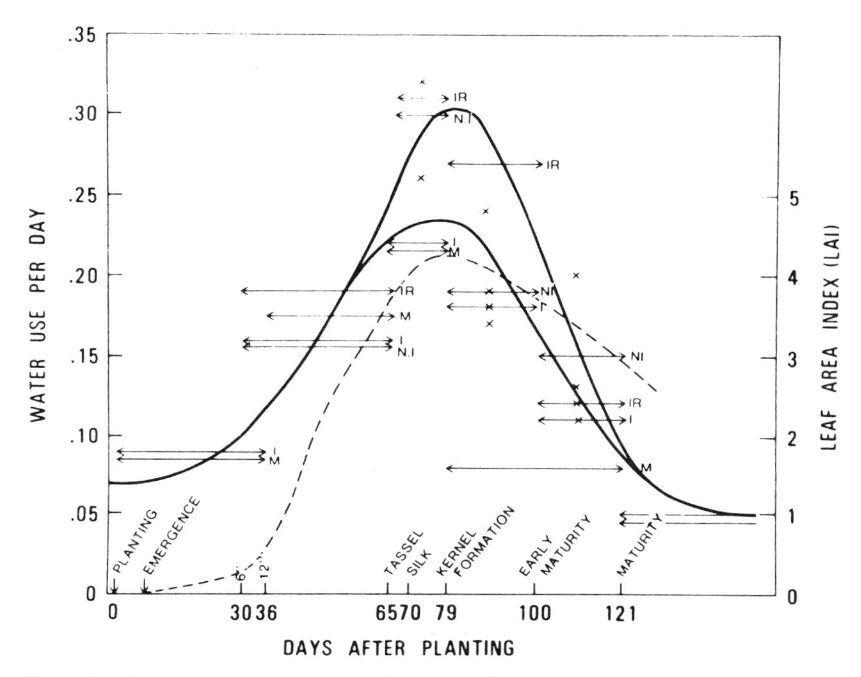


**Figure 2. Water use per day for maize (Shaw, 1977). I = Iowa; M = Minnesota; NI = west Tennessee, not irrigated; IR = west Tennessee, irrigated; x with no line = Washington. Dashed line is Leaf-Area-Index for typical Iowa data.**

the critical tasseling-silking period. The water use during this period varies widely from 5 to 7.5 mm/day (0.2 to 0.3 in./day), depending upon the atmospheric demand, and generally far exceeds that received by rainfall. By the time kernels are forming, water use is decreasing, primarily because of a decreasing demand by the atmosphere.

It is important that the weather factors be considered in terms of normal weather conditions for an area and those conditions necessary to provide optimum plant development and grain production (Figure 3). This point is so obvious that it sometimes is forgotten. Although increased temperatures, for example, may be beneficial in an area where temperatures are too cool, these increased temperatures will be detrimental in an area already at or above the optimum.

## EFFECT OF WEATHER DURING SPECIFIC DEVELOPMENT STAGES ON GRAIN YIELD

### Before Planting

Even before planting, there are factors occurring that may be important in determining final yields. Of particular importance in the western part of

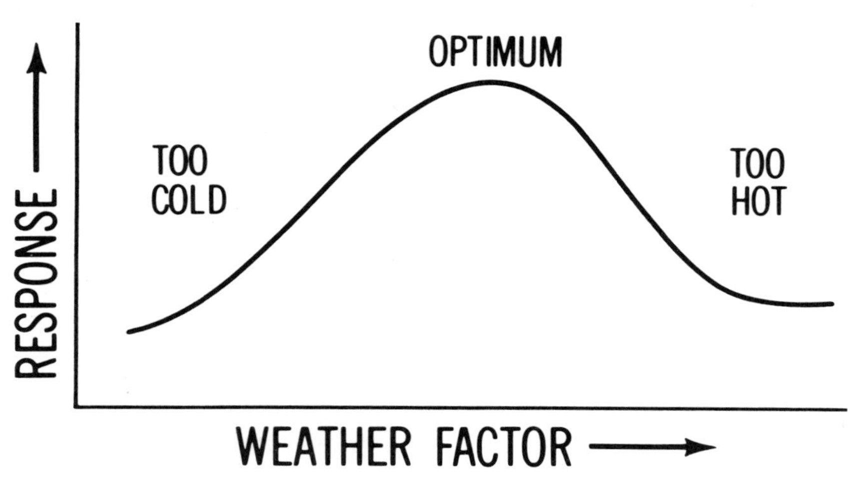


**Figure 3. Response to environmental factors.**

the Corn Belt is the soil-moisture reserve. During the high-use period in July and early August, normal rainfall is not sufficient to meet the needs of the plant, and the moisture reserve is drawn upon. Although from Illinois to the east, field capacity or greater in the spring is the general rule, and the moisture reserve for a given soil relates little to the final yield, in Iowa under conditions that were moderately dry in the summer, the author has measured yield increases of 600 to 1000 kg/ha for each additional 25 mm of soil moisture available in mid-April. Wet conditions during the spring, however, may delay planting. In many areas, delayed planting can reduce yields, as shown for Iowa in Table 1.

### Planting to Emergence

Weather is a major factor in determining the date of planting and the time to emergence. The bulk of corn planting is done when the average air temperature reaches 14 to 16 C. The average soil temperature at planting depth for corn will be near this same value, but over much of the Corn Belt, temperatures generally will tend toward "too cool" rather than "too warm."

**Table 1. Corn yield for various planting dates in Iowa as a percentage of optimum yield.**

| | Planting Date | | | | | | |
|---|---|---|---|---|---|---|---|
| | May 1 | May 10 | May 20 | June 1 | June 10 | June 20 | July 1 |
| Yields as % of optimum | 100 | 98 | 92 | 84 | 71 | 55 | 33 |

Weather conditions can have a significant effect on stand, but other than that, little direct relation to yield is evident.

**Early Vegetative Growth Period**

Relatively little information is available on freezing injury to corn as it is related to final grain yield. Early in the season when the growing point is below the surface, there is little effect from a freeze. The author has observed plants that were killed above ground, yet by silking time could not be separated from plants that emerged in the same field after the freeze. As plants get larger, more damage may be done. Once the growing point is above ground, cold temperatures can kill the plant. When leaves are partly killed by a freeze, the damage may approximate that due to removal of leaf area by hail. Arny and Upper (1973) examined a field where injury occurred when corn plants were 90-cm high. Plants with six or more damaged leaves due to a freeze on 23 June, yielded only 6000 kg/ha, or about 71% of the yield of 8500 kg/ha for undamaged plants. They estimated that comparable hail damage would cause a 10 to 20% yield reduction. This was not a replicated experiment, however, and observations were made only in a damaged field of corn. The author has observed damage to corn resulting from a freeze near silking time. Corn at the silking stage was severely damaged, much of the leaf area was killed, and little yield would have been produced. In an adjacent field where the ears were well formed at the time of the freeze, almost no visible injury to leaves occurred. Because freezes seldom occur at this period of growth, little information is available, but these casual observations indicate that the plant is very susceptible to freezing injury at certain stages of development.

There is little relation between moisture stress during the early vegetative growth period and final yield unless the weather becomes so severe that the crop is killed. If the weather is relatively dry during this period, better root development will occur. Any detrimental effects of early-season stress may be offset later by the better root development that makes more soil moisture available. Salter and Goode (1967) cited data from Russian workers that indicated little or no effect of stress during the early vegetative stage on final yield, possibly because of this deeper rooting. In the current soil-moisture program used in Iowa, moisture extraction is predicted to occur to a depth of 100 cm in a year with a wet May and June, 150 cm in a normal year, and 210 cm in a year with a dry May and June. Soil-moisture measurements substantiate this pattern. Because of the importance of midsummer weather, however, no direct yield relationships are evident for this period.

During the planting and early vegetative growth periods, temperatures are below optimum over most of the Corn Belt. The expansion of young leaves is very sensitive to both water stress and low temperatures for air and soil. Leaf elongation is more sensitive than photosynthesis (Barlow and Boersma, 1976); Barlow et al., 1977; Lehenbauer, 1914), but the effect on grain yield is not very clear.

Willis (1956) found that growth of corn plants in a greenhouse increased linearly with increasing soil temperature from 16 to 24 C. When heating cables were used in the field to warm the soil, he found that yield increased with increasing soil temperatures up to 23C at a depth of 10 cm and then decreased with higher temperatures. A bare, unheated soil with an average temperature of 23 C yielded 7300 kg/ha while a bare, heated soil with an average temperature of 27 C yielded 6800 kg/ha. In this instance, the heating was detrimental. An unheated mulched area with an average temperature of 22 C yielded 7200 kg/ha while a mulched area that was heated to 23.5 C yielded 8000 kg/ha. Burrows (1959), in a mulch-heating study, found that, while mulch created a cooler temperature that was detrimental in Iowa, it would create a beneficial effect later by conserving moisture. He found no change in yield between unheated, bare soil with an average temperature of 20.4 C and a bare soil heated all season to an average temperature of 22.5 C. But he obtained a higher yield on a mulched soil that was not heated than one that was heated, probably because of greater moisture conservation for the unheated area. He pointed out, however, that if moisture is not critical, then effects of soil temperature could be important.

Rykbost et al. (1975) simulated soil heating with electrical cables. The experiment was conducted over four years near Corvallis, Oregon, and heating was for the entire season. Soil temperature was increased 1 C at 5 cm, 4 C at 25 cm, and 6.5 C at 45 cm. The average daily temperature of the soil profile to a depth of 220 cm increased about 10 C. Air temperatures at nearby Eugene, Oregon, average only 16.3 C for the month of June, so temperatures are definitely below optimum. Over the four years of the experiment, grain yield was increased from 8.2 metric tons/ha to 10.5 metric tons/ha, but for the years in which high yields were obtained without heating, there was little or no additional yield response to heating. I interpret this to mean that heating was not needed in warmer years.

Adams (1970) reported an increase in corn yields of 1000 kg/ha at Temple, Texas, by use of a clear plastic mulch. Although the increase was not statistically significant, the effect was believed to result from a faster growth during the first four to six weeks. Moisture conservation effects also may have been present. Average air temperature near planting time was 17.3 C.

Van Wijk et al. (1959) examined the effect of mulching on early-season growth of corn in Iowa, Minnesota, Ohio, and South Carolina. Where the early-season temperatures were below optimum (Iowa, Minnesota, Ohio), mulch reduced the growth rate, but in South Carolina, where early-season temperatures are much warmer, the cooling due to the mulch had little effect. Early season temperatures, therefore, must be interpreted in terms of the normal temperatures for the area.

Correlation studies provide another source of information, but one should recognize that they are indicative of statistical relations, not necessarily cause and effect. Correlations with early season weather and grain yield are generally low. Wallace (1920) estimated that an average May temperature of 15.6 C in central Iowa resulted in average yields. Rose (1936) found that, in areas of the Corn Belt where May temperature averaged below 15 C, temperature was positively correlated with yield, but where May temperature averaged above 16.1 C, temperature was negatively correlated with yield. In the western part of the Corn Belt, rainfall usually is positively correlated with yield, but Wallace (1920) estimated that May rainfall above 12.7 cm caused decreased yields.

The average response of corn yields to summer temperatures in the Corn Belt is shown in Figure 4 (Thompson, 1963). Normal rainfall during June is assumed. The optimum June temperatures of 21 to 23 C were similar

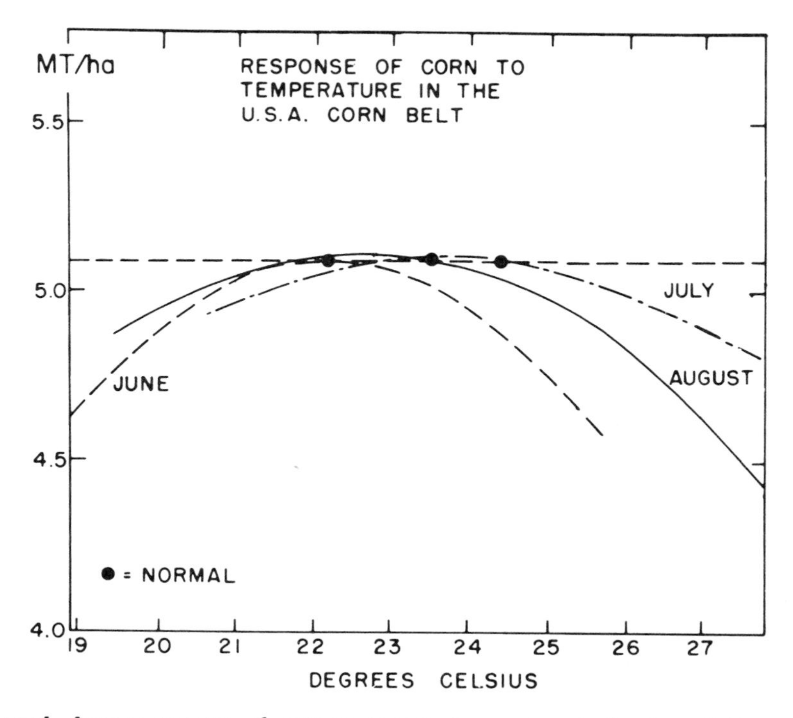


**Figure 4. Average response of corn to summer temperature in five Corn Belt states (Thompson, 1963).**

to those cited by Wallace (1920) and similar to the average temperatures that actually occur for June. The responses for rainfall obtained by Thompson (1966) are shown in Figure 5. Yield response is only slight for rainfall during early vegetative growth in June with a small yield increase for amounts below normal and a small decrease for amounts above normal. In individual years, however, these relationships depend upon the amount of soil moisture present at the start of the season and on the weather occurring after June. Sopher et al. (1973) also found that excess moisture and cool temperatures early in the season were important factors in reducing yields in soils of the South Atlantic Coastal Plains. Shaw (1974) has developed a stress index which is weighted according to the stage of plant development. The range in yields is wide when the index has a low value; i.e., when little stress has occurred. In examining these values, Shaw (1974) found that yields were consistently lower in years when percolation occurred through the 5-foot profile in May and June than in years when no percolation occurred. The years with excess moisture are designated in Figure 6 for a station in central Iowa. This indicates that excess moisture can be limiting to yields in many years, although the

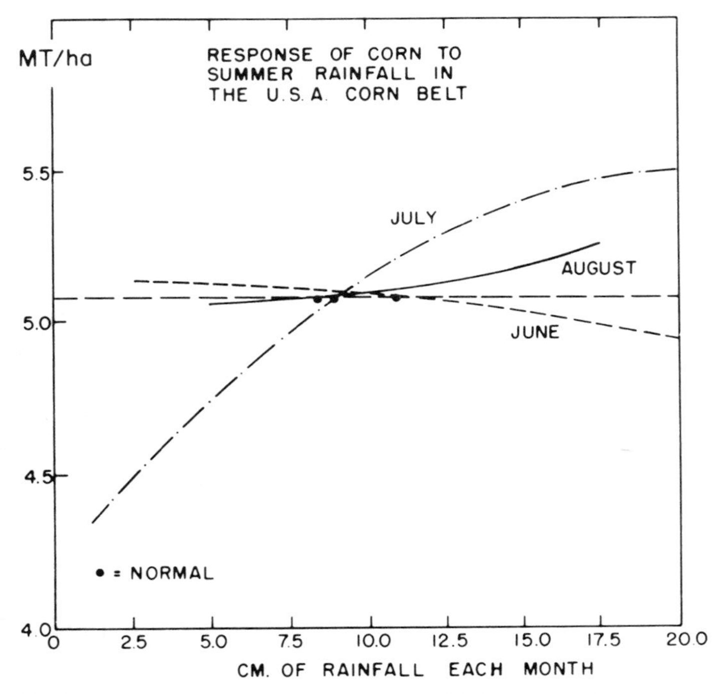


**Figure 5. Average response of corn to summer rainfall in five Corn Belt States (Thompson, 1966).**

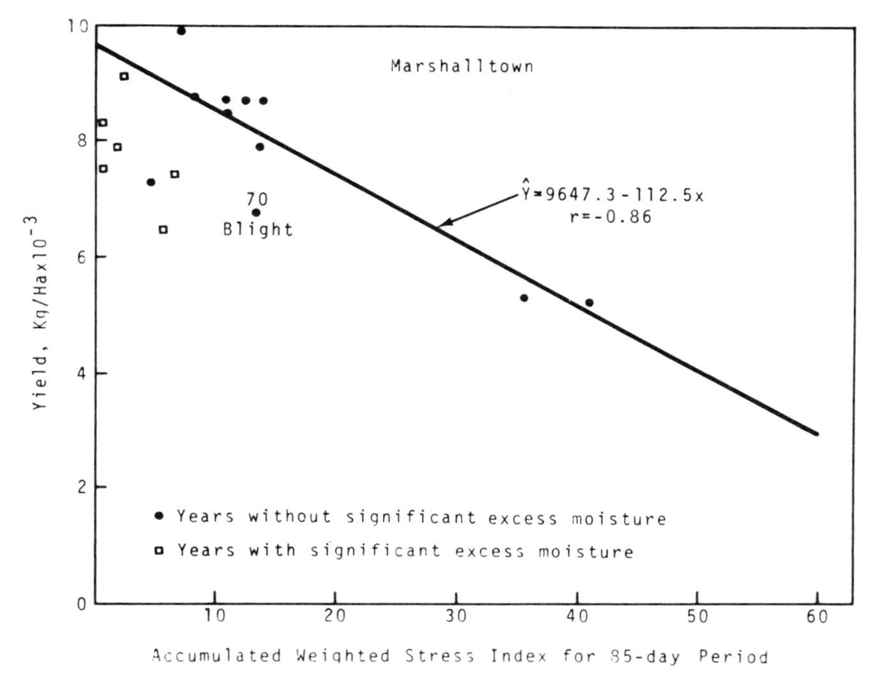


**Figure 6. Relation between weighted stress index and corn yield near Marshalltown, Iowa.**

major yield reductions are attributable to moisture deficits. The reduction associated with excess moisture could be a result of several factors, such as late planting, poor rooting, nitrogen deficiency, etc.

### Late Vegetative Growth Up to Tasseling

In much of the Corn Belt, the late vegetative growth period would include most of July, although the early vegetative growth period extends into early July in some areas. Tasseling and silking may occur in mid- to late July, depending upon the season, and in a very late season silking will extend into August. July temperatures generally are negatively correlated with yield. Thompson (1962, 1963, and 1966) found the average July temperature (assuming July rainfall and other months' temperature and rainfall normal) of about 24 C (Figure 4) to be somewhat above optimum. Temperatures above normal reduce the yield sharply. That the optimum July temperature is particularly affected by the July rainfall (Figure 7) emphasizes the interaction between rainfall and temperature present in most periods. With a dry July, temperatures much below normal are optimum, while with a wet July,

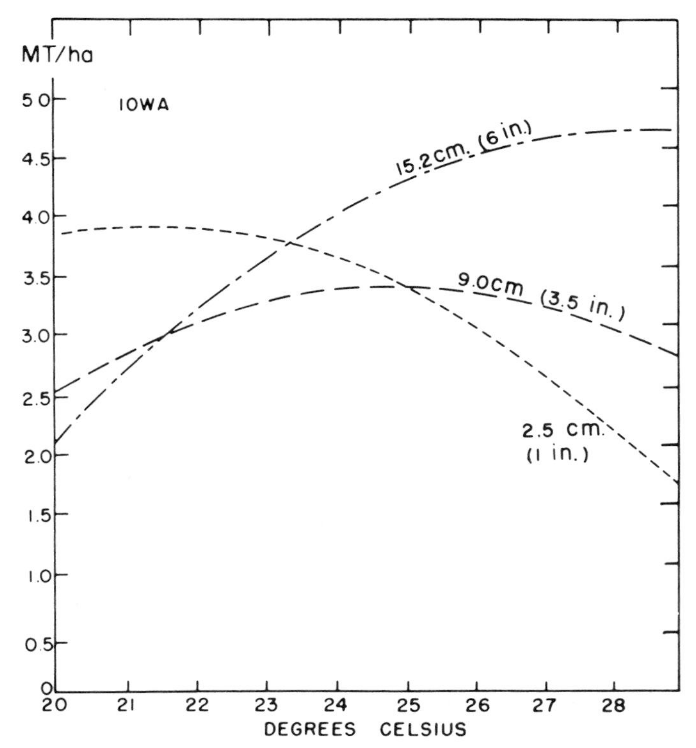


**Figure 7. Relation of corn yield to July temperature with different levels of rainfall in Iowa (Thompson, 1963).**

temperatures above optimum do not depress the yield. Statistically, one can examine the effects of rainfall and temperature separately, but it is almost impossible to separate them under field conditions.

Conducting an experiment with irrigation does not necessarily remove all moisture stress. I mentioned this recently to a plant physiologist who readily agreed. He had conducted his Ph.D. research in North Carolina in greenhouses in the summer. His statement was, "My plants growing in nutrient cultures wilted almost every day." Under conditions of very high temperature, some moisture stress occurs frequently because of the high transpiration demands. Very high temperatures as experienced in Kansas in 1980, no doubt combined with excessively high evapotranspiration rates, would not be counterbalanced even with irrigation.

Moisture stress during the period of late vegetative growth to tasseling will cause yield reductions. A number of researchers (Claassen and Shaw, 1970a; Denmead and Shaw, 1960; Mallett, 1972; Robins and Domingo, 1953;

Wilson, 1968) have worked on this problem. Typically, corn has been grown with a restricted rooting area and kept well watered except when stress was imposed. Their results are summarized in Figure 8 (Shaw 1977). The shaded area covers the range of most yield reductions reported and the line through the shaded area represents the average yield reduction. This figure is applicable to periods of severe stress of a few days duration. The stress imposed by different researchers varied in intensity, but generally was of four to six days in duration. The intensity of the stress depended to some extent upon the atmospheric demand for evapotranspiration occurring during the experiment. Our research has shown that it is very easy to impose a high degree of stress when atmospheric demand is high, but under low-demand conditions severe stress is difficult to impose. The degree of stress also is affected by the fertility level. If both moisture and fertility stresses occur, as is typical of field conditions, the yield reductions are greater. If fertility stress does not occur, which may happen in experiments with short-duration stress periods and limited rooting volume, the yield reduction will be less. Under natural field conditions, a moisture stress usually is combined with some degree of fertility stress because moisture usually is limiting at the shallower soil depths where plant nutrients often are more concentrated. During the late vegetative stage, a moderately severe stress may reduce yield near 3% per day. Shaw (1978) has found indications from his soil-moisture program that very severe stress for several consecutive 5-day periods seems to have an increasingly detrimental effect and, if it occurs just before silking, can introduce greater yield reductions by interfering with pollination.

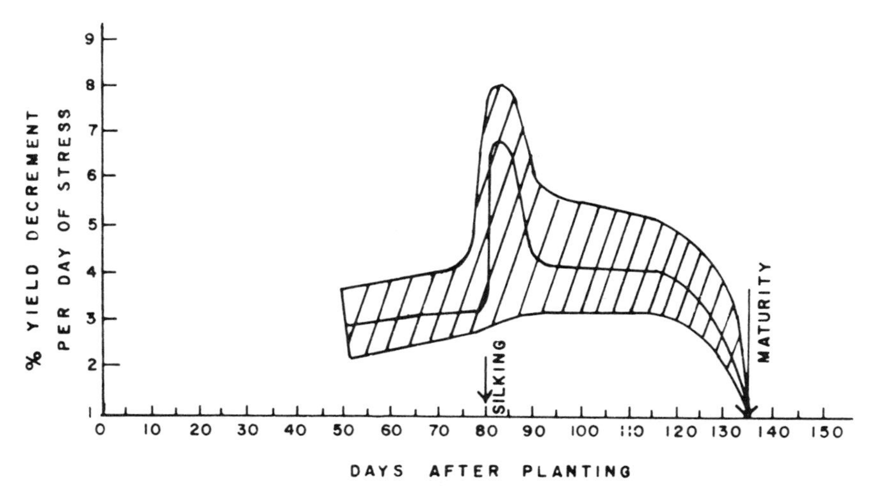


**Figure 8. Schematic diagram of relationship between age of crop and percentage yield decrement due to one day of moisture stress (Shaw, 1977).**

Excess moisture also can affect yield, but the reduction is closely tied to the nitrogen level in the soil. Ritter and Beer (1969) surface-flooded an area of corn with several inches of water when the corn was 76-cm high. The soil involved was one in which permeability increased with depth. At a low level of nitrogen fertilization, 24 hours of flooding reduced yields 14%, and the reduction increased to 30% with 96 hours of flooding. With a high level of nitrogen fertilization, little yield reduction occurred even with 96 hours of flooding. When flooded near the time of silking, no reduction in yield occurred with the high level of nitrogen fertilization, but a 16% reduction in yield occurred with 96 hours of flooding at the low nitrogen level.

**Tasseling, Silking, and Pollination Period**

The tasseling, silking, and pollination period, when number of fertilized ovules is being determined, is the most sensitive period to stress. The yield reduction depends upon the intensity of the stress and the stage of development at which it occurs. If stress occurring just before tasseling is severe enough, the tassels will not emerge, and pollination will not take place. Claassen and Shaw (1970b) found that if a water stress was imposed when 6% of the plants had silked, yield was reduced only 3% per day but, if applied when 75% of the plants had silked, the reduction was 7% per day. On the basis of Figure 8, this seems to be typical for conditions in which moisture stress and fertility stress occur. Shaw (1949), Du Plessis and Dijkhuis (1967), and Berbecel and Eftimescu (1973) found that stress before and during flowering caused the time between pollen shed and silking to be delayed. With severe stress, silking may be delayed until after all, or much, of the pollen has shed, increasing the number of barren stalks and poorly filled ears. My work has indicated that very severe stress during a 10-day period centered around silking will result in a complete crop failure. Barbecel and Eftimescu (1973) found that maximum temperatures above 32 C around tasseling and silking resulted in higher rates of kernel abortion. This may well have been a moisture-stress condition. If too many kernels are aborted, the total sink size may limit yield, but corn has a much greater potential (size x kernel number) than usually is reached.

**Fertilization to Physiological Maturity**

Significant yield reductions still can occur during the grain-filling stage. Yield reductions of more than 4% per day can occur for three to four weeks after pollination, but then, as the crop moves closer to maturity, the yield reduction will decrease (Figure 8). Greater reductions will occur if a fertility stress is confounded with a moisture stress as typically occurs under field conditions.

In the Corn Belt, the first part of the grain-filling period is late July or early August. Average temperatures during August are higher than those associated with optimum yields in the Corn Belt (Wallace, 1920; Rose, 1936; Davis and Harrell, 1941; Kiesselbach, 1950; Thompson, 1963). Thompson (Figure 4) estimated that the optimum temperature is about 1 C below the normal temperature, provided that other factors are normal. Bondavalli et al. (1970) found rainfall to be more important than temperature during the first half of August, but temperature to be more important during the second half. Peters et al. (1971), in an unreplicated experiment, found that a night air temperature of 29.4 C for the period from flowering to maturity reduced corn yield almost 40% compared with the cooler temperature of 16.6 C. The higher temperature may have induced a water stress. Normal nighttime minimum temperatures in August are 17 to 18 C in central Iowa. Nighttime temperatures definitely may be too high for optimum corn yields during warm, humid periods in the Corn Belt. Thompson's analysis (1963) showed relatively little effect of rainfall in August. In a dry year, with low soil moisture reserves, considerable stress can occur unless there are adequate rains. In a wet year, or if soil moisture reserves are high, too much rain during August is not beneficial, and although it may not reduce yield directly, it could reduce yield by contributing to harvesting problems. This factor may be particularly important on poorly drained soils. Cooler temperatures in late July and August will tend to reduce the atmospheric demand on the crop. Less stress will occur, and higher yields will result. In some respects, cool temperatures can substitute for some lack of rainfall or soil moisture. The hot temperatures in August generally cause a high demand period for water and high temperatures typically indicate a stress period. Although regression analyses often show temperature in August as more important than rainfall, yield reductions often occur in August due to moisture stress.

In September, the crop is approaching maturity, and by the end of the month, much of the corn is physiologically mature. In a dry year, moisture stress in September still can reduce yields. A wet September may have two effects. First, it can cause a significant increase in the soil moisture reserves for the next season. Second, if September is too wet, harvesting problems may result. Once the crop is physiologically mature, weather will affect yield only if it increases harvesting losses.

A freeze before physiological maturity can reduce the final yield by killing the leaves so that no further photosynthesis can take place. The extent of the reduction in yield will depend upon when the freeze occurs relative to physiological maturity, or stage 10 (Figure 1). Some dry matter still can be translocated from the stalk into the grain as maturity is approached (Figure 1). An early freeze or a late maturing crop may result in freeze-damage to corn. The late maturation is largely associated with weather events before silking.

## SEASONAL EFFECTS OF WEATHER

Obviously, the total effect of weather on yield is an integration of what has happened over the entire season. The major effects at specific stages of growth have been covered in earlier sections. Additional data involve total seasonal effects.

Thompson (personal communication) found that accumulated degrees above 32.2 C were related to yield. He accumulated the degrees above 32.2 C for maximum temperature and found that for each 5.6 C accumulated, yields of corn were reduced by 62.7 kg/ha (1 bu/acre). Schwab et al. (1958) found a high negative relationship between the number of days with temperatures of 32.2 C or higher and corn yields on either irrigated or nonirrigated sandy soils. The effect may well have resulted more from moisture stress than temperature. Recent results by Doyle Peters (University of Illinois, personal communication) may help to explain the temperature and moisture-stress relation. In work that he is conducting on photosynthesis, he has found that stomates close either with moisture stress or with high temperatures of about 35 to 38 C under certain conditions. Under conditions of high temperature and cloudiness, the stomates close even when the leaf is not under moisture stress. Under these cloudy conditions, leaf and air temperature are approximately the same because of poor radiational cooling. With clear skies and a dry climate, however, there is little or no closing of the stomates because the higher radiational cooling keeps the leaf temperature several degrees below air temperature.

Duncan et al. (1973) grew corn at Davis and Greenfield, California, and at Lexington, Kentucky, all of which are on about the same latitude. The corn was irrigated as necessary to provide adequate moisture. Yields were highest at Davis, which had the greatest solar radiation, the highest daytime temperatures, and the intermediate night temperatures. Yields were lowest at Lexington, which had the least solar radiation, moderately high daytime temperatures, and the highest nighttime temperatures. One could infer that both high solar radiation and low nighttime temperatures were beneficial. High daytime temperatures were not detrimental under the clear-sky conditions at Davis.

In controlled experiments, radiation usually shows positive effects. Pendleton et al. (1967) found that increased radiation increased corn yields, and Duncan and Hesketh (1968) found that shading reduced yields. Regression analyses of yield and climatic data frequently have shown negative effects of increased radiation. Kiesselbach (1950) found that a seasonal 1% increase in sunshine reduced grain yield by 96 kg/ha while an increase of 1 g · cal in the seasonal mean daily total radiation reduced the yield by 24 kg/ha. Radiation is confounded with rainfall and temperature, and the net effect will depend upon the levels of these factors. If high radiation is related to greater moisture stress, a negative correlation should be expected.

## NOTES

R. H. Shaw, Curtiss Distinguished Professor, Agricultural Climatology, Agronomy Department, Iowa State University, Ames, Iowa 50011.

Journal Paper No. J-10020 of the Iowa Agriculture and Home Economics Experiment Station, Ames, Iowa 50011. Project No. 2290.

## LITERATURE CITED

Adams, J. E. 1970. Effects of mulches and bed configuration. II. Soil temperature and yield responses of grain sorghum and corn. Agron. J. 62: 785-790.

Arny, D. C., and C. D. Upper. 1973. Example of the effects of early season frost damage on yield of corn. Crop Sci. 13: 760-761.

Barlow, E. W. R., and L. Boersma. 1976. Interaction between leaf elongation, photosynthesis, and carbohydrate levels of water-stressed corn seedlings. Agron. J. 68: 923-926.

Barlow, E. W. R., L. Boersma, and J. L. Young. 1977. Photosynthesis, transpiration, and leaf elongation in corn seedlings at suboptimal soil temperatures. Agron. J. 69: 95-100.

Berbecel, O., and M. Eftimescu. 1973. Effect of agrometeorological conditions on maize growth and development. Institute of Meteorology and Hydrology, Bucharest, Romania. p. 10-31. (English translation).

Bondavalli, B., D. Colyer, and E. M. Kroth. 1970. Effects of weather, nitrogen and population on corn yield response. Agron. J. 62: 669-672.

Burrows, W. C. 1959. Mulch influence on soil temperature and corn growth. Ph.D. Dissertation. Iowa State University. Univ. Microfilms, Ann Arbor, Michigan.

Claassen, M. M., and R. H. Shaw. 1970a. Water deficit effects on corn. I. Vegetative components. Agron. J. 62: 649-652.

Claassen, M. M., and R. H. Shaw. 1970b. Water deficit effects on corn. II. Grain components. Agron. J. 62: 652-655.

Davis, F. E., and G. D. Harrell. 1941. Relation of weather and its distribution to corn yields. U.S. Dept. Agric. Tech. Bull. 806.

Daynard, T. B., and W. G. Duncan. 1969. The black layer and grain maturity in corn. Crop Sci. 9: 473-476.

Denmead, O. T., and R. H. Shaw. 1960. The effects of soil moisture stress at different stages of growth on the development and yield of corn. Agron. J. 52: 272-274.

Duncan, W. G., and J. D. Hesketh. 1968. Net photosynthetic rates, relative leaf growth rates and leaf numbers of 22 races of maize grown at eight temperatures. Crop Sci. 8: 370-374.

Duncan, W. G., D. L. Shaver, and W. A. Williams. 1973. Insolation and temperature effects on maize growth and yield. Crop Sci. 13: 187-191.

Du Plessis, D. P., and F. J. Dinkhuis. 1967. The influence of the time lag between pollen shedding and silking on the yield of maize. S. Afr. J. Agric. Sci. 10: 667-674.

Hanway, D. G. 1966. Irrigation. p. 155-176. In W. H. Pierre, S. A. Aldrich, and W. P. Martin (eds.) Advances in corn production: Principles and practices. Iowa State Univ. Press, Ames.

Hanway, J. J. 1971. How a corn plant develops. Iowa Coop Ext. Serv., Spec. Rep. 48 (rev.).

Kiesselbach, T. A. 1949. The structure and reproduction of corn. Nebraska Agric. Exp. Stn. Res. Bull. 161.

Kiesselbach, T. A. 1950. Progressive development and seasonal variation of the corn crop. Nebraska Agric. Exp. Stn. Res. Bull. 166.

Lehenbauer, P. A. 1914. Growth of maize seedlings in relation to temperature. Physiol. Res. 1: 247-288.

Mallett, J. B. 1972. The use of climatic data for maize yield predictions. Ph.D. Dissertation. Dept. of Crop Sci., Univ. of Natal, Pietermaritzburg, South Africa.

Pendleton, J. W., D. B. Egli, and D. B. Peters. 1967. Response of *Zea mays* L. to a "light rich" field environment. Agron. J. 59: 395-397.

Peters, D. B., J. W. Pendleton, R. H. Hageman, and C. M. Brown. 1971. Effect of night air temperature on grain yield of corn, wheat and soybeans. Agron. J. 63: 809.

Rench, W. E., and R. H. Shaw. 1971. Black layer development in corn. Agron. J. 63: 303-305.

Ritter, W. F., and C. E. Beer. 1969. Yield reduction by controlled flooding of corn. Trans. ASAE 12: 46-50.

Robins, J. S., and C. E. Domingo. 1953. Some effects of severe soil moisture deficits at specific growth stages in corn. Agron. J. 45: 618-621.

Rose, J. K. 1936. Corn yield and climate in the Corn Belt. Geogr. Rev. 26: 88-102.

Rykbost, K. A., L. Boersma, H. J. Mach, and W. E. Schmisseur. 1975. Yield response to soil warming. Agronomic crops. Agron. J. 67: 733-738.

Salter, P. J., and J. E. Goode. 1967. Crop responses to water at different stages of growth. Commonwealth Agric. Bur. Res. Rev. 2.

Schwab, G. D., W. D. Shrader, P. R. Nixon, and R. H. Shaw. 1958. Research on irrigation of corn and soybeans at Conesville and Ankeny, Iowa, 1951-1955. Iowa Agric. Home Econ. Exp. Stn. Res. Bull. 458.

Shaw, R. H. 1949. Studies on corn phenology and maturity in Iowa. Ph.D. Dissertation. Iowa State Univ., Ames.

Shaw, R. H. 1974. A weighted moisture-stress index for corn in Iowa. Iowa State J. Res. 48: 101-114.

Shaw, R. H. 1977. Water use and requirements of maize. A review. p. 119-134. In Agrometeorology of the maize crop. Publication 481. World Meteorological Organization, Geneva.

Shaw, R. H. 1978. Calculation of soil moisture and stress conditions in 1976 and 1977. Iowa State J. Res. 53: 119-127.

Shaw, R. H., and W. E. Loomis. 1950. Bases for the prediction of corn yields. Plant Physiol. 25: 225-244.

Sopher, C. D., R. J. McCracken, and D. D. Mason. 1973. Relationships between drouth and corn yields on selected south Atlantic coastal plain soils. Agron. J. 65: 351-354.

Thompson, L. M. 1962. An evaluation of weather factors in the production of corn. Cen. Agric. Econ. Adjust. Rep. 12T, Iowa State Univ., Ames.

Thompson, L. M. 1963. Weather and technology in the production of corn and soybeans. Cen. Agric. Econ. Dev. Rep. 17, Iowa State Univ., Ames.

Thompson, L. M. 1966. Weather variability and the need for a food reserve. Cen. Agric. Econ. Dev. Rep. 26, Iowa State Univ., Ames.

Van Wijk, W. R., W. E. Larson, and W. C. Burrows. 1959. Soil temperature and the early growth of corn from mulches and unmulched soil. Soil Sci. Soc. Am. Proc. 23: 428-434.

Wallace, H. A. 1920. Mathematical inquiry into the effect of weather on corn. Monthly Weather Rev. 43: 439-446.

Willis, W. O. 1956. Soil temperature, mulches and corn growth. Ph.D. Dissertation. Iowa State Univ., Ames.

Wilson, J. H. 1968. Water relations of maize. Pt. 1. Effects of severe soil moisture stress imposed at different stages of growth on grain yields of maize. Rhodesian J. Agric. Res. 6: 103-105.

# 6

# REDUCTION IN YIELD OF FOREST AND FRUIT TREES BY WATER AND TEMPERATURE STRESS

T. T. Kozlowski

Yields of forest and fruit trees rarely attain their maximum because of the limitations imposed on plant growth by a variety of environmental stresses. It is difficult, however, to determine precisely how much a single environmental stress factor reduces yield. Ultimate yield, whether measured as production of wood, seeds, or fruits, reflects an integrated response to changes in physiological processes that are regulated over a long time span by a complex of fluctuating and interacting environmental factors. The relative importance of individual environmental stress factors on plant growth changes with time. For example, correlations of cambial growth of *Quercus ellipsoidalis* with air temperature decreased late in the growing season as soil moisture was progressively depleted and growth was limited largely by plant water deficits (Kozlowski et al., 1962).

Growth is influenced by both the duration and severity of various environmental stresses. Sometimes the effects of environmental stress on yield are evident immediately; at other times they may not be apparent for a very long time. The lag in response to stress varies greatly with the aspect of yield measured, the species, and the stage of plant growth at which the stress occurs. The impact of environmental stress on growth is also complicated by environmental preconditioning (Rowe, 1964).

While recognizing the large number of stresses affecting the yield of wood plants, this chapter will emphasize the importance of water and temperature stresses which together account for major reductions in yield of forest and fruit trees.

## WATER STRESS

Tree growth is reduced by either too little or too much water, but mostly by the former. Arid conditions, which prevail over about a third of the world's land area, severely inhibit tree growth and, over most of the remaining land area, growth is reduced by periodic droughts. The amounts of such growth losses usually are not realized because few data are available to show how much more growth would occur if trees had favorable water supplies throughout the growing season. Metabolism and growth of plants in drying soil are influenced by even mild water deficits in plant tissues and such deficits usually occur long before drying soil approaches the permanent wilting percentage. Growth limitations caused by internal water deficits in plants frequently are overlooked because growth reduction is attributed to other factors such as plant competition, disease, or insect pests. For example, reduced growth resulting from weed competition commonly involves desiccation caused by competition for water and minerals. Root diseases and insect injury to roots often interfere with absorption of water and cause desiccation of shoots. The drying of plant tops following xylem plugging associated with vascular wilt disease often plays a major role in foliage wilt and the ultimate death of infected plants (Kozlowski et al., 1962). Water deficits may also predispose host plants to attacks by certain fungus pathogens and insects (Vite′, 1961; West, 1979).

Trees produce highest yields in regions of abundant and uniformly distributed rainfall, but even there they often develop midday water deficits caused by absorption lagging behind transpiration. This is shown by midday decreases in leaf moisture content and in leaf water potential ($\psi$), stomatal closure, and shrinkage of leaves, stems, roots, and reproductive structures (Kramer and Kozlowski, 1979). These transient water deficits in leaves may not reduce yield appreciably because the leaves usually rehydrate at night. However, persistent plant water deficits during rainless periods greatly decrease yield. Water deficits often are especially severe in recently transplanted trees. Even in the relatively humid climate of Japan, death of seedlings occurs in sandy soil and the relative growth rate and net assimilation rate of cultivated mulberry are reduced by summer droughts (Tazaki et al., 1980).

### Water Stress and Seed Germination

Water stress often limits seed germination and seedling establishment. Most seeds must imbibe two to three times their dry weight to initiate germination. Thereafter, the young seedlings require a continuously increasing supply of water as their transpiring capacity increases. Seedlings in the cotyledon stage of development are especially sensitive to water stress and droughts often lead to spectacular losses of seedlings. For example, germination of *Eucalyptus* seeds under favorable conditions in Australia may exceed 80%, but desiccation often reduces survival to less than 1% (Jacobs, 1955).

Most viable seeds with permeable seed coats can germinate in soil at field moisture capacity, but as soils dry both the rate of germination and final germination percentage decline rapidly. For example, 88% of citrus seeds germinated with no soil water deficit (0 bars) but only 23 and 3% germinated at -2.3 and -4.7 bars, respectively (Kaufmann, 1969). According to Satoo (1966) germination of *Chamaecyparis obtusa* seed decreased about 6% for each decrease of 1 bar in soil water potential, germination of *Pinus densiflora* 25%, and that of *P. thunbergii* 1.5%.

### Water Stress and Vegetative Growth

The first and most direct effect of water stress on growth is through reduction in cell enlargement. The smaller leaf area and the closure of stomata combine to reduce the amount of photosynthate available for growth. In many temperate zone species the rate of photosynthesis decreases when leaf $\psi$ decreases to minus a few bars and becomes negligible at zero turgor. When droughted trees are irrigated, photosynthesis does not always return to normal (Figure 1). Following irrigation, transpiration rates usually recover faster than photosynthetic rates, indicating a lag in recovery of the photosynthetic machinery. Failure of photosynthesis to recover rapidly in previously droughted plants may indicate injury to chloroplasts, damage to stomata, and/or death of roots. Some stomata may be permanently injured by drought and will not reopen even if full turgor is regained after irrigation.

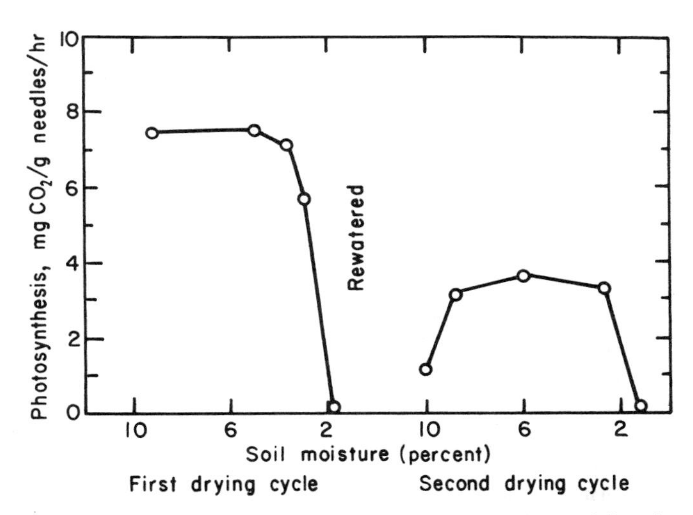


**Figure 1. Effect of two soil drying cycles on net photosynthesis of *Pseudotsuga menziesii* seedlings. From Zavitkovski and Ferrell (1970).**

Some trees respond to drought by early leaf abscission following changes in: (1) amounts and balances of several growth hormones, and (2) synthesis and activity of enzymes that hydrolyze the middle lamella between cells of the abscission layer. In other trees the leaves simply wither and dry during drought. Shedding of leaves during dry summers also is well known (Kozlowski, 1973).

**Shoot Growth**

Plant water deficits decrease growth of shoots by inhibiting initiation of leaf primordia, expansion of leaves, and internodal elongation. The amount of reduction in growth caused by water deficits varies among species and with severity, duration, and time of occurrence of droughts.

Many temperate zone trees have "fixed" growth (e.g. northern pines, *Picea, Fagus*). The winter buds contain all of the leaves that will appear during the following growing season. In this type the growing season is short (Figure 2) and growth in a given season is strongly influenced by rainfall during bud formation in the preceding year. In species with "free" growth

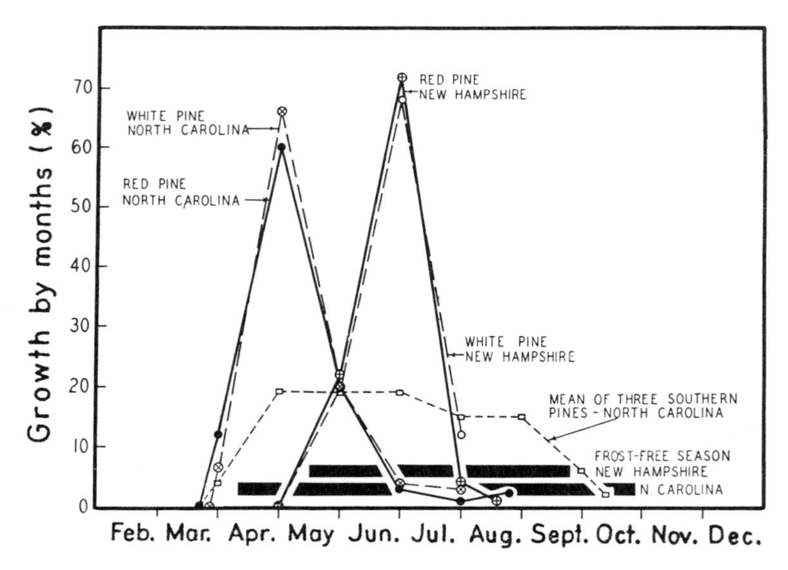


**Figure 2. Variations in seasonal duration of shoot elongation of northern pines (red and eastern white pines) in North Carolina and New Hampshire and of three southern pines (loblolly, shortleaf, and slash pines) in North Carolina. The two northern pines have preformed shoots (fixed growth) and exhibit only one annual growth flush whereas the southern pines grow in recurrent flushes. From Kramer (1943).**

(e.g. *Populus, Betula, Liriodendron*) part of the leaves are preformed, but new leaves are added during the growing season. Thus, growth may be affected by rainfall during both the preceding and the current year. In a third type, characteristic of our southern pines and many tropical trees, shoot expansion occurs in a series of "flushes." For example, in *Pinus taeda* the winter buds expand rapidly in the spring, then another bud forms at the stem tip and opens and expands. This may occur from two or three to five or six times during the growing season. The long growing season of trees with recurrent flushing makes them particularly susceptible to late summer droughts.

In species with fixed growth favorable water supplies during the time of bud formation result in large buds with numerous leaves. In the following year such buds expand into much longer shoots with many more leaves than do small buds formed under drought conditions. In 20-year-old *Pinus strobus* trees mild droughts during midsummer reduced the number of preformed needles by 40%. A drought during the next year resulted in smaller needles and shorter internodes (Lotan and Zahner, 1963). In *Pinus resinosa* shoot length was highly correlated with the size of the winter bud (Kozlowski et al., 1973). Clements (1970) showed that irrgation of young *Pinus resinosa* late in the summer induced the formation of large buds, which expanded into long shoots bearing many needles the following year. Irrigation in the spring of that year had little effect on shoot length. Some studies indicate, however, that shoot growth of species with fixed growth is influenced by water deficits during both the year of bud formation and the year of bud expansion (Zahner and Stage, 1966).

Droughts in late summer often have little effect on current-year shoot elongation of species having fixed growth because their shoot elongation is completed by mid-summer. Conversely, shoot expansion of species exhibiting free growth or recurrent flushing patterns usually is greatly reduced by droughts in late summer. As mentioned earlier, in the latter two groups shoot expansion and production of new leaves continue late into the summer. Differences in effects of water deficits on shoot growth of pines are shown by studies of *Pinus resinosa* having fixed growth (Lotan and Zahner, 1963) and *Pinus taeda* with recurrently flushing growth (Zahner, 1962). Height growth of *Pinus resinosa* was affected little in the same year as a late summer drought whereas height growth of *Pinus taeda* was reduced about 50% (Figure 3) and the number of growth flushes was reduced from four to two.

**Cambial Activity**

Activity of the cambium is very sensitive to water stress. In gymnosperms up to 90% of the variation in xylem increment has been attributed to water deficits in arid climates and up to 80% in humid climates (Zahner, 1968).

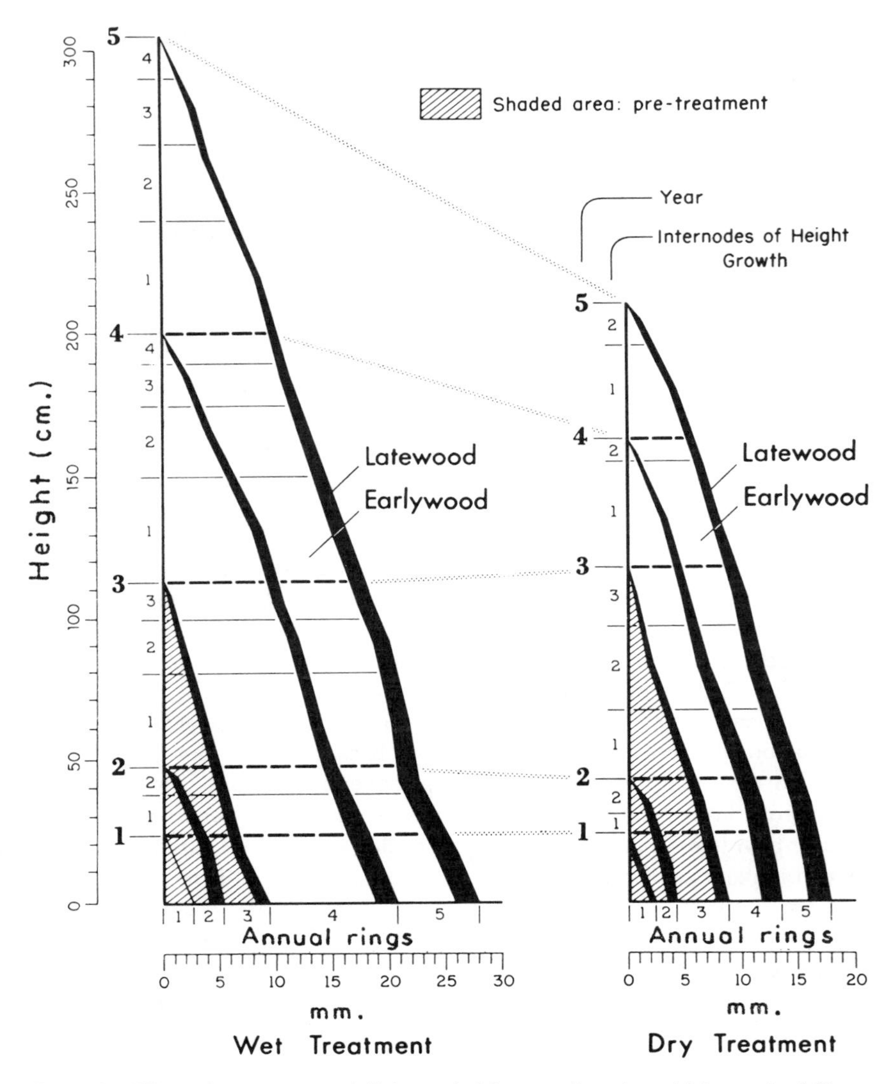


**Figure 3.** Effect of severe water deficits on height growth and cambial growth of *Pinus taeda* seedlings. Wet treatment trees were irrigated throughout the year by surface watering of soil to field capacity. Dry treatment trees received no water from May 1 to the end of October. From Zahner (1962).

The manner by which cambial activity is affected is complex and appears to involve both indirect and direct effects (Kramer and Kozlowski, 1979). Cambial activity is inhibited directly by water deficits when they become severe enough to dehydrate tissues containing mother cells and undifferentiated cambial derivatives and cause loss of turgor (Whitmore and Zahner, 1967). The indirect effects of water deficits on cambial activity are exerted by reduced synthesis and downward transport in the stem of carbohydrates and hormonal growth regulators.

Year to year variation in water supply of temperate-zone trees seldom plays an important role in initiation of cambial activity because soil water is seldom limiting in the spring. However, later in the growing season the water balance of a tree has an important regulatory role on cambial activity and wood production. The amount of xylem increment, its distribution along the stem, seasonal duration of wood production, time of initiation of latewood, and duration of latewood production are very responsive to availability of water at various times during the growing season (Kramer and Kozlowski, 1979).

As internal water deficits develop in trees, wood production slows or stops, but it usually accelerates or resumes with the next rain. In Canada diameter growth of some, but not all, tree species was about half as great in dry years as in wet ones (Figure 4). The difference in growth reduction caused by drought was related to soil water supply on different sites. The importance of water supply to wood production was emphasized by Zahner (1962) who showed that annual rings of irrigated *Pinus taeda* trees were more than twice as wide as rings of trees undergoing drought (Figure 3).

Seasonal cambial activity continues for a much longer time in wet years than in dry ones. For example, during a dry summer in Arkansas, diameter growth of *Pinus taeda* stopped by August. However, it resumed during September which was rainy, with the result that about a third of the annual ring was produced during September and October (Zahner, 1958). The seasonal duration of cambial activity is influenced by crown development as well as by water supply (Table 1). When several droughts occur during a single growing season, multiple growth rings often form in trees. Severe water deficits may also prevent any xylem from forming in the lower stem (Kramer and Kozlowski, 1979).

In addition to inhibiting cambial activity while a drought is in progress, water deficits have carry-over effects to the subsequent years. Such lag effects result from the influence of water supply on crown development and its physiological activity. The rate of cambial activity, which depends on a downward flow of carbohydrates and hormonal growth regulators from the crown, varies with leaf area. Since leaf development of species with fixed growth depends to a considerable extent on the weather during the previous year

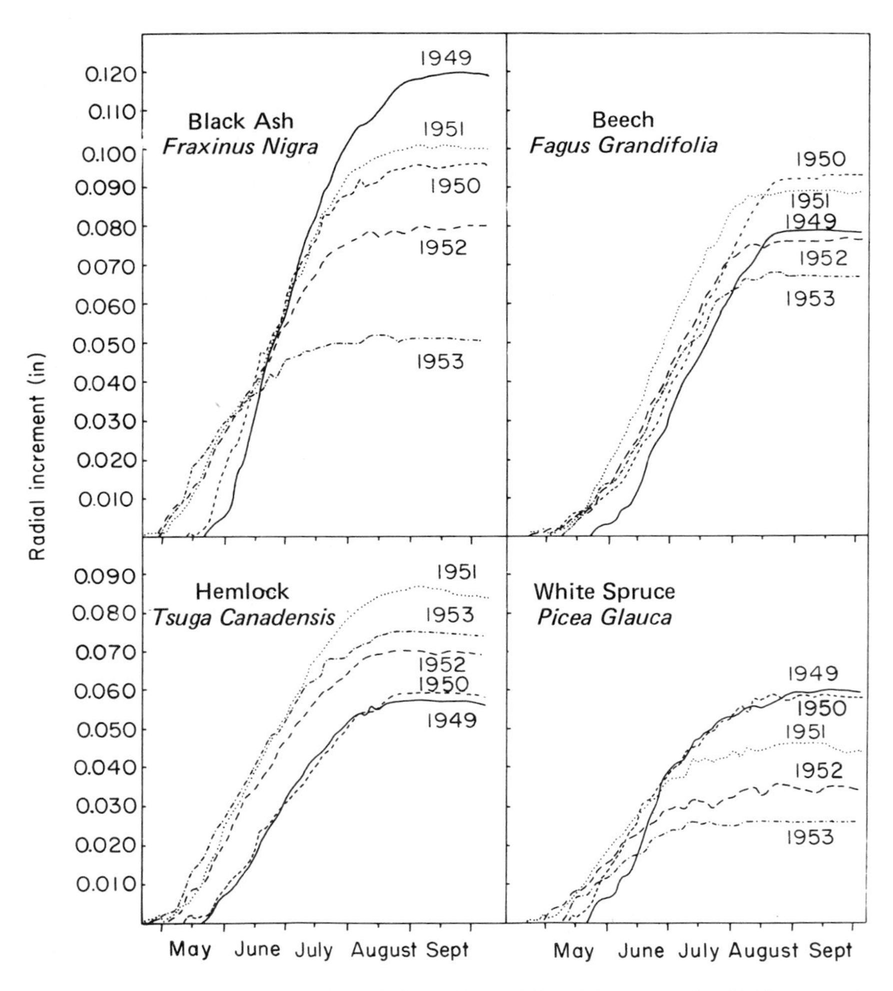


**Figure 4. Variations in time of initiation of cambial activity, rate of radial increment, and duration of growth of four species of forest trees during 5 growing seasons in Ontario, Canada. Note variations in growth characteristics during a wet year (1949) and a dry one (1953). Growth of hemlock was not reduced during the dry year because it was on a wet site. From Fraser (1956).**

when buds were forming, cambial activity is influenced by water supply of both the current and previous year. Zahner and Stage (1966) found that water deficits during June of both years inhibited cambial activity of *Pinus monticola* in Idaho.

**Table 1. Variations in duration of circumferential increase of different crown classes of 30-year old *Pinus taeda* trees at various times within the growing season.[a] From Bassett (1966).**

| | % of Growth Period During Which Trees Increased in Circumference | | | | | |
|---|---|---|---|---|---|---|
| **Crown Class** | **March 1 - May 31** | **June 1 - Aug. 16** | **Aug. 16 - Oct. 31** | **March 1 - Oct. 31** | **During Wet Year (1961)** | **During Dry Year (1963)** |
| Dominant | 96 | 79 | 66 | 80 | 96 | 58 |
| Codominant | 90 | 66 | 53 | 70 | 92 | 46 |
| Intermediate | 77 | 43 | 38 | 53 | 79 | 28 |
| Suppressed | 51 | 15 | 17 | 28 | 49 | 8 |

[a]Data are averages for 5 growing seasons (1960 to 1964) including a wet year (1961) and a dry year (1963).

In addition to controlling the amount of wood produced, water deficits influence the relative amounts of earlywood and latewood which, in turn, affect the quality of wood. Drought triggers early formation of latewood and sustained drought thereafter shortens the period of latewood formation, as shown in Figure 3 (Kraus and Spurr, 1961; Howe, 1968).

### Root Growth

Tree roots are very responsive to water deficits. Retardation and even cessation of root growth during drought, and acceleration or resumption of growth after rain or irrigation, are well known. In arid regions soil penetration by roots usually is restricted to the depth to which soil is wetted. *Pinus taeda* seedlings maintained in well-watered soil developed large, extensively branched root systems but, when grown in soil undergoing drying cycles before irrigation, developed small, unbranched root systems (Kozlowski, 1949). Kaufmann (1968) subjected *Pinus taeda* and *P. strobus* seedlings to three successive drying cycles. When the soil $\psi$ decreased to near -6 or -7 bars the rate of root growth decreased to about 25% of that of trees growing in soil maintained near field moisture capacity. Root growth was inhibited more during the second and third drying cycles than during the first one. Water deficits not only inhibit root growth but also lead to suberization of root tips, thereby reducing their capacity to absorb water and minerals even after the soil is rewetted (Kramer, 1969).

Because water in soil not penetrated by roots is largely unavailable, small and sparsely branched root systems do not absorb sufficient water to meet transpiration requirements. As a result shoot water deficits develop, leading to stomatal closure as well as reduced synthesis and downward transport of carbohydrates and hormonal growth regulators. This sequence of events eventually causes a reduction in amount of wood produced (Kozlowski, 1969).

### Water Stress and Reproductive Growth

Yield of fruit and seed may be reduced by water stress at any stage of reproductive growth. However, water stresses during bud initiation, fruit set, and fruit enlargement are likely to be particularly injurious. Severe droughts during the period of flower bud initiation can result in reduction or failure of the fruit or seed crop the following year. Although it has been claimed that, at least in apples, mild water stresses can cause a larger proportion of the buds to develop flower primordia (Teskey and Shoemaker, 1978), the evidence is not conclusive. Water stress after flowering reduced fruit set in apples to about 35% of that on irrigated trees, and the latter also retained more of the fruit (Table 2). An adequate water supply during fruit expansion also is critical for high yield (Morris et al., 1962; Crane and Uriu, 1965) and nearly all kinds of fruits show shrinkage when subjected to water stress. Water stress during maturation results in unfilled seeds and shriveled kernels in nuts such as pecans and walnuts. It also causes preharvest drop of apples and some other fruits.

### Effects of Irrigation on Yield

In arid and semiarid regions irrigation is essential for the production of fruit crops and it is beginning to be utilized to increase yield in relatively humid areas. It is necessary in California and the Yakima Valley of Washington, but it also is proving profitable for citrus fruits in Florida where the annual rainfall is about 125 cm. Even in humid England irrigation increases the yield of apples (Table 3). However, in some years frequent irrigation can reduce yield (Table 3), probably by saturating the soil. Many experiments have shown quantitative reductions in growth and yield with increase in water stress (Beutel, 1964; Suzuki and Kaneko, 1970; Maotani, 1972; Proebsting and Middleton, 1980; Stolzy et al., 1963). In dry regions and on sandy soils irrigation might be profitable for forest tree seed orchards.

As mentioned earlier there often is a lag between the occurrence of water stress and its effects (Uriu, 1964). There likewise can be a lag between

**Table 2. Effect of water deficits on various aspects of reproductive growth in Cox's Orange Pippin apple trees. From Powell (1974).**

| | | Treatment | | |
|---|---|---|---|---|
| | **Date** | **Droughted** | **Control** | **Irrigated** |
| Flowers per tree | May 12 | 3228 | 3431 | 2464 |
| Fruitlets per tree | June 9 | 275 | 540 | 591 |
| Fruit set as percentage of flower number | June 9 | 8.5 | 15.7 | 24.0 |
| Fruit clusters per tree | June 9 | 144 | 252 | 219 |

**Table 3. Effect of irrigation during 1953 to 1960 on yield (kg/tree) of apples during bearing years. From Goode and Hyrycz (1964).**

| | Treatment | | | |
|---|---|---|---|---|
| Year | Frequent Irrigation | Medium Irrigation | Infrequent Irrigation | Unwatered Control |
| 1954 | 24.5 | 32.3 | 30.9 | 26.5 |
| 1956 | 76.1 | 76.1 | 57.0 | 61.1 |
| 1958 | 113.2 | 113.4 | 98.5 | 78.4 |
| 1960 | 171.1 | 183.5 | 164.9 | 122.0 |
| Total | 384.9 | 405.3 | 351.3 | 288.0 |

irrigation and its effects, especially when it affects the formation of flower buds which do not open until the following year.

In the long term, tree breeders can improve woody plants with respect to one or more desiccation-avoiding characters such as high leaf diffusion resistance, abundant leaf waxes, early shedding of leaves during drought, small leaves, and high root-shoot ratios. Drought tolerance of trees generally represents response to various degrees of specific desiccation-avoiding adaptations. The contributions of desiccation-avoiding characters to drought tolerance vary both between and within species. Hence, tree breeders must assess the need for improving some characters more than others in their plant materials. They also should recognize that some characters contributing to desiccation avoidance, such as early stomatal closure and leaf shedding reduce photosynthesis as well as transpiration. Therefore, some growth will be lost in exchange for tree survival and decreased injury by drought.

Desiccation avoidance might be increased by selection or breeding for high stomatal diffusion resistance (small stomata, few stomata, and stomata that close early during drought). Desiccation avoidance in deciduous species might also be achieved by developing plants that will shed leaves early during developing droughts (Kozlowski, 1973), but this also reduces photosynthesis. Another approach might be to improve water use efficiency (the ratio of dry matter produced to water used in evapotranspiration). Some species, such as *Acer saccharum,* with small stomata and low transpiration rates per unit of leaf area (Kozlowski et al., 1974) undergo water stress when they develop extensive leaf areas (Pereira and Kozlowski, 1977). The possibility of using dwarfing rootstocks with such species to increase avoidance of desiccation by changing root-shoot ratios might be explored. The best direction to take in breeding and selection programs depends on climatic conditions and the type of crop. It is necessary to avoid severe stress in fruit crops to obtain good yields, and mere survival is much less important in orchards than in shade or forest trees.

## TEMPERATURE STRESS

In general, vegetative growth increases as temperature is raised and attains a maximum in the range of 20 to 35 C depending on species and seed source. As temperature is increased further, growth decreases rapidly. This pattern is related to the effects of temperature on enzyme activity.

Reduced growth at temperature extremes involves subtle interrelations among food, water, and hormones, Because roots depend on shoots for carbohydrates, and shoots depend on roots for water and inorganic nutrients, the growth of the two is closely coordinated. A specific temperature-induced reduction in function of either shoots or roots therefore may be expected to inhibit development of both. Carbohydrates produced by shoots are used preferentially for shoot growth, with roots receiving the surplus not used in the shoots. Rapid consumption of photosynthate by respiration under high temperatures, or reduced production of photosynthate under low temperatures, therefore tends to reduce root growth more than shoot growth (Kozlowski and Keller, 1966). Excessive transpiration leading to stomatal closure at high temperatures also has adverse effects on availability of carbohydrates for root growth.

Growth inhibition at low temperatures may involve complicated interactions of mineral and water relations. Soil temperature can alter the concentration of soluble nutrients in the soil and also affect the capacity of plants to absorb and use minerals. Absorption of soil minerals is closely related to the size and activity of root systems. Therefore, restriction of root growth tends to reduce absorption of minerals unless they are very abundant (Nielsen and Humphries, 1966).

Low soil temperature reduces absorption of water largely by increasing resistance of water movement across the living root cells. Other factors, such as decrease in rate of root growth and in metabolism, also contribute to decreased water absorption. In addition to reducing passive mineral uptake, low soil temperature inhibits active uptake of ions (Kramer, 1969).

### Temperature Stress and Vegetative Growth

*Shoot Growth.* Several investigators have shown that height growth of trees decreases with increase in altitude, and that such decreases reflect both reduction in length of growing season and in rate of growth (Figures 5 and 6) with decreasing temperature. When potted *Larix decidua* seedlings, which have long shoots exhibiting free growth, were placed at three altitudes, the beginning of annual shoot growth was delayed with increase in altitude, but shoot elongation ended at about the same time at all altitudes. Shoot elongation of *Picea abies,* which has fixed growth also started progressively later with increase in altitude (Figure 5). On Mt. Washington in Nevada, height

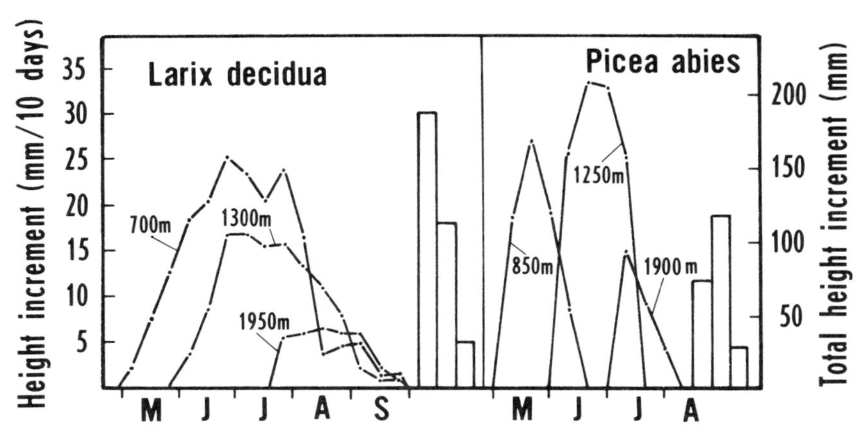


Figure 5. Effect of altitude on the amount and seasonal duration of height growth of young *Larix decidua* and *Picea abies* trees in Austria. Adapted from Tranquillini and Unterholzner (1968) and Oberarzbacher (1977).

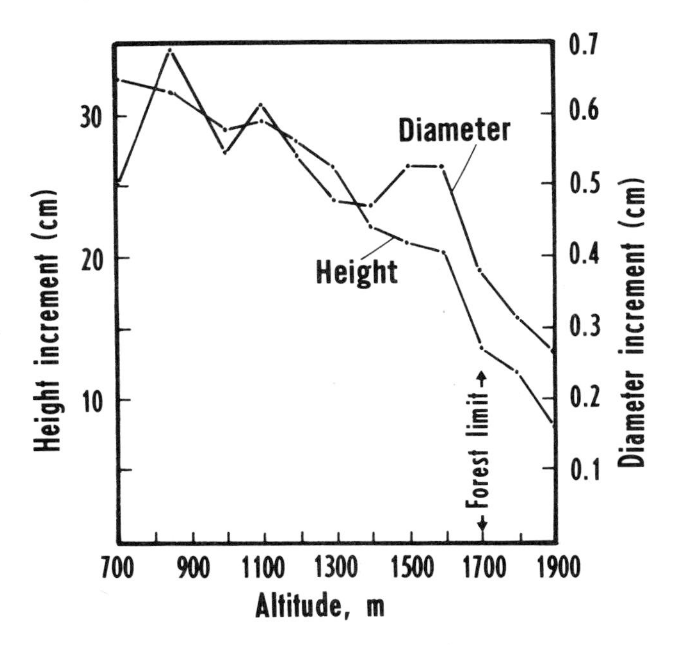


Figure 6. Effect of altitude on height and diameter growth of mature (70 to 140-year old) *Picea abies* trees in the Austrian alps. From Holzer (1973).

growth of *Pinus longaeva* dropped very rapidly as timberline was approached. Tree height averaged 6 m at 3400 m and only 2 m near timberline (LaMarche and Mooney, 1972). Size of leaves also generally decreases with increase in elevation. Needles of *Pinus cembra* were 7.6 cm long on trees at 1300 m, but they averaged only 5.2 cm in length at 2000 m near timberline (Tranquillini, 1965).

*Cambial Activity.* In a given year, temperature influences cambial activity and wood production through its effects on the time of initiation of cambial activity as well as on its rate and duration. Cambial activity begins in the spring after some critical minimal temperature is attained and usually is closely correlated with temperature early in the season (Kozlowski et al., 1962). Delayed responses of cambial activity to temperature are associated with the influence of temperature on leaf production. In New Zealand, *Pinus radiata* on a warm site had many more needle primordia in buds, produced more growth flushes, and had longer shoots than trees of the same clone on a cool site (Bollman and Sweet, 1979). The larger crowns of the trees on the warm site may be expected to account for more wood than would be produced by the smaller crowned trees on the cool site.

Effects of latitude on cambial activity that are mediated largely through temperature are well known. In a wide range of species the cambium is active for a longer time with an increase in the length of the frost-free season. For example, on the east coast of Hudson Bay, annual cambial growth of *Picea abies* continues for less than 8 weeks but further south at Chalk River, Ontario, it lasts for about 12 weeks. Long term variations in temperature also affect tree growth. For example, Mikola (1962) stated that the decrease in cambial growth observed in northern Finland in the early 1900s and the increase in later years were closely correlated with mean July temperatures.

Some idea of the effect of temperature on wood production of forest trees can be gained from studies of uniform material planted at various altitudes. In the Austrian Alps the diameter increment of *Picea abies* was 6 mm at low and moderate altitudes but only 3 mm near timberline (Figure 6). The reduced cambial growth at high altitudes reflected a short growing period associated with a long delay in beginning of cambial activity as well as reduced physiological activity. Seasonal cambial activity of *Larix decidua* near the Austrian timberline (1950 m) began 10 weeks later than at 700 m, and 2 weeks later than at 1300 m. Differentiation of cambial derivatives was correspondingly delayed at higher altitudes (Tranquillini and Unterholzner, 1968).

*Root Growth.* Roots usually begin to grow shortly after the soil becomes free of frost. Hence, seasonal root growth may be expected to begin much earlier at low than at high altitudes. Low soil temperature inhibits both root regeneration and growth of existing roots. The root regenerating potential of *Larix* and *Picea* seedlings was very low when the soil temperature was 4 C.

It increased at 12 C and was optimal at 20 C (Tranquillini, 1973). At high altitudes root growth usually was inhibited more than shoot growth, leading to lower root-shoot ratios (Tranquillini, 1979). Some root growth occurs at temperatures as low as 4 or 5 C in cool climate species, but probably not below 10 or 12 in warm climate species. Root growth of trees has been observed in every month of the year in the southeastern United States. The optimum temperature for root growth probably is lower than for shoot growth, and the high temperatures found in exposed surface soil sometimes injure or kill roots.

*Dry Weight Increment.* Increase in dry weight of trees is progressively reduced with increase in altitude. In Austria, dry weight production on seedlings at timberline (1950 m) was reduced by 42% over that at 650 m in *Pinus mugo,* 54% in *Picea abies,* and 73% in *Nothofagus solandri* (Benecke, 1972). Mature trees growing at high altitudes show small growth increments. For example, total above-ground dry matter production of fully stocked *Nothofagus solandri* stands near timberline (1340 m) in New Zealand was 5 t/ha as against 7 t/ha at 900 m (Wardle, 1970).

Growth of seedlings in the cotyledon stage of development is particularly sensitive to extremes of temperature (Kozlowski and Borger, 1971). Over a 7-week period, dry weight increment of young *Pinus resinosa* seedlings was reduced by temperatures above and below 20 C, with root growth reduced more than shoot growth. When compared with growth at 20 C, shoot growth was reduced by about 26% at 10 C, 5% at 15 C, 34% at 25 C, and 35% at 30 C. Root growth was reduced by 61% at 10 C, 25% at 15 C, 46% at 25 C, and 56% at 30 C (Kozlowski, 1967).

**Temperature Stress and Reproductive Growth**

Yield of fruits and seeds is influenced by temperature through effects on floral initiation, release of bud dormancy, anthesis, fruit set, and growth of fruits. According to Pereira (1975), both yield and quality of apple crops are determined by temperature at three critical periods of reproductive growth. Most important is the absence of late spring frosts. Secondly, the temperature at pollination must be adequate. The third critical period is the month following full bloom. Warm bright weather in the first half of that month generally increases crop yield, whereas in the second half it often decreases it.

*Breaking of Bud Dormancy.* In the temperate zone southward extension of many northern species of trees is limited by lack of enough low temperature to break bud dormancy. In the southern United States yield of fruit trees sometimes is greatly reduced when winters are too mild to break dormancy of flower buds. When the amount of chilling is inadequate, peach flower buds near branch tips may develop into blossoms while basal buds remain dormant. Inadequate chilling results in small and deformed blossoms as well as low germination of pollen. The stigma and style fail to grow while the rest of the flower achieves normal size. Such abnormal flowers fail to set fruit. Leaf growth also is inhibited and overall vigor of the tree is greatly reduced. The amount of chilling required for breaking of dormancy of flower buds varies greatly among cultivars (Table 4).

*Freezing Injury.* Growth and yield are reduced when woody plants do not develop enough cold hardiness, do not harden fast enough to survive early freezes, or lose cold hardiness too rapidly. Both the nature and amount of injury vary with the duration of a freeze and the organs affected. Early autumn frosts injure inadequately hardened shoots, especially those of late-season growth flushes. Late application of fertilizers and irrigation often prolong shoot growth and increase the possibility of frost injury. Injury to the cambium by freezing may product "frost" rings characterized by death of differentiating cambial derivatives. Freezing of steams may also induce frost cracks, lesions, and sunken cankers that can become sites for invasion by fungi and insects.

When the soil freezes, many small, physiologically active roots are killed. The amount of injury varies with species, rootstock, and soil type. Frost

**Table 4. February 15th chilling requirements of flower buds of peach cultivars. From Teskey and Shoemaker (1978).**

| Cultivars | Hours of Chilling Required at 7.2 C (45 F) |
|---|---|
| Mayflower | 1150 |
| Raritan Rose, Dixired, Fairhaven | 950 |
| Sullivan Early Elberta, Trigem, Elberta, Dixired, Georgia Belle, Halehaven, Redhaven, Candor, Rio Oso Gem, Golden Jubilee, Shippers Late Red | 850 |
| Afterflow, July Elberta (Burbank), Hiland, Hiley, Redcap, Redskin | 750 |
| Maygold, Bonanza (dwarf), Suwannee, June Gold, Springtime | 650 |
| Flordaqueen | 550 |
| Bonita | 500 |
| Rochon | 450 |
| Flordahome (double flowers) | 400 |
| Early Amber | 350 |
| Jewel, White Knight No. 1, Sunred Nectarine, Flordasun | 300 |
| White Knight No. 2 | 250 |
| Flordawon | 200 |
| Flordabelle | 150 |
| Okinawa | 100 |
| Ceylon | 50-100 |

heaving of seedlings in frozen soil may result in loss of nursery stock because roots are broken and the shoots and exposed roots become desiccated.

*Chilling Injury.* Both vegetative and reproductive growth of many tropical and subtropical trees, and a few temperate zone trees, are adversely influenced by low temperatures above freezing. Chilling injury of fruits is of particular interest because they often are stored and shipped at low temperatures. Symptoms of chilling injury to fruits include surface pitting, lesions, discoloration, susceptibility to decay organisms, and shortening of storage life. Furthermore, fruits subjected to chilling temperature do not ripen normally (Lyons, 1973).

The critical temperature at which chilling injury occurs is near 10 to 12 C for many tropical fruits, but the precise temperature varies appreciably. For example, the lower temperature limit is 10 to 12 C for many varieties of banana; 8 to 12 C for citrus, avocado, and mango; and 0 to 4 C for temperate zone fruits such as apple. Chilling injury also is affected by duration of exposure. For example, exposure of many varieties of banana for a few hours to temperatures between -1 and 7 C lowers the quality of the fruit, and 12 hours of exposure results in such extensive injury that the fruits are not marketable.

Levitt (1980), McWilliam (1982) and Wilson (1982) treat chilling injury as direct, indirect, or the result of secondary stress. Rapid development of injury symptoms usually reflects direct injury whereas slow development indicates indirect injury. The latter, which may require days or weeks of exposure to the low temperature stress before it appears, may be associated with increase in membrane permeability and leakage of solutes, starvation, abnormal respiration, accumulation of toxins, and biochemical lesions.

*Winter Desiccation Injury.* Much winter injury in gymnosperms and broadleaved evergreens results from shoot desiccation rather than from direct thermal injury. When the injury is severe all the leaves and buds, and often the trees, are killed. More commonly only the leaves are killed and the trees survive. Winter desiccation appears to be one of the most important factors limiting the range of gymnosperms in the temperate zone (Weiser, 1970).

Winter desiccation injury occurs when absorption of water cannot keep up with transpirational losses. In many parts of the temperate zone, appreciable transpiration occurs as the air warms sufficiently during sunny winter or spring days and increases the vapor pressure gradient between the leaves and surrounding air. Because the soil is cold or frozen, water cannot be absorbed through the roots rapidly enough to replace transpirational losses; hence, the shoots become desiccated.

### Selection and Breeding for Cold Hardiness

Because cold hardiness is a polygenic property it is difficult to breed directly for cold hardiness. However, trees can be screened for traits that are correlated with cold hardiness (Namkoong et al., 1980). During acclimation of plant tissues to cold, changes occur in the concentrations of sugars, proteins, amino acids, nucleic acids, and lipids (Kramer and Kozlowski, 1979). However, such data have been less useful in breeding programs than has information on seasonal shoot growth characteristics. The ideal cold hardy plant begins shoot growth late in the spring, grows rapidly during the summer, accumulates large amounts of reserve carbohydrates, and stops shoot elongation early in the autumn. Because the time of initiation and end of seasonal shoot growth are strongly hereditary (Thielges and Beck, 1976), much can be done by tree breeders to produce cold hardy varieties. This is especially important because of widespread interest in introducing rapid-growing exotic forest trees into many countries.

In addition to the need for trees that have cold hardiness, three additional aspects of growth should be considered in a breeding program (Glerum, 1976). These are the time of year when cold hardiness begins and the rate of hardening, the degree of maximum cold hardiness and whether or not it is high enough for a given site, and the time of beginning of dehardening and the rate of dehardening.

There is considerable within-species variation in these characteristics. As a result of natural selection, climatic races have become adapted to particular environments. For example, northern provenances often are more resistant than southern provenances to cold. Examples are *Acer saccharum* (Kriebel, 1957), *Pinus ponderosa* (Squillace and Silen, 1962), *Quercus rubra* (Flint, 1972), and *Cornus stolonifera* (Smithberg and Weiser, 1968). Campbell and Sorenson (1973) demonstrated a close relationship between early bud set and cold tolerance of *Pseudotsuga menziesii* provenances. For each additional week by which bud set preceded frost, the proportion of frost damaged seedlings decreased by 25%. A complication for tree breeders is that some species do not show marked provenance variation in cold hardiness (Sakai and Weiser, 1973).

In the USDA breeding program for scion hardiness of citrus, the mandarins have been shown to be cold tolerant and many hybrids involving them have been tested. Mandarin hybrids are more hardy than hybrids involving pummelo, grapefruit, or lemon. In Japan *Poncirus* has been the most widely used rootstock, and cultivars of the early-ripening and cold-hardy Satsuma are the most important scion types. In the USSR several new frost resistant hybrids have been obtained by breeding (Soost and Cameron, 1975).

### Effects of High Temperature

Relatively high temperatures often reduce growth and injure woody plants. The amount of injury sustained varies with species, age of tree, and particularly with the duration of exposure to high temperature. Injuries induced by temperatures in the 15 to 40 C range usually are considered indirect; those caused by temperatures of 45 C or higher are classed as direct. Indirect injury occurs slowly, often hours or days after exposure to high temperature, whereas direct injury occurs during heating or very shortly thereafter.

High temperatures sometimes reduce growth indirectly because apparent photosynthesis declines after a critical high temperature is reached while respiration continues to increase above that temperature, leading to depletion of carbohydrates. Indirect injury may also be associated with formation of toxic compounds, biochemical lesions, breakdown of protoplasmic proteins, and desiccation resulting from high transpiration rates.

Direct heat injury is less common than indirect injury. However, direct heat injury sometimes occurs and may be expressed as lesions on fleshy fruits and stem lesions often described as "sunscald" and "bark scorch." The mechanisms of direct heat injury may involve damage to cellular membranes, protein denaturation, and lipid liquefaction. As emphasized by Levitt (1980), the mechanisms of heat injury are complex and attempts to identify a single mechanism seem futile.

## NOTES

T. T. Kozlowski, Dept. of Forestry, University of Wisconsin, Madison, WI 53705.

## LITERATURE CITED

Bassett, J. R. 1966. Seasonal diameter growth of loblolly pines. J. For. 64: 674-676.

Benecke, U. 1972. Wachstum, $CO_2$-Gaswechsel und Pigmentgehalt einiger Baumarten nach Ausbringung in verschiedene Höhenlagen. Angew. Bot. 46: 117-135.

Beutel, J. A. 1964. Soil moisture, weather and fruit growth. Calif. Citrograph 49: 372.

Bollman, M. P., and G. B. Sweet. 1979. Bud morphogenesis of *Pinus radiata* in New Zealand. II. The seasonal shoot growth pattern of seven clones at four sites. N.Z. J. For. Sci. 9:153-165.

Campbell, R. K., and F. C. Sorensen. 1973. Cold-acclimation in seedling Douglas-fir related to phenology and provenance. Ecology 54: 1148-1151.

Clements, J. R. 1970. Shoot responses of young red pine to watering applied over two seasons. Can. J. Bot. 48: 75-80.

Crane, J. C., and K. Uriu. 1965. The effect of irrigation on response of apricot fruits to 2,4,5-T application. Proc. Amer. Soc. Hort. Sci. 86: 88-94.

Fraser, D. A. 1956. Ecological studies of forest trees at Chalk River, Ontario, Canada. II. Ecological conditions and radial increment. Ecology 37: 777-789.

Flint, H. L. 1972. Cold hardiness of twigs of *Quercus rubra* L. as a function of geographic origin. Ecology 53: 1163-1170.

Glerum, C. 1976. Frost hardiness of forest trees. p. 403-420. In M. G. R. Cannell and F. T. Last (eds.) Tree physiology and yield improvement. Academic Press, London.

Goode, J. E., and K. J. Hyrycz. 1964. The response of Laxton's Superb to different soil moisture conditions. J. Hort. Sci. 39: 254-276.

Holzer, K. 1973. Die Vererbung von physiologischen und morphologischen Eigenschaften der Fichte. II. Mutterbaummerkmali Unveroffentiches Manuskript.

Howe, J. B. 1968. The influence of irrigation on wood formed in ponderosa pine. For. Prod. J. 18: 84-93.

Jacobs, M. R. 1955. Growth habits of the eucalypts. Austr. Forest. Timber Bur. 1-262.

Kaufmann, M. R. 1968. Water relations of pine seedlings in relation to root and shoot growth. Plant Physiol. 43: 281-288.

Kaufmann, M. R. 1969. Effects of water potential on germination of lettuce, sunflower, and citrus seeds. Can. J. Bot. 47: 1761-1764.

Kozlowski, T. T. 1949. Light and water in relation to growth and competition of Piedmont forest tree species. Ecol. Monogr. 19: 207-231.

Kozlowski, T. T. 1967. Growth and development of *Pinus resinosa* seedlings under controlled temperatures. Adv. Front. Plant Sci. 19: 17-27.

Kozlowski, T. T. 1969. Tree physiology and forest pests. J. For. 69: 118-122.

Kozlowski, T. T. (ed.). 1973. Shedding of plant parts. Academic Press, New York.

Kozlowski, T. T. 1976. Water supply and leaf shedding. p. 191-231. In T. T. Kozlowski (ed.) Water deficits and plant growth. Vol. IV. Soil water measurement, plant responses, and breeding for drought resistance. Academic Press, New York.

Kozlowski, T. T. 1979. Tree growth and environmental stresses. Univ. of Washington Press, Seattle.

Kozlowski, T. T., and G. A. Borger. 1971. Effect of temperature and light intensity early in ontogeny on growth of *Pinus resinosa* seedlings. Can. J. For. Res. 1: 57-65.

Kozlowski, T. T., W. H. Davies, and S. D. Carlson. 1974. Transpiration rates of *Fraxinus americana* and *Acer saccharum* leaves. Can. J. For. Res. 4: 259-267.

Kozlowski, T. T., and T. Keller. 1966. Food relations of woody plants. Bot. Rev. 32: 293-382.

Kozlowski, T. T., J. E. Kuntz, and C. H. Winget. 1962. Effect of oak wilt on cambial activity. J. For. 60: 558-561.

Kozlowski, T. T., R. H. Torrie, and P. E. Marshall. 1973. Predictability of shoot length from bud size in *Pinus resinosa.* Can. J. For. Res. 3: 34-38.

Kozlowski, T. T., C. H. Winget, and J. H. Torrie. 1962. Daily radial growth of oak in relation to maximum and minimum temperature. Bot. Gaz. 124: 9-17.

Kramer, P. J. 1969. Plant and soil water relationships: A modern synthesis. McGraw-Hill, New York.

Kramer, P. J., and T. T. Kozlowski. 1979. Physiology of woody plants. Academic Press, New York.

Kramer, P. J. 1943. Amount and duration of growth of various species of tree seedlings. Plant Physiol. 18: 239-251.

Kraus, J. F., and S. H. Spurr. 1961. Relationship of soil moisture to the springwood-summerwood transition in southern Michigan red pine. J. For. 50: 510-511.

Kriebel, H. B. 1957. Patterns of genetic variation in sugar maple. Ohio Agr. Expt. Sta. Res. Bull. 791.

Lamarche, V. C., and H. A. Mooney. 1972. Recent climatic change and development of the bristlecone pine *(Pinus longaeva* Bailey) krummholz zone. Mt. Washington, Nevada. Arct. Alp. Res. 4: 61-72.

Levitt, J. 1980. Responses of plants to environmental stresses. Vol. I. Chilling, freezing, and high temperature stresses. Academic Press, New York.

Lotan, J. E., and R. Zahner. 1963. Shoot and needle responses of 20-year-old red pine to current soil moisture regimes. For. Sci. 9: 497-506.

Lyons, J. M. 1973. Chilling injury in plants. Annu. Rev. Plant Physiol. 24: 445-466.

Maotani, T., Y. Machida, and K. Yamatsu. 1977. Studies on leaf water stress in fruit trees. VI. Effects of leaf water potential on growth of satsuma trees. J. Jap. Soc. Hort. Sci. 45: 329-334.

McWilliam, J. R. 1982. Physiological basis for chilling stress and the consequences for crop production. p. 113-132. In C. D. Raper, Jr., and P. J. Kramer (eds.) Crop reactions to water and temperature stresses in humid, temperate climates. Westview Press, Boulder, Colorado.

Mikola, P. 1962. Temperature and tree growth near the northern timber line. p. 265-274. In T. T. Kozlowski (ed.) Tree growth. Ronald Press, New York.

Morris, J. R., A. A. Kattan, and E. H. Arrington. 1962. Response of Elberta peaches to the interactive effects of irrigation, pruning, and thinning. Proc. Amer. Soc. Hort. Sci. 80: 177-189.

Namkoong, G., R. D. Barnes, and J. Burley. 1980. Screening for yield in forest tree breeding. Commonw. For. Rev. 59: 61-68.

Nielsen, K. I., and E. C. Humphries. 1966. Effects of root temperature on plant growth. Soils and Fertilizers 29: 1-7.

Oberarzbacher, P. 1977. Beitrage zur physiologische Analyze des Hohenzuwachses von verschiedenen Fichtenklonen entlang eines Hohenprofils im Wipptal (Tirol) und in Klimakammern. Diss. Univ. Innsbruck.

Pereira, H. C. 1975. Climate and the orchard. Commonwealth Bur. Hort. Plant. Crops (G.B.), Res. Rev. 5.

Pereira, J. S., and T. T. Kozlowski. 1977. Influence of light intensity, temperature, and leaf area on stomatal aperture and water potential of woody plants. Can. J. For. Res. 7: 143-153.

Powell, D. B. B. 1974. Some effects of water stress in late spring on apple trees. J. Hort. Sci. 49: 257-272.

Proebsting, E. L., Jr., and J. E. Middleton. 1980. The behavior of peach and pear trees under extreme drought stress. J. Amer. Soc. Hort. Sci. 105: 380-385.

Rowe, J. S. 1964. Environmental preconditioning with special reference to forestry. Ecology 45: 399-403.

Sakai, A., and C. J. Weiser. 1973. Freezing resistance of trees in North America with reference to tree regions. Ecology 54: 118-126.

Satoo, T. 1966. Variation in response of conifer seed germination to soil moisture conditions. Tokyo Univ. For. Misc. Inf. 16:17-20.

Smithberg, M. H., and C. J. Weiser. 1968. Patterns of variation among climatic races of red-osier dogwood. Ecology 49: 495-505.

Soost, R. K., and J. W. Cameron. 1975. Citrus. p. 508-540. In J. Janick and J. N. Moore (eds.) Advances in fruit breeding. Purdue University Press, Lafayette, Indiana.

Squillace, A. E., and R. R. Silen. 1962. Racial variation in ponderosa pine. For. Sci. Monogr. 2.

Stolzy, L. H., O. C. Taylor, M. J. Garver, and P. B. Lonbard. 1963. Previous irrigation treatments as factors in subsequent irrigation level studies in orange production. Proc. Amer. Soc. Hort. Sci. 82: 199-203.

Suzuki, T., and M. Kaneko. 1970. The effect of suction pressure in the soil solution during summer on growth and fruiting of young satsuma orange trees. J. Jap. Soc. Hort. Sci. 39: 99-106.

Tazaki, T., K. Ishihara, and T. Ushijima. 1980. Influence of water stress on the photosynthesis and productivity of plants in humid areas. p. 309-321. In N. C. Turner and P. J. Kramer (eds.) Adaptation of plants to water and high temperature stress. John Wiley & Sons, New York.

Teskey, B. J., and J. S. Shoemaker. 1978. Tree fruit production. Avi, Westport, Connecticut.

Thielges, B. A., and R. C. Beck. 1976. Control of bud break and its inheritance in *Populus deltoides.* p. 253-259. In M. G. R. Cannell and F. T. Last (eds.) Tree physiology and yield improvement. Academic Press, New York.

Tranquillini, W. 1965. Über den Zusammenhang zwischen Entwicklungszustand und Dürreresistenz junger Zirben *(Pinus cembra* L.) im Pflanzengarten. Mitt. Forstl. Bundesversuchsanst. Mariabrunn 66: 241-271.

Tranquillini, W. 1979. Physiological ecology of the Alpine timberline. Springer-Verlag, Berlin.

Tranquillini, W. 1973. Der Wasserhaushalt junger Forstpflanzen nach dem Versetzen und seine Beeinflusbarkeit. Zentralbl. Gesamt. Forstwes. 90: 46-52.

Tranquillini, W., and R. Unterholzner. 1968. Das Wachstum Zweijähriger Lärchen einheitlicher Herkunft in verschiedener Seehöhe. Zentralbl. Gesamt. Forstwes. 85: 43-59.

Uriu, K. 1964. Effect of post-harvest soil moisture depletion on subsequent yield of apricots. Proc. Amer. Soc. Hort. Sci. 84: 93-97.

Vité, J. P. 1961. The influence of water supply on oleoresin exudation pressure and resistance to bark beetle attack in *Pinus ponderosa.* Contrib. Boyce Thompson Inst. 21: 37-66.

Wardle, J. A. 1970. Ecology of *Nothofagus solandri.* N. Z. J. Bot. 8: 494-646.

Weiser, C. J. 1970. Cold resistance and injury in woody plants. Science 169: 1269-1278.

West, P. W. 1979. Date of onset of regrowth dieback and its relation to summer drought in eucalypt forests of southern Tasmania. Ann. Appl. Biol. 93: 337-350.

Whitmore, F. W., and R. Zahner. 1967. Evidence for a direct effect of water stress in the metabolism of cell walls in *Pinus.* For. Sci. 13: 397-400.

Wilson, J. M. 1982. Interaction of chilling and water stress. p. 133-148. In C. D. Raper, Jr. and P. J. Kramer (eds.) Crop reactions to water and temperature stresses in humid, temperate climates. Westview Press, Boulder, Colorado.

Zahner, R. 1958. September rains bring pine growth gains. Southern Forest Expt. Sta., U.S. Forest Service Note 113.

Zahner, R. 1962. Terminal growth and wood formation by juvenile loblolly pine under two soil moisture regimes. For. Sci. 8: 345-352.

Zahner, R. 1968. Water deficits and growth of trees. p. 191-254. In T. T. Kozlowski (ed.) Water deficits and plant growth. Vol. II. Plant water consumption and response. Academic Press, New York.

Zahner, R., and A. R. Stage. 1966. A procedure for calculating daily moisture stress and its utility in regressions of tree growth on weather. Ecology 47: 64-74.

Zahner, R., and F. W. Whitmore. 1960. Early growth of radically thinned loblolly pine. J. For. 58: 628-634.

Zavitkovski, J., and W. K. Ferrell. 1970. Effect of drought upon rates of photosynthesis, respiration, and transpiration of seedlings of two ecotypes of Douglas-fir. II. Two-year-old seedlings. Photosynthetica 4: 58-67.

SECTION III

# PHYSIOLOGICAL BASIS OF INJURY AND TOLERANCE

*The third section deals with the physiological effects of temperature and water stress. Eastin et al. caution that efforts to relate losses in grain yield of sorghum to short-term effects of water and temperature stress on photosynthesis or other single physiological processes often are unsuccessful because yield is not controlled by any single physiological process. For example, while brief stress periods during differentiation of pistil and stamen primordia can irreversibly reduce seed number and yield of sorghum, photosynthesis may return to normal following relief from stress. Also, root growth, a potentially competing sink to reproductive growth, is rapid at the time that seeds are being set in the inflorescence. Thus, the lower respiration rate in roots of a drought-tolerant sorghum hybrid than in roots of a normal hybrid could explain the greater stability of the former in seed number and yield. Understanding such limiting effects of stress on growth and differentiation is critical to successful genetic and cultural manipulation for yield stability. Another process limiting yield of stressed corn and sorghum is accelerated senescence in the placento-chalazal pad of the seeds. This reduces assimilate translocation and seed growth even though the leaves may remain green and capable of photosynthesis.*

*According to McWilliam, chilling (exposure to temperatures below 12 or 15 C) often produces significant stress in the many tropical and subtropical crops that are important as summer crops in the temperate zone. Effects on seed germination and chlorophyll development are particularly severe. Chilling causes a serious increase in flower sterility in rice and a reduced yield of chilling-sensitive crops such as bean, cotton, and rice. Chilling often causes wilting because of slow closure of stomata and reduced permeability of the roots to water. Plants grown from seeds matured at low temperature often are more tolerant of chilling. Water stress also increases tolerance of chilling. The possibility of breeding for chilling tolerance has been demonstrated for sorghum and tomato, and selection for tolerance might be made on cell cultures. Wilson continued the discussion of chilling injury and agreed that it usually is caused by water stress, resulting from decreased*

*root permeability and failure of the stomata to close. Stomatal closure can be brought about by spraying the leaves with ABA. However, for some plants such as Episcia, injury is a direct effect of chilling. Final death after prolonged chilling involves disturbance of metabolism such as changes in lipids and ATP, photo-oxidation, and decrease in photosynthesis and translocation.*

# 7

# PHYSIOLOGICAL ASPECTS OF HIGH TEMPERATURE AND WATER STRESS

**J. D. Eastin, R. M. Castleberry, T. J. Gerik, J. H. Hultquist, V. Mahalakshmi, V. B. Ogunlela, and J. R. Rice**

The literature on environmental stress is replete with references on the effects of water and temperature stress on specific physiological processes, especially photosynthesis. Fewer reports, but nonetheless a significant body of information, are available on stress effects on respiration and many associated synthetic processes. Both logic and intuition support the general inclination prevalent in recent decades to concentrate a great deal of research effort on the influences of environmental stresses on primary essential physiological processes such as photosynthesis, respiration, translocation, etc. These are obviously plant processes essential to the production of grain, and yet rarely can significant, positive correlations between grain yield and any of these processes be found in the literature. Figure 1 (Murata, 1981) illustrates significant positive correlations between mean crop growth rate (total dry matter production) and maximum photosynthetic rate over a wide range of both $C_3$ and $C_4$ crops, but no such correlation appears evident when a grain yield function is substituted for a total dry matter function. Apparently, essential physiological processes, when considered individually, are not the first grain yield-limiting processes. Developmental limitations appear more critical in grain production. This is not to suggest that research on essential physiological processes such as photosynthesis, respiration, translocation, etc., should be altered necessarily, but rather that a broader perspective should be developed simultaneously to detail the order of yield limiting plant developmental processes. Perhaps, then, a consideration of how essential physiological process limitations relate to developmental limitations would

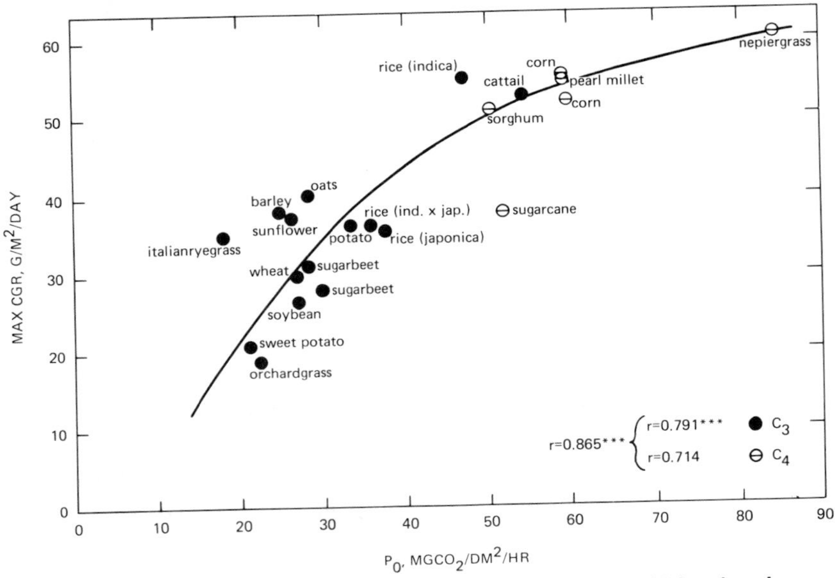


**Figure 1. Relationships of leaf photosynthetic activity ($P_0$) with high values in max. crop growth rate (MaxCGR) among crop species. (Courtesy of S. Yoshida, 1980).**

be more productive.

Considerably less effort has been expended in experimenting on environmental influences (chiefly water and temperature) on the complex integrated effects of all essential physiological and developmental processes on grain than on physiological processes individually. Such is not surprising partly because of the difficulty in controlling water and temperature levels for extended periods in field, greenhouse and growth room environments to span discrete developmental periods. Analyzing related physiological and developmental processes simultaneously often seems to range from difficult to impossible if the number of processes being evaluated is very large. Recent developmental investigations (Ogunlela, 1979, for example) suggest that lengths of the developmental time periods which impact heavily on grain yield can be quite short. Therefore, the task of better quantifying temperature and water stress effects simultaneously on important developmental and related physiological processes to better explain grain yield limitations may not be as difficult as once suspected.

One of the most effective means of sharpening appreciation of yield-limiting plant processes is simply to focus initially on the seed number and seed size components of yield. It long has been clear, often through the efforts of plant breeders (Kambal and Webster, 1966; Stickler et al., 1961;

Blum, 1967 and 1970; Quinby, 1963; Doggett, 1967; Beil and Atkins, 1967; Fischer and Wilson, 1975; Warrington, et al., 1977; Ahad, 1979; Yoshida, 1980; Fischer, 1980; Eastin and Sullivan, 1974; Eastin, 1980), that the seed-number component of yield usually is correlated more positively with yield than is the seed-size (weight) component. The seed-size component obviously is important but yield increases at reasonable yield levels relate primarily to increases in seed number per unit land area. There are, however, good reasons to concentrate on the seed-size component, as will be illustrated later.

We have selected a rather narrow range of references characterizing the most yield-sensitive developmental periods that point toward some of the gaps in our knowledge of factors limiting grain yields. Sorghum will be the principal illustrative crop used. Literature citations by no means are exhaustive, even for sorghum.

## DEVELOPMENTAL STAGES

What, then, are the developmental factors relating to the seed-size and seed-number components of yield? A simple, phenological description used for grain sorghum (Eastin, 1972) and wheat (Warrington et al., 1977) can be used for discussion purposes:

$GS_1$ (vegetative) – planting to panicle initiation (PI)
$GS_2$ (inflorescence development) – PI to bloom
$GS_3$ (grain fill) – bloom to kernel dark layer.

More detailed phenological descriptions are available for cereals in general, but this simple system suffices for this discussion. Obviously, seed-number potential is set during $GS_2$ and seed-size considerations relate mostly to $GS_3$ events. $GS_1$ is the least sensitive period and will not be included in this discussion.

### $GS_2$ Vegetative-Floral Competition.

Existing evidence suggests that competition between simultaneously expanding vegetative and floral parts during $GS_2$ can limit seed-number potential. We conducted an irrigated experiment at Mead, Nebraska, in 1970, utilizing three RS 626 grain sorghum hybrids which were isogenic except for height (Table 1). Grain yields of the three hybrids did not differ significantly but stover yields did. The shorter 3x3-dwarf and 4x3-dwarf (normal commercial hybrid heights) hybrids produced 16 and 19% less total dry matter than the 2x3-dwarf. Seed weight, however, was about 30% higher in the tall sorghum. Since grain yields were the same, seed number per unit land area was 30% higher in the 3x3-dwarfs and 4x3-dwarfs compared to the 2x3-dwarf hybrids. Reason for the differential seed-number and seed-size reactions in the isogenic tall and short hybrids lies in the fact that there are no height differences among the hybrids at panicle initiation and little vegetative growth

**Table 1.** Comparative production data for tall (2x3-dwarf), normal (3x3-dwarf), and short (4x3-dwarf) RS 626 grown under irrigation at Mead, Nebraska (1970). Germination to maturity was 106 days. Dry matter is adjusted to 14% moisture. (J. D. Eastin, unpublished data.)

| | | Dry Weight | | | Dry Matter Accumulation Rate | | | Seed Weight | |
|---|---|---|---|---|---|---|---|---|---|
| Hybrid Height | Grain to Stover Ratio[a] | Grain | Total | % of 2x3-dwarf total | Grain | Total | % of 2x3-dwarf total | Total | % of 2x3-dwarf over others |
| | | *– kg/ha –* | | *– % –* | *– kg/ha/day –* | | *– % –* | *g/1000 seed* | *– % –* |
| 2x3-dwarf | 0.76 | 7741 | 17,940 | 100 | 73.3 | 169 | 100 | 29.3 | – |
| 3x3-dwarf | 1.14 | 7988 | 15,080 | 84 | 75.4 | 142 | 84 | 22.7 | 29 |
| 4x3-dwarf | 1.08 | 7528 | 14,501 | 81 | 71.0 | 137 | 81 | 22.5 | 30 |

[a]Ratio is higher than normal because of leaf loss in a heavy, unseasonal snow before final harvest.

occurs after bloom. Therefore, the extra 16 to 19% dry matter production in the tall hybrid occurred during $GS_2$ when seed number potential was being set. Since post-bloom floret abortion was not apparent, competition for assimilates between simultaneously expanding vegetative and floral parts presumably reduced seed number 30%. A second lesson accruing from the experiment is that seed size capacity is at least 30% larger than is normally realized in many sorghum hybrids. Numerous grain-removal experiments have demonstrated potential seed size increases of up to 40 and 50%. These results suggest (1) that seed size is a critical yield-compensating component when environmental rigors reduce seed-number set, and (2) that seed-size potential is adequate to permit conservatively a 20 to 30% increase in grain yield if the assimilate source can be increased by a faster photosynthetic rate or maintenance of current rates over a longer grain-filling period.

Research on dwarf wheat cultivars (Fischer, 1980; Ahad, 1979) strongly suggests that yield advantages observed in certain environments generally result from higher seed number per unit area. However, seed size is a more important factor in some genotypes than others.

**Sensitivity within $GS_2$.**

Castleberry (1973) tested sensitivity of sorghum to an environmental light variable by thinning a stand of 300,000 plants/ha to 180,000 at approximately weekly intervals beginning 13 days after planting. The 180,000-plant stand was still thick enough to permit nearly complete interception of available photosynthetically active radiation (PAR) by mid to late $GS_2$. Differences in PAR reception in the two stands were very small but the amount of PAR per plant was on the order of 40% greater in the thinner stand. Figure 2 illustrates a sharp drop off in yield from thinning at mid $GS_2$ which occurred because this is the approximate time floret components (particularly stamen and pistil primordia) are formed. Beyond this time, no compensation in seed number per plant is possible, and seed-size increases did not make up for seed-number losses due to thinning.

Eastin (unpublished, 1971) fed $^{14}CO_2$ photosynthetically to the last expanded leaf during $GS_2$ and evaluated $^{14}C$ distribution in the plant including the developing panicle and the juvenile stem tissue (generally unelongated internodes) supporting the developing panicle. Figure 3 illustrates the dynamic nature of panicle development when spikelet primordia are developing (10 to 14 days after PI), followed by floret components (pistil, stamen, glumes, etc.) at about 14 to 21 days. If the specific activity ratio of panicle to juvenile stem can be considered a crude indicator of relative sink strengths of the two (and it is questionable), it appears that the panicle is a comparatively strong sink at PI but weakens by the critical floret differentiation stage when seed-number potential is being set. The most critical point illustrated in

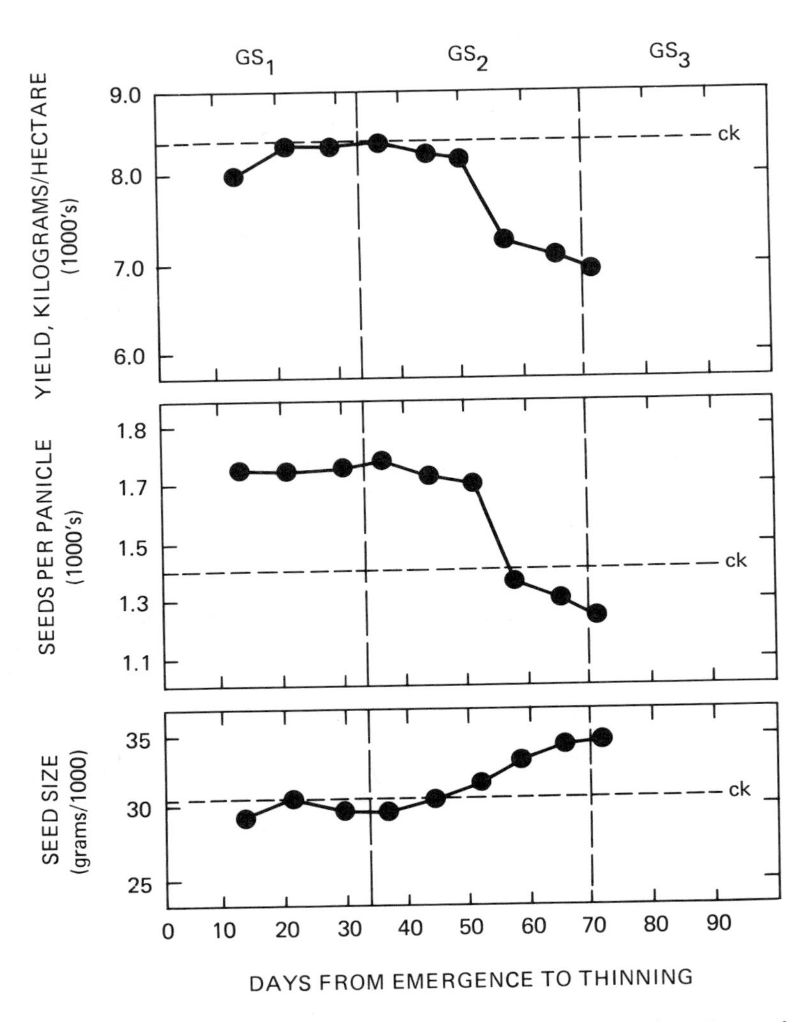


**Figure 2. The effects of thinning treatments on EH 101 yield, seed number, and seed size in Experiment 2. (Castleberry, 1973).**

Figure 3 is the very high growth rate of a very small panicle at floret differentiation. Assimilate supply to the developing panicle may be a limiting factor. However, since the panicle weight of 637 mg at floret differentiation (FD) is very small compared to the total plant weight, ability to polarize assimilate transport to the developing panicle and subsequently utilize it metabolically may be more limiting than is immediate photoassimilate supply *per se.*

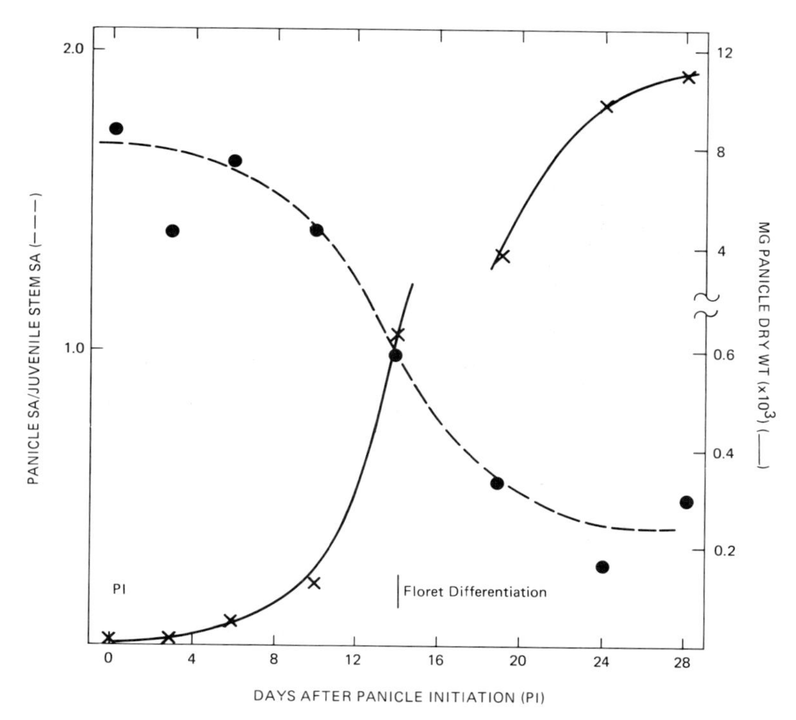


**Figure 3. Changes in the ratio of panicle specific activity (SA) to juvenile stem tissue (unelongated internodes) SA during panicle development prior to bloom in RS 671 grain sorghum. (J.D. Eastin, unpublished data.)**

## $GS_2$ Water Effects

Hultquist (1973) compared drought-stress effects on developmental and physiological characteristics in a drought-resistant (DeKalb C42-Y) and a normal sorghum hybrid (RS 626). Stressing the hybrids severely at peduncle elongation (about FD time) resulted in drastically different responses. The normal hybrid could suffer damage (death) over half the panicle and would exert a partial panicle when stress was relieved. If the drought-resistant hybrid was stressed severely enough to damage any portion of the panicle, the whole plant usually died. A stress level sufficient to do that was very severe. Leaf feeding of $^{14}CO_2$ to both hybrids under stress showed markedly different assimilate-distribution characteristics. C42-Y generally retained a little higher proportion of its fixed carbon in the fed leaf and distributed less of the total exported assimilate to the lower nodes and roots so that a greater portion of exported $^{14}C$ could be distributed to the developing panicle (Figure 4). The same generally was true upon relief of stress (Figures 5 and

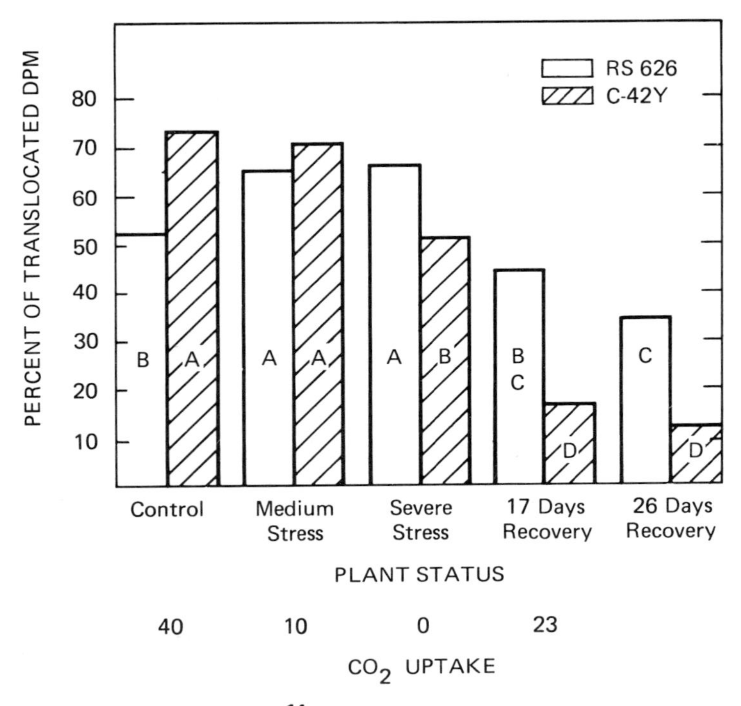


**Figure 4. Percent of translocated $^{14}C$ (DPM) in lower nodes and roots. Carbon dioxide uptake is expressed as mg $CO_2/dm^2/hr$. Unlike letters indicate significant differences at the 5% level. (Hultquist, 1973).**

6). Also during stress the root hairs of the stress-tolerant hybrid were a more active assimilate sink than in the normal hybrid even though less total assimilate was transported to the root (Figure 7).

The interpretation of drought resistance in this situation bears comment. Under intermittent water stress conditions in the Great Plains of the United States from Central Nebraska south into Kansas and in the High Plains of Texas, C42-Y generally yields higher than RS 626. The severe stress imposed in Hultquist's (1973) pot experiments (sufficient to kill C42-Y plants) seldom occurs in fields of those areas. Apparently the tendency of C42-Y to polarize a greater portion of its assimilates to the developing panicle, both during panicle development and grain fill, relates to its generally superior yield. Under more severe stress conditions, RS 626 may persist longer by allotting assimilate to the root at the expense of panicle development and then starting regrowth from tillers when stress is relieved. RS 626 characteristics relate more to persistence under stress than to maximum grain production. The desirable type or level of "drought resistance" for one production area may

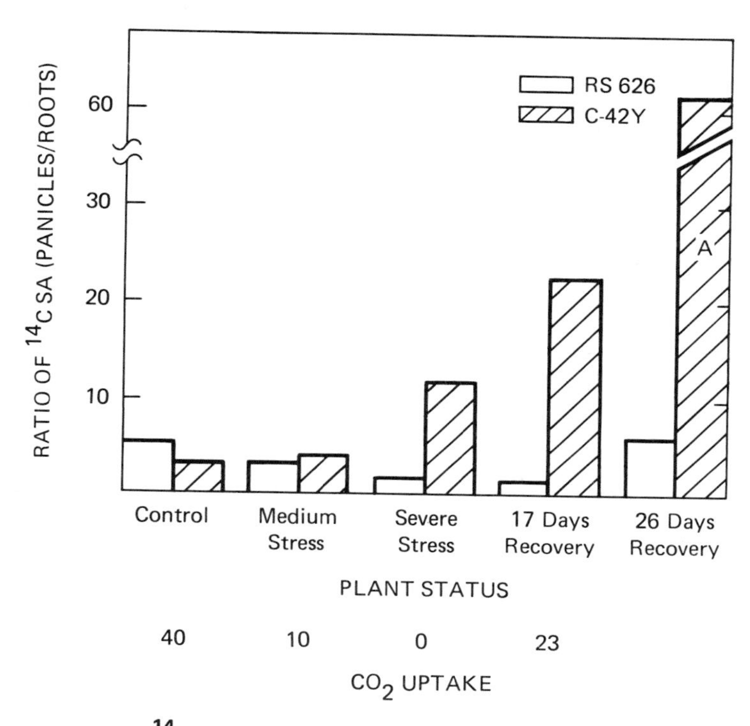


**Figure 5. Ratio of $^{14}C$ specific activity (SA) panicles/roots. Carbon dioxide uptake is expressed as mg $CO_2/dm^2/hr$. Unlike letters indicate significant at the 5% level. (Hultquist, 1973.)**

be different from another area depending upon the average level of drought and heat stress prevalent in the area.

Bennett (1979) applied stress treatments to sorghum by using osmotica in nutrient solutions. A -6.5 bar osmotic stress for 12 days during panicle development caused a reduction in leaf water potential of 2 bars associated with a 20% reduction in photosynthesis and a 42% loss in yield attributable primarily to a reduction in seed number.

## $GS_2$ Temperature Effects

Eastin et al. (1976) and Eastin (1976) tested post PI day/night temperature effects on grain yield in four sorghum hybrids. Temperature combinattions noted in Table 2 were applied at PI and held to maturity. Major effects occurred during $GS_2$ because seed number was heavily influenced and floret abortion during $GS_3$ was not obvious.

Hybrids range from temperate to cool tolerant types. Several points are clear. First, a 5 C elevation in night temperature above the apparent optimum

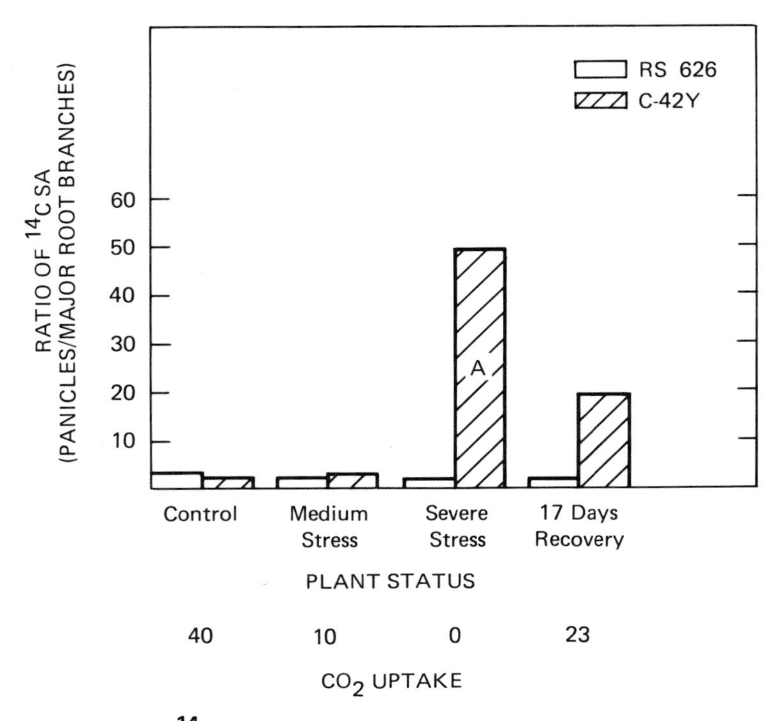


**Figure 6. Ratio of $^{14}C$ specific activity (SA) panicles/major root branches. Carbon dioxide uptake is expressed as mg $CO_2/dm^2/hr$. The letter "A" indicates a significant difference at the 5% level. (Hultquist, 1973.)**

for any of the hybrids reduced yield/head from about 25 to 36%. An increase of 10 C above optimum reduced yields about half. Second, hybrids definitely have different temperature optima. Third, metabolic or production efficiency, as expressed as grams of grain per $GS_3$ day, was definitely influenced by temperature. For example, respective grams of grain per $GS_3$ day for SJ7x156 at 29/17, 29/22, and 29/27 C are 1.30, 1.23 and 0.93. Similarly, grams of grain per $GS_3$ day for RS 610 changes from 1.19 at 29/22 C (optimum) to 0.96 at 29/27 C. Whether these production efficiency differences arise from reduction of seed number alone or involve lower efficiencies in metabolic pathways is not known.

Growth-room data on temperature-induced yield reductions were quite striking. Therefore, a system was set up to control night temperature in the field relative to ambient temperature. Ogunlela (1979) collected data with this system, part of which is reported in Table 3. Temperature treatment (ambient + 5 C) was applied at weekly intervals beginning at PI rather than over all of $GS_2$. The most sensitive period was floret differentiation (FD)

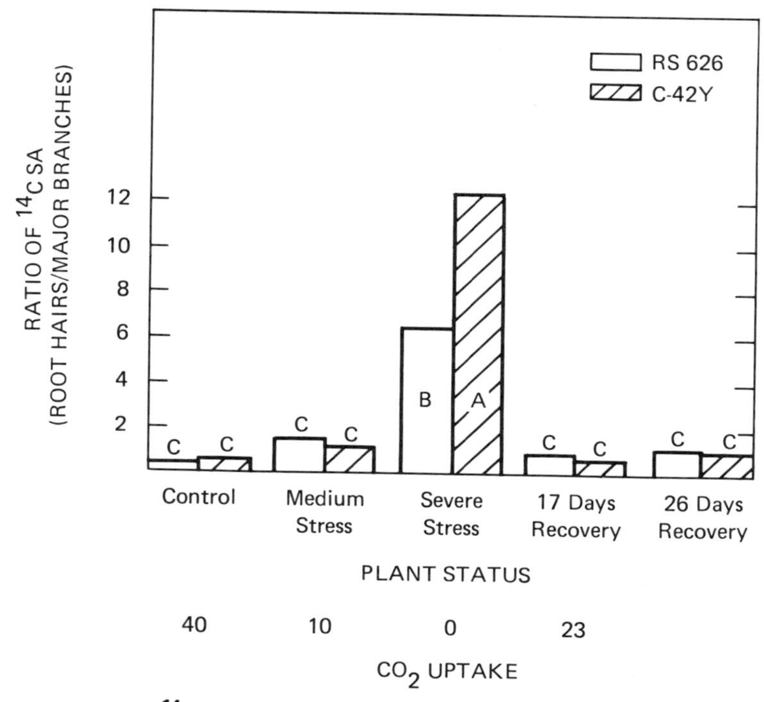


**Figure 7. Ratio of $^{14}C$ specific activity (SA) in root hairs and minor branches/major branches. Carbon dioxide uptake is expressed as mg $CO_2/dm^2/hr$. Unlike letters indicate significant differences at the 5% level. (Hultquist, 1973.)**

and development. The one week of temperature elevation reduced yields 28%. No water stress in the plots was evident.

A very basic problem that needs thorough investigation is why most grain crops are highly sensitive to damage from stresses imposed during floret differentiation and development. Photosynthate supply may be limiting or capacity of the inflorescence to compete with simultaneously expanding vegetative parts may be limiting even though a fairly generous assimilate supply may be available. Data in Figure 3 were cited earlier in reference to this problem. Perhaps stress factors upset growth regulator balances which alter capacity of the inflorescence to utilize assimilates or inhibit adequate assimilate transport to the inflorescence. These could be separate or combined influences if, in fact, they are pertinent at all. The one clear fact is that a good deal of basic research is necessary to characterize what is happening in terms of developmental control.

Note that all yield reductions were paralleled by nearly identical seed-number reductions and further that the reduction in grams of grain per $GS_3$

**Table 2. Temperature influence on sorghum grain yield in growth-room plants. Percent reduction from maximum yield is given in parenthesis. (Eastin et al., 1976.)**

| Genotype | Day/Night Temperature, C[a] | | | |
|---|---|---|---|---|
| | 29/17 | 29/22 | 29/27 | 34/22 |
| | | *– grain yield (g/head) –* | | |
| RS 610[b] (CK60 x 7078) | 36.8 (14%) | 42.6 — | 31.8 (25%) | 27.6 (35%) |
| CK60 x 60[c] | 32.8 ( 4%) | 34.0 — | 21.7 (36%) | 30.0 (12%) |
| 33[d] x 606 | 62.3 — | 42.1 (32%) | 28.3 (55%) | 31.9 (49%) |
| 606 | 18.8 — | 18.0 ( 4%) | 8.9 (53%) | 12.3 (35%) |
| SJ7[c] x 156[c] | 50.7 — | 38.3 (25%) | 26.8 (47%) | 30.9 (39%) |

[a]Temperature regimes imposed from panicle initiation to physiological maturity.
[b]Temperature hydrid.
[c]Cool-tolerant genotype.
[d]Intermediate temperature genotype.

day per plant followed similar trends. Production efficiency or metabolic efficiency appears to relate very closely to sink capacity (seed number). This may be through a sink influence on photosynthesis, but such an effect is not well understood particularly in biochemical terms.

Assimilate partitioning relates to seed number and is a problem when water stress is involved as demonstrated by Hultquist (1973). Rice (1979) investigated temperature x water interaction effects on root respiration during panicle development since roots are a potential competitive sink for assimilates when seed-number potential is being set. He grew DeKalb C-46 (stress-tolerant sorghum hybrid) and RS 671 (normal hybrid) in nutrient solution and applied osmotica to create water stress at different temperatures. Root respiration was monitored at all combinations of temperature and water status. Respiration rate is presumably an indicator of assimilate requirement. Figure 8 illustrates an average of 31% higher root respiration rates in RS 671 than C-46 over the range of temperature and water stress combinations. That much higher assimilate requirement during panicle development may be significant in terms of simultaneously setting a high seed-number potential.

## Grain Filling

Eastin et al. (1973) demonstrated the functional effectiveness of the

**Table 3.** Influence of night temperature on yield and other characteristics of RS 671 grain sorghum at Lincoln, Nebraska. Night temperatures in the field were regulated at 5 C above ambient. Values in parenthesis are percent reduction from the control except for weight per 1000 seeds which is change from control. (Ogunlela, 1974.)

| Time of Temperature Treatment | Grain Yield | Seed Number | Weight of Seed | Rate of Dry Matter Accumulation in Grain |
|---|---|---|---|---|
| | *– g/plant –* | *– no./plant –* | *– g/1000 seed –* | *- g grain/$GS_3$ day/plant -* |
| Control | 66.9 | 2659 | 26.6 | 2.09 |
| $PI_1$ to $PI_7$ [a] | 59.3 (11%) | 2333 (12%) | 27.2 ( 2%) | 1.85 (11%) |
| $PI_8$ to $FD_1$ [b] | 53.4 (20%) | 2174 (18%) | 27.8 ( 5%) | 1.71 (18%) |
| $FD_1$ to $FD_7$ | 48.0 (28%) | 1855 (30%) | 29.7 (12%) | 1.49 (29%) |
| $FD_8$ to $BL_1$ [c] | 52.7 (21%) | 2176 (18%) | 27.8 ( 5%) | 1.66 (21%) |
| $BL_1$ to $BL_7$ | 55.9 (16%) | 2223 (16%) | 25.5 (-4%) | 1.80 (14%) |

[a] PI is panicle initiation (subscripts are days). [b] FD is floret differentiation (stamen and pistil primordia). [c] BL is bloom.

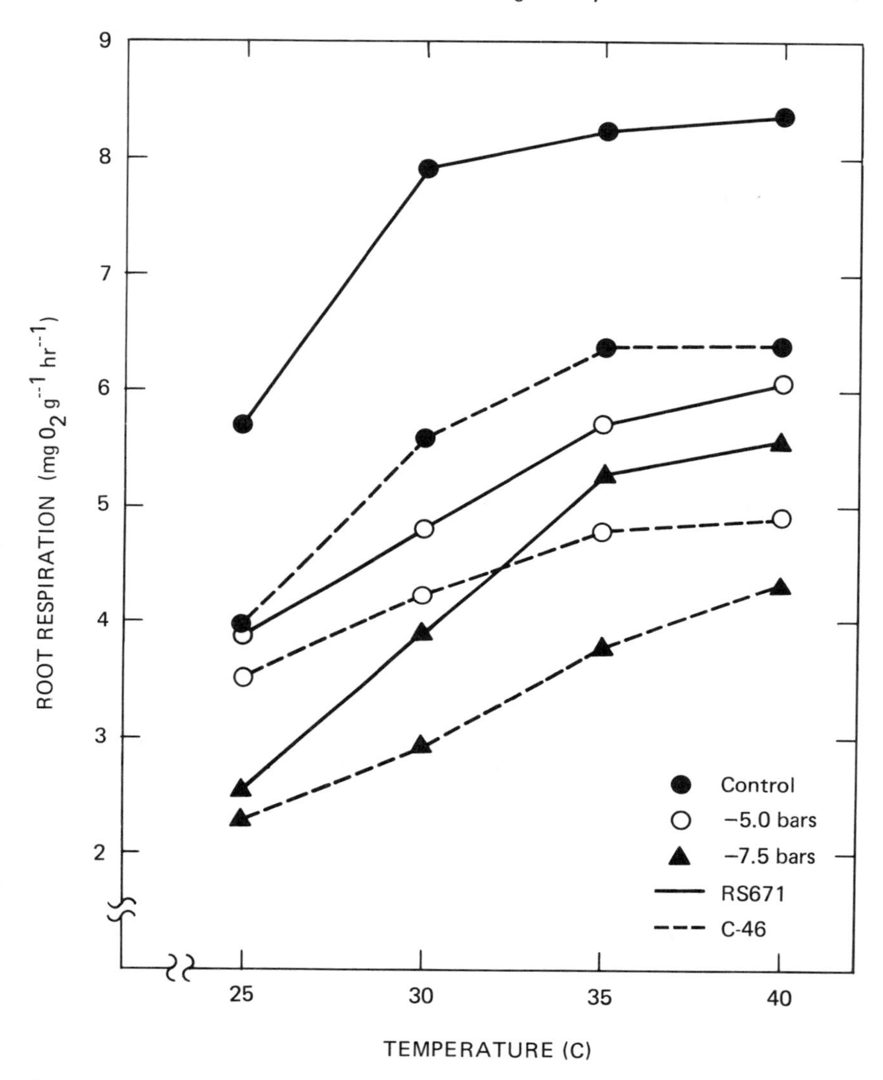


**Figure 8. The effect of temperature (25, 30, and 40 C) on root respiration in RS671 and Dekalb C-46 grain sorghum, under control and water stress conditions, during the boot stage of development. (Rice, 1979.)**

dark layer or black layer in signaling physiological maturity in grain by $^{14}C$ labeling of assimilate and noting its movement from leaf to grain. Subsequent determination of bloom and dark layer dates permitted precise definition of grain-fill length and its relation to yield.

Both temperature and water influence length of grain fill but neither effect is well documented over a wide range of genotypes and environments. We do have considerable dryland and irrigated field data but comparative values do not differ widely because the environment was relatively favorable. Temperature effects from the yield data in Table 2 are pronounced. An increase of 5 C above optimum night temperature reduced $GS_3$ length from 8 to 20%. With grain-fill rates under good conditions in the field commonly being on the order of 150 to 190 kg/ha/day, the importance of extending the grain fill period is obvious (Eastin, 1980).

Temperature effects on length of grain fill in sorghum, corn, and wheat (Warrington et al., 1977) are probably sizeable and economically important in humid, temperate regions. Under moderately good conditions, correlation coefficient between grain yield and $GS_3$ length in sorghum is about 0.7. Also, a comparison of corn versus sorghum under good production conditions shows that corn has about a 35% longer grain fill period and yields about 30% more (Eastin, unpublished). Both crops have similar photosynthetic rates. The essential difference lies in developmental characteristics which permit corn to fill grain longer.

Substantial variability exists for length of grain fill in sorghum (Eastin, 1972) and in corn (Fischer, 1980). Selection for length of grain fill is feasible. Also, consideration should be given to investigation of factors that result in black layer formation in cereals.

Kiesselbach and Walker (1952) were among the earliest to recognize the existence of a dark layer area, although the complete significance of it was not stated. They recognized passage of all solutes from the mother plant to the seed through specialized conducting tissue in that area but did not concern themselves with its use as an indicator of physiological maturity. Figure 9 (Giles et al., 1975) illustrates tissue arrangements associated with conducting the phloem solutes that are necessary for kernel growth and development. The phloem terminates in phloem parenchyma tissue just beyond the upper tip of the pedicel. The phloem parenchyma lies adjacent to the placento-chalazal pad (Kiesselbach and Walker, 1952) which arises from fusion of the ovule with the ovary wall. The closing layer, or dark layer at maturity, lies just outside this area. All solutes to the kernel must pass through this area to be absorbed and further transported through the transfer cells which arise from basal endosperm tissue.

About two weeks before maturity in corn, the closing layer cells begin to shrink as cell content diminishes. Also, cytoplasm and nuclei losses occur in basal endosperm tissue as they are crushed and absorbed by an expanding scutellum as the embryo grows in corn. Darkening in the closing tissue area gives rise to the dark layer used to determine physiological maturity.

According to Giles et al. (1975) the appearance of phenolic compounds in the phloem parenchyma is associated with formation of the dark layer

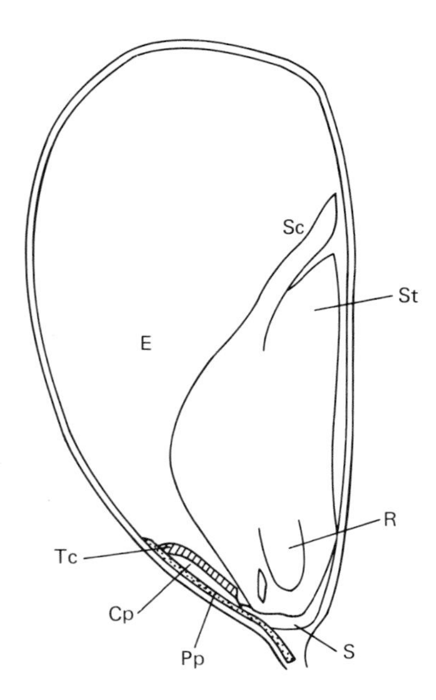


**Figure 9. Cross-section of a sorghum kernel illustrating specialized conducting tissues in the assimilate translocation pathway (Giles et al., 1975). Placento-chalazal pod (cp), endosperm (E), phloem parenchyma (Pp), radicle (R), seed coat (S), scutellum (Sc), shoot (St), and transfer cells (Tc).**

adjacent to the transfer cells in sorghum. Senescence of phloem tissue is suspected (in forcing cessation of transport) due to blockage by slime plugs, mucilage and pectic compounds. Neither the transfer cells nor the meristematic area outside the placento-chalazal pad are crushed or compressed at maturity in sorghum.

L. K. Fussell and D. M. Dwarte (personal communication) evaluated dark-layer function and anatomy in *Pennisetum americanum* in which the anatomy of the region is similar to sorghum. However, pigmented compounds causing dark color accumulated in the placento-chalazal pad instead of the phloem parenchyma as in both corn and sorghum. Also, the scutellum appears to digest and absorb endosperm cell layers, including transfer cells, as in corn.

The sorghum, corn, and millet examples of apparently contrasting mechanisms of cutting off assimilate flow to developing kernels bear investigation as they relate to seed-size potential in cereals. There is appreciable potential for manipulating yield upward through maintaining a functional transport system to the kernel in order to extend grain fill length. For example, as pointed out earlier, the seed-size potential in sorghum is 25 to 50% larger than is normally realized in the field. Since sorghum has a perennial growth habit and the leaves remain green long after kernel dark-layer formation, photoassimilates cannot be directed to grain, the economic product.

Investigations are needed on the course of events preceding and during assimilate cutoff to grain of corn, sorghum, and pearl millet. How can embryo development be arrested to avoid scutellar growth and absorption of kernel transfer cells? How can loss of cell contents in the basal endosperm and closing-layer tissues be delayed beginning about two weeks before maturity? Does the accumulation of pectins, mucilages, phenols, etc., in phloem parenchyma relate to a general metabolic slow-down in sorghum kernels? How can a dynamic state be maintained and general senescence be avoided? How is growth regulation in developing seeds mediated? Our knowledge is so sparse that research is needed to determine whether these are even pertinent questions.

Several investigators have detailed the nature of the grain-fill curve in sorghum but knowledge of that period remains sparse. Dickinson (1976) applied high temperatures in the field to sorghum panicles and found the most sensitive period to be seven to nine days after anthesis. This is probably about the time endosperm cell expansion was ceasing and starch synthesis was beginning as reported in wheat by Evers (1970) and Wardlaw (1970) at 16 days. However, M. P. Cochrane and C. M. Duffus (personal communication) recently demonstrated division of endosperm cells in barley as late as 28 to 30 days after anthesis. Obviously more complete definition of cell division and cell enlargement events would be useful in determining the extent to which one could expect to increase yield through the seed-size component.

**Metabolic Considerations**

Speculation regarding metabolic considerations associated with pace of developmental sequences is risky considering present knowledge. Nonetheless, some questions appear to be in order. The pace of development (presumably linked to metabolic pace) during $GS_2$ influences seed-number potential and yield. Data in addition to that in Table 2 confirm that as $GS_2$ length declines (at adapted temperatures) total seed number declines. Similarly, as $GS_3$ length declines yield declines and, as pointed out from data in Tables 2 and 3, production or metabolic efficiency (g grain/plant/day) likewise declines. Chowdhurry and Wardlaw (1978) illustrated that $GS_3$ declines with elevated temperatures in wheat, sorghum and rice. Part of the declines in yield and production efficiency is related to seed number or sink strength. The question also must be posed as to whether or not general metabolic efficiency declines with increase in metabolic pace beyond some level.

Since yield losses apparently are related partly to night effects, the possibility of implicating dark reactions exists. We chose to test effects of temperature (which influences metabolic pace) on the efficiency of converting seed dry matter to new growth by germinating seed of hybrids and lines at 20 C and splitting seed lots to grow for 4 days at 25 C, 30 C, 35 C and 40 C. New growth then was separated from the seed to get both the remaining seed weight and new growth. New growth of hybrids and lines was the same at 30 C, but at all other temperatures there appeared to be greater hybrid growth than line growth. At the same time the percentage of seed weight lost to new growth tended to be lower for hybrids.

Calculation of g new growth/g seed weight lost gives an estimate, though far from perfect, of growth efficiency or metabolic efficiency of the growth process. Grams new growth/g seed weight lost was the same for lines and hybrids at 30 C (suggesting near optimal conditions) but hybrids tended toward higher efficiency levels both above and below 30 C. It appears that metabolic pace dictated by temperature (and other environmental variables) may influence metabolic efficiency. This is another area which merits attention. Evaluation of factors involved will include at least considerations of sink strength, growth respiration, and maintenance respiration.

Since metabolic pace may bear on metabolic efficiency and does, without doubt, bear on yield by way of influencing length of $GS_2$ and $GS_3$, estimates of temperature influence on metabolic pace are of interest. We attempted to measure temperature effect on metabolic pace by measuring temperature effect on respiration rate of sorghum plants at PI (Table 4). Respiration was chosen since it is coupled to many essential synthetic processes. The temperatures used were 10, 15, 21 and 27 C. The average 15.3% increase in respiration per degree C is appreciable. Since changes in respiration, and presumably metabolic pace, are large for small changes in temperature, it is

**Table 4. Influence of temperature on respiration rate of sorghum plants at panicle initiation. (J. D. Eastin, unpublished data.)**

| | Genotype | | |
|---|---|---|---|
| **Temperature Range** | **RS 610** | **SJ7 x 156** | **Mean** |
| *– C –* | *– % increase in respiration/degree C –* | | |
| 10 to 15 | 18.1 | 15.8 | 17.0 |
| 15 to 21 | 14.6 | 15.0 | 14.8 |
| 21 to 27 | 14.1 | 14.3 | 14.2 |
| Mean | 15.6 | 15.0 | 15.3 |

of interest to know what is the range of response of different genotypes to different temperatures (environments). Average respiration rates of RS 610 (temperature) and SJ7 x 156 (cool tolerant) (Tables 2 and 4) are 1.41 and 1.78 mg $CO_2$ evolved/g dry weight/hour over the 10 to 27 C range. SJ7 x 156 respires 26% faster (1.78/1.41) on the average which must be related to its ability to grow satisfactorily in cooler, high altitude areas.

Gerik (1979) evaluated three random mating populations of sorghum for variable panicle respiration response in the field at three temperature levels (3 times a day). Mean, minimum, and maximum panicle respiration rates (mg $CO_2$/g/hr) were 1.91, .58 and 5.22, respectively. The wide respiration rate range of 4.64 mg $CO_2$/g/hr should be sufficient to permit selecting genotypes to best fit a particular environment. Potential gains from such a selection scheme are unknown.

Mahalakshmi (1978) and Gerik (1979) found panicle respiration to be much less responsive than vegetation respiration to temperature. Mahalakshmi (1978) found response to temperature of starch synthetase activity in grain to be low as was grain respiration response.

Photosynthesis has been neglected, partly in anticipation of it being covered elsewhere. Norcio's work (1976) is cited to demonstrate ability to select for heat tolerance. Sullivan et al. (1977) reported on Sullivan's heat tolerance test used to isolate heat tolerant sorghums which showed superior yielding ability (Ogunlela, 1974) during the extraordinarily hot Nebraska summer of 1973. Photosynthesis declined in RS 626 at 40 C but remained active in 4104, a heat tolerant selection, at 43 C. Norcio (1976) further showed with tests on a hybrid and its parent lines that heat tolerance for photosynthesis can be manipulated genetically.

Heavy emphasis has been given to developmental processes (growth type processes) at the expense of individual physiological processes because other workshop participants have covered them. Also growth processes are usually far more sensitive to environmental stresses than are photosynthesis, respiration, etc. (Hsiao, 1973). Research on yield limitations in cereals should emphasize environmental effects on cell division, cell elongation, differentiation, and senescence processes.

## SUMMARY

Analysis of yield-limiting problems for crops in humid, temperate areas should include careful evaluation of water and temperature effects on developmental sequences that influence the seed-number and seed-size components of yield. In regard to seed number, particular attention should be given to problems of assimilate partitioning during inflorescence development when vegetative and floral structures are being expanded simultaneously. Cereals are very sensitive to environmental stresses at this stage. A mechanistic explanation of why differentiation and expansion of floret components is more sensitive than vegetative components to environmental stresses might be useful in trying to select for tolerant genotypes. The efficiency of root function and metabolism, as a competing sink during inflorescence development, appears important.

Investigations of grain fill are lacking even more than those on problems of inflorescence development. If seed-size potential in cereals is to be exploited, much work will be needed on the nature of senescence in grain endosperm transfer cells, the placento-chalazal pad and adjacent phloem parenchyma cells.

Temperature effects of yield appear more subtle and sizeable than might be suspected. Investigations on temperature x water interactions are wanting. Genotypes obviously have different temperature optima. Temperature is a prime controller of metabolic pace and respiration rate. If respiration can be considered a crude measurement of metabolic pace, adequate genetic variability appears to be available to select sorghums for a wide range of temperature conditions. The closeness of a metabolic pace and respiration association remains to be defined.

Generally more effort has been expended on effects of water and temperature stress on photosynthesis, respiration, translocation, etc., than on developmental processes. Such work should be continued. Developmental processes that correlate more closely with yield and yield components, however, merit more effort. Developmental processes depend upon photosynthesis, respiration, etc., but also involve the processes of cell division, differentiation, and growth that generally are more sensitive to environmental stresses than are the primary physiological processes.

## NOTES

J. D. Eastin, Professor, Department of Agronomy, University of Nebraska, Lincoln, Nebraska 68582. R. M. Castleberry, T. J. Gerik, J. H. Hultquist, V. Mahalakshmi, V. B. Ogunlela, and J. R. Rice are former graduate students.

Contribution of the Nebraska Agricultural Experiment Station. Published as Paper No. 6852, Journal Series, Nebraska Agricultural Experiment Station, Lincoln, Nebraska 68583.

## LITERATURE CITED

Ahad, A. A. 1979. Yield component study in winter wheat. M.Sc. Thesis. University of Nebraska, Lincoln, Nebraska.

Beil, G. M., and R. E. Atkins. 1967. Estimates of general and specific combining ability in $F_1$ hybrids for grain yield and its components in grain sorghum, *Sorghum vulgare.* Crop Sci. 7: 225-228.

Bennett, J. M. 1979. Responses of grain sorghum [*Sorghum bicolor* (L.) Monech] to osmotic stresses imposed at various growth stages. Ph.D. Thesis. University of Nebraska, Lincoln, Nebraska.

Blum, A. 1967. Effect of soil fertility and plant competition on grain sorghum panicle morphology and panicle weight components. Agron. J. 59: 400-403.

Blum. A. 1970. Heterosis in grain production by sorghum. Crop Sci. 10: 28-31.

Castleberry, R. M. 1973. Effects of thinning at different growth stages on morphology and yield of grain sorghum [*Sorghum bicolor* (L.) Moench]. Ph.D. Thesis, University of Nebraska, Lincoln, Nebraska.

Chowdhurry, S. I., and I. F. Wardlaw. 1978. The effect of temperature on kernel development in cereals. Aust. J. Agric. Res. 29: 205-223.

Dickinson, T. E. 1976. Caryopsis development and the effect of induced high temperature in [*Sorghum bicolor* (L.) Moench]. M.S. Thesis. University of Nebraska, Lincoln, Nebraska.

Doggett, H. 1967. Yield increase from sorghum hybrids. Nature 216: 798-799.

Eastin, J. D. 1972. Efficiency of grain dry matter accumulation in grain sorghum. p. 7-17. In J. I. Sutherland and R. J. Falasca (eds.) Proceedings of the twenty-seventh annual corn and sorghum research conference. American Seed Trade Association.

Eastin, J. D. 1976. Temperature influence on sorghum yield. p. 19-23. In H. P. Loden and D. Wilkinson (eds.) Proceedings of the thirty-first annual corn and sorghum research conference. American Seed Trade Association.

Eastin J. D. 1980. Sorghum development and yield. In press. In S. Yoshida (ed.) Proceedings of symposium on potential productivity of field crops under different environments. International Rice Research Institute, Los Banos, Laguna, Phillipines.

Eastin, J. D., I. Brooking, and S. O. Taylor. 1976. Influence of temperature on sorghum respiration and yield. Agron. Abstracts, p. 71. American Society of Agronomy.

Eastin, J. D., J. H. Hultquist, and C. Y. Sullivan. 1973. Physiologic maturity in grain sorghum. Crop Sci. 13: 175-178.

Eastin, J. D., and C. Y. Sullivan. 1974. Yield considerations in selected cereals. p. 871-877. In L. Bieleski, A. R. Ferguson, and M. M. Creswell (eds). Mechanisms of regulation of plant growth. Bulletin 12, Roy Soc. of N.Z., Wellington.

Evers, A. D. 1970. Development of endosperm of wheat. Ann. Bot. 34: 547-555.

Fischer, R. A. 1980. Wheat. In press. In S. Yoshida (ed.) Proceedings of symposium on potential productivity of field crops under different environments. International Rice Research Institute, Los Banos, Laguna, Philippines.

Fischer, K. S., and G. L. Wilson. 1975. Studies of grain production in [*Sorghum bicolor* (L.) Moench.] V. Effect of planting density on growth and yield. Aust. J. Agric. Res. 26: 31-41.

Gerik, T. J. 1979. The relation of photosynthesis and dark respiration in grain sorghum [*Sorghum bicolor* (L.) Moench] to yield, yield components and temperature. Ph.D. Thesis, University of Nebraska, Lincoln, Nebraska.

Giles, K. L., H. C. M. Bassett, and J. D. Eastin. 1975. The structure and ontogeny of the hilum region in *Sorghum bicolor.* Aust. J. Bot. 23: 795-802.

Hsiao, T. C. 1973. Plant responses to water stress. Ann. Rev. Plant Physiol. 24: 519-570.

Hultquist, J. H. 1973. Physiologic and morphologic investigation of sorghum [*Sorghum bicolor* (L.) Moench] I. Vascularization. II. Response to internal drought stress. Ph.D. Thesis. University of Nebraska, Lincoln, Nebraska.

Kambal, A. E., and O. J. Webster. 1966. Manifestation of hybrid vigor in grain sorghum and the relations among the components of yield, weight per bushel and height. Crop Sci. 6: 513-515.

Kiesselbach, T. A., and E. R. Walker. 1952. Structure of certain specialized tissues in the kernel of corn. Amer. J. Bot. 39: 561-569.

Mahalakshmi, V. 1978. Temperature influence on respiration and starch synthetase in [*Sorghum bicolor* (L.) Moench]. Ph.D. Thesis, University of Nebraska, Lincoln, Nebraska.

Norcio, N. V. 1976. The effects of high temperatures and moisture stress on photosynthetic and respiration rates of grain sorghum. Ph.D. Thesis, University of Nebraska, Lincoln, Nebraska.

Ogunlela, V. B. 1974. A field study of heat and drought tolerance of grain sorghum [*Sorghum bicolor* (L.) Moench] as an approach to genetic improvement. M.S. Thesis, University of Nebraska, Lincoln, Nebraska.

Ogunlela, V. B. 1979. Physiological and agronomic responses of a grain sorghum [*Sorghum bicolor* (L.) Moench] hybrid to elevated night temperatures. Ph.D. Thesis, University of Nebraska, Lincoln, Nebraska.

Quinby, J. R. 1963. Manifestation of hybrid vigor in sorghum. Crop Sci. 3: 288-291.

Rice, J. R. 1979. Physiological investigations of grain sorghum [*Sorghum bicolor* (L.) Moench] subjected to water stress conditions. Ph.D. Thesis, University of Nebraska, Lincoln, Nebraska.

Stickler, F. C., A. W. Pauli, H. H. Laude, H. D. Wilkins, and J. L. Mings. 1961. Row width and plant population studies with grain sorghum at Manhattan, Kansas. Crop Sci. 4: 297-300.

Sullivan, C. Y., N. V. Norcio, and J. D. Eastin. 1977. Plant responses to high temperature. p. 301-317. In A. Muhammed, R. Aksel, and R. C. von Barstel (eds.) Genetic diversity in plants. Plenum Press, New York.

Wardlaw, I. F. 1970. The early stages of grain development in wheat: Response to light and temperature in a single variety. Aust. J. Biol. Sci. 23: 765-774.

Warrington, I. J., R. L. Dunstone, and L. M. Green. 1977. Temperature effects at three development stages on the yield of the wheat ear. Aust. J. Agric. Res. 28: 11-27.

Yoshida, S. 1980. Rice. In Press. In S. Yoshida (ed.) Proceedings of symposium on potential productivity of field crops under different environments. International Rice Research Institute, Los Banos, Laguna, Philippines.

8

# PHYSIOLOGICAL BASIS FOR CHILLING STRESS AND THE CONSEQUENCES FOR CROP PRODUCTION

**J. R. McWilliam**

Many crops of tropical and subtropical origin, including corn, cotton, soybean, sorghum, rice, and tomato, suffer chilling injury when subjected to non-freezing temperatures below about 10 to 15 C. This response is in sharp contrast to the behavior of temperate species which are tolerant of low temperatures in this range and in some cases may have temperature optima close to those in the upper chilling range.

Common symptoms of chilling injury include poor establishment, chlorosis, retarded growth and development, and reduced yield. Chilling injury is not restricted to crops growing under field conditions and also is responsible for symptoms of surface pitting, necrosis, and discoloration of a wide range of tropical fruits and vegetables during cold storage (Lyons, 1973).

Chilling stress is of little or no consequence to crops growing in tropical latitudes, but can become a serious problem when tropical species are grown as summer crops in temperate environments. With these crops, chilling injury occurs most commonly in the spring and early summer when there is a greater probability of experiencing low soil and air temperatures. Chilling injury also is observed during latter stages of crop growth and particularly during the sensitive stages of flowering and fruit or grain development.

In this chapter I will restrict discussion to a consideration of the primary events involved in chilling injury and the nature and significance of the secondary biochemical and physiological dysfunctions that are entrained as a consequence of these. Also, I propose to look at the short and long term consequences of these physiological dysfunctions on important aspects of growth and development of crops, the nature and extent of the variation observed for low temperature tolerance, and possible avenues for ameliorating or minimizing low temperature lesions in these chilling-sensitive crops. (Low temperature as used in this review refers to temperatures in the chilling range for sensitive crops.)

## THE MECHANISM OF CHILLING INJURY

A hypothesis to explain both the non-linear temperature dependency of Arrhenius plots of many reactions in the tissues of chilling-sensitive plants when exposed to low temperature and the presence of a 'break' around a postulated 'critical temperature' for chilling injury was based on the concept of a single primary event. This event was thought to involve a physical phase transition in the membrane, from a normal flexible liquid-crystalline to a solid gel structure as the temperature fell below the critical level for the particular species (Lyons, 1973; Raison et al., 1971).

There is evidence that chilling resistance in plants is associated with a high concentration of unsaturated fatty acids, particularly linolenic acid (18:3) (Lyons, 1973), and that inhibition of linolenic acid biosynthesis in membranes of plants increases their sensitivity to chilling injury and prevents cold-hardening (St. John, 1976; St. John and Hilton, 1976). However, lipid analyses of bulk membranes suggest that there is no basis for a phase transition or separation at physiological temperatures about 0 C (Bishop et al., 1979), although the possibility that phase transitions may occur in smaller discrete domains within critical regions of the membrane has not been ruled out.

The effect of low temperature on the activity of a number of soluble enzyme proteins has also been reported by Duke et al. (1977) and Dowton and Hawker (1975); however, there is some evidence that the sharp breaks in the Arrhenius plots for these enzymes at low temperatures may be due to a conformational change induced by a physical change in associated lipid or in the lipid component of an associated membrane. Studies with the enzyme PEP-carboxylase (not membrane-bound) from chilling-sensitive $C_3$ and $C_4$ plants indicate that chilling temperatures below about 10 C have a direct effect on the enzyme activity, possibly through altered substrate affinity (Graham et al., 1979; McWilliam and Ferrar, 1974). These results suggest that the low temperature, in addition to effects on membranes, may have direct effect on the structure and hence activity of enzyme proteins.

Another primary effect of chilling in sensitive species is a decrease in membrane permeability. Kramer (1942) was one of the first to observe this decreased hydraulic conductivity of the roots of chill-sensitive species. More recently, Markhart et al. (1979a) have demonstrated a similar decline in the conductivity of soybean roots following chilling, and my own evidence (Unpublished data) for the slow closure of stomates of chilled cotton leaves may be attributable, at least in part, to the reduced flux of water out of the guard cells at low temperature (Figure 1).

Based on this and other evidence (Lyons et al., 1979b), a hypothesis is proposed to explain the effects of chilling temperatures on sensitive plants. Low temperature invokes one or a number of primary events involving a change in the molecular ordering, or state, of cell membranes. This change leads to reduced permeability and possibly polymeric separation or conformational changes in free or membrane-associated enzymes. Any or all of these primary events lead quite rapidly to the inhibition of metabolic processes and the accumulation of toxic products, an increase in the $Q_{10}$'s of reaction rates leading to physiological dysfunction and, if sustained, to a range of visible symptoms and ultimately death (Figure 2).

## PHYSIOLOGICAL BASIS OF CHILLING INJURY AND CONSEQUENCES FOR CROP PRODUCTION

The extent and severity of chilling injury is a function of the temperature and duration of the chilling stress, the species, and the condition and stage of plant development. Although the effects of chilling injury can be observed at any stage, it tends to be more serious during germination and early seedling growth when plants are more vulnerable to the metabolic and physiological dysfunctions that follow exposure to chilling temperatures.

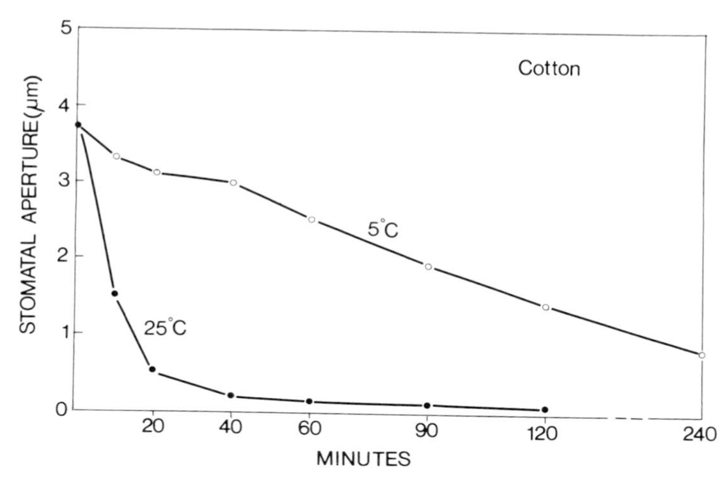


**Figure 1. Change in stomatal aperture with time in epidermal strips from cotton *(Gossypium hirsutum)* leaves floated on polyethylene glycol (–12 bars) at 5 and 25 C. (J. R. McWilliam, unpublished data.)**

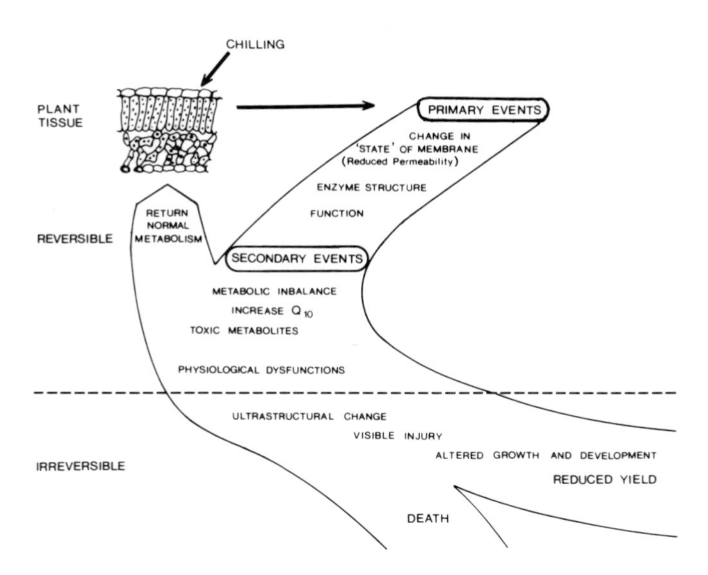


**Figure 2. Schematic pathway of the sequential events that follow the primary lesions induced by the exposure of chilling-sensitive plants to low temperatures in the chilling range.**

Chilling symptoms are expressed at low temperatures, especially if prolonged, but appear more rapidly after plants are returned to an elevated temperature. Although a wide range of chilling symptoms are observed, the underlying primary biophysical mechanisms discussed earlier appear to be common to chilling-sensitive species and are responsible for the secondary symptoms which are ultimately expressed. Descriptions of some of the more important of these follow.

## Biochemical and Physiological Responses

The biochemical and physiological events that follow as a consequence of the primary chilling events in chilling-sensitive plants have been reviewed recently (Levitt, 1972; Lyons, 1973; Lyons et al., 1979a; and Raison, 1974). These events include a reduction in the rate of many reactions as reflected in increased $Q_{10}$'s, an imbalance in metabolites, and a disproportionate decrease in respiration rates of both isolated mitochondrion and intact tissue. They also include reduced capacity for oxidative phosphorylation, inhibition of protein synthesis, impairment or cessation of protoplasmic streaming, accumulation of toxic products and ethylene, increase in permeability of cell membranes, and the loss of electrolytes from a wide range of plant tissues. These dysfunctions are associated with ultrastructural changes including the progressive deterioration of cellular membranes and the loss of cytoplasmic structure possibly through direct effects of low temperature on the cytoskeleton. These early symptoms of chilling injury lead directly to the impairment of other vital plant functions, including cell division and cell extension, tissue water relations, chlorophyll synthesis, photosynthesis, and ultimately to dry matter production and yield.

**Growth and Development**

The primary and secondary temperature lesions described in this review have important consequences for the growth and development of chilling-sensitive crop plants under field conditions. Some of the more important of these consequences follow.

*Germination and Seedling Emergence.* Seeds of chilling-sensitive crop plants, such as corn, rice, soybean, cotton, and sorghum, are injured and often fail to germinate at temperatures much below about 10 to 12 C which are common in field soils during late spring. Many seeds are particularly sensitive to chilling injury during the first few minues of imbibition and the injury can be reduced if the seeds are imbibed before exposure to low temperature (Christiansen, 1967). The injury is thought to be due to the impairment of membrane reorganization during hydration of the cells, as evidenced by the high rate of leakage of solutes (Bramlage et al., 1979; Leopold and Musgrave, 1979; Simon, 1974).

Arrhenius plots of germination rate for two chilling-sensitive sorghum species, a commercial U.S. hybrid *(S. bicolor)* and a wild tropical species *(S. verticilliflorum),* are illustrated in Figure 3a. A consistent feature of the germination response of such species is the sudden increase in the $Q_{10}$ below a certain temperature range, which in this case was higher (14 to 16 C) for the more sensitive tropical species than for the commercial U.S. hybrid (11 to 12 C). A similar difference has been observed in the case of indica rice varieties (Nishiyama, 1976). Changes in the $Q_{10}$'s for the respiration of the intact seed of these two species followed closely the pattern for germination (McWilliam et al., 1979).

Emergence of seedlings following germination involves the elongation of the mesocotyl/hypocotyl and the radicle. In chilling-sensitive mungbean *(Vigna radiata)* the temperature coefficient for the growth of the hypocotyl and radicle increases abruptly below about 15 C (Raison and Chapman, 1976). A similar response has been observed for the elongation of the mesocotyl of sorghum seedlings (Figure 3b). There are also numerous reports of injury and visible necrosis of the developing radicle of many chilling-sensitive crops such as corn and cotton (Christiansen, 1968; Creencia and Bramlage, 1971).

Slow germination of seed of sensitive species, such as cotton, sorghum,

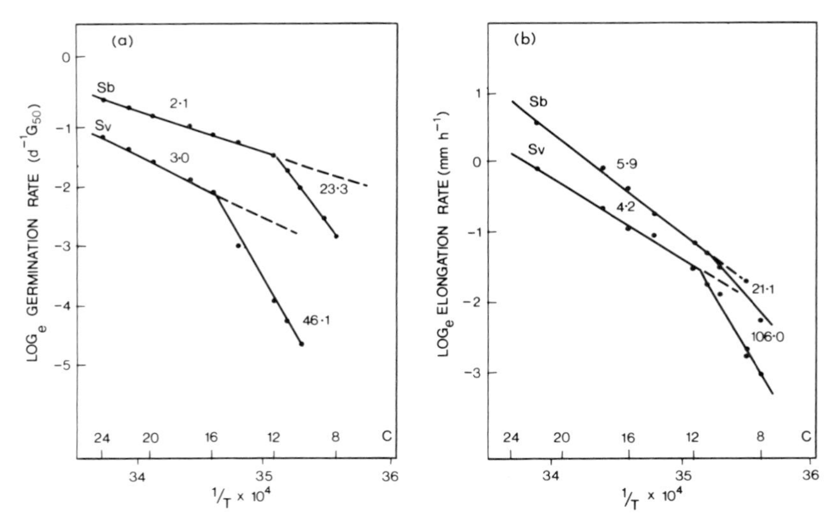


**Figure 3. Arrhenius plots for (a) germination rate (reciprocal of days to 50% germination) and (b) mesocotyl elongation rate (mm hour$^{-1}$) for two sorghum species (Sb., *S. bicolor,* commercial hybrid S.v., *S. verticilliflorum,* tropical wild sorghum). $Q_{10}$ values derived from regressions are indicated for each slope. (Adapted from McWilliam et al., 1979.)**

and soybean, at low soil temperatures often predisposes seedlings to attack by fungal and other soil borne pathogens and can cause a reduction in the growth and development of plants when returned to a favorable temperature regime (Christiansen, 1964; McWilliam et al., 1979; Obendorf and Hobbs, 1970). Chilling also can have the longer term effects of reducing growth, delaying flowering, and influencing the quality of the final product (Christiansen and Thomas, 1969).

*Chlorophyll Synthesis.* Another common symptom of chilling injury during the early growth of chilling-sensitive crops is the slow growth and chlorosis of the first formed seedling leaves. Chlorophyll synthesis is severely depressed at low temperature in many chilling-sensitive species (McWilliam and Ferrar, 1974; van Hasselt and Strikwerda, 1976). At chilling temperatures, the ability of plants in the light to accumulate chlorophyll is determined by the difference between synthesis and photo-oxidation, the latter being a function of the incident light energy flux. At 15 C at levels of irradiance as low as 15% of full sunlight, there is no net synthesis of chlorophyll in etiolated leaf tissue of corn and sorghum (Figure 4), and if the exposure is prolonged, the injury is irreversible (McWilliam et al., 1979; Millerd and McWilliam, 1968). Evidence from electron micrographs of leaf tissue (McWilliam et al. 1979) suggests that the failure to develop chlorophyll under these conditions is associated with the arrested development of the thylakoid membrane

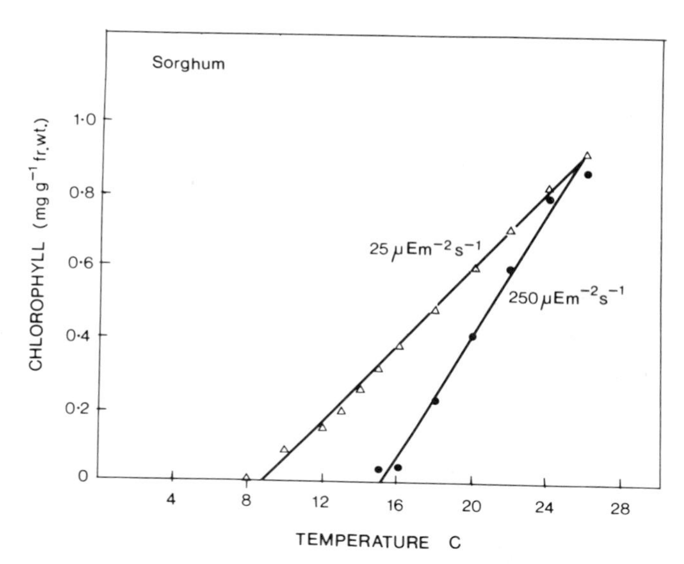


**Figure 4. Chlorophyll accumulation in etiolated leaves of *Sorghum bicolor* seedlings when exposed to a range of temperatures from 8 to 26 C at irradiances of 25 and 250 $\mu$E m$^{-2}$sec$^{-1}$. (Adapted from McWilliam et al., 1979.)**

system of the developing plastids (Figure 5).

The effects of chilling on chlorophyll synthesis are not restricted to seedlings. Leaves of chilling-sensitive tropical plants often develop chlorotic bands (Faris bands) when exposed to diurnal temperatures with a minimum around 10 C (Taylor et al., 1975). Also Bagnall (1979) and Muramoto et al. (1971) have reported loss of chlorophyll and chlorosis in leaves of adult sorghum and cotton plants, respectively, during extended periods at low temperature in the light.

*Photosynthesis.* Before there is any evidence of chlorophyll breakdown in green tissue at chilling temperatures, a reduction in photosynthesis can be observed within minutes of exposure to temperatures below 10 C in chilling-sensitive species such as sorghum (Bagnall, 1979; McWilliam and Ferrar, 1974). This decline in $CO_2$ uptake follows the decline in the activity ($V_{max}$) of the primary carboxylating enzyme PEP carboxylase in this species (McWilliam and Ferrar, 1974). In addition to the direct effect of low temperature on the photosynthetic process, the injury to the leaf incurred during chilling can affect the photosynthetic capacity of the leaf when returned to an elevated temperature. This has been reported for corn (Teeri et al., 1977) and cotton (Patterson and Flint, 1979), and data for sorghum are presented in Figure 6.

Another reason for depressed photosynthesis in $C_4$ chilling-sensitive

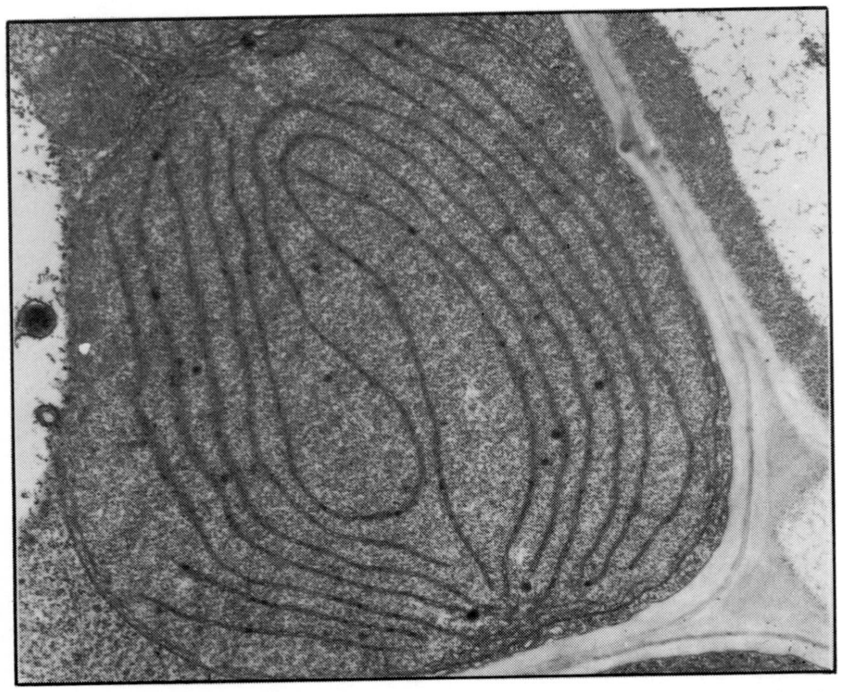

**Figure 5. Arrested development of a plastid from an etiolated leaf of *Sorghum bicolor* containing only primary lamella layers and no chlorophyll, chilled for 48 hours at 15 C at an irradiance of 250 $\mu$E m$^{-2}$sec$^{-1}$.**

species grown under conditions of low night (5 to 10 C) and warm day (20 C) temperatures is the accumulation of starch in the mesophyll and bundle sheath chloroplasts attributable to the depressed amylolytic activity at low night temperatures (Ford et al., 1975; Hillard and West, 1970). Under such conditions, the excessive build-up in starch can depress photosynthesis by impairing the physical integrity of the chloroplast membranes (West, 1970).

All these effects on photosynthesis during and subsequent to chilling stress are expressed as differences in dry matter production as shown by Teeri et al. (1977) for corn.

*Water Relations.* One of the earliest visible symptoms of low temperature stress in chilling-sensitive species is the wilting and subsequent dehydration of leaf tissue which becomes visible within hours of the onset of chilling in sensitive species such as cotton, bean and cucumber (Crookston et al., 1974; Rikin et al., 1976; Wilson, 1976; Wright, 1974). The reduction in leaf water potential occurs when whole seedlings or roots alone are chilled, and it tends to be more pronounced in cotton when the roots are chilled (5 C)

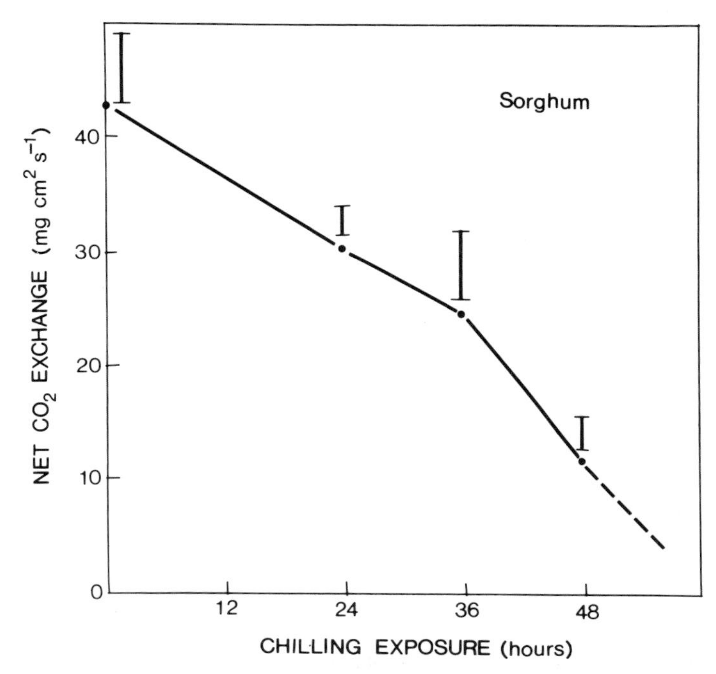


**Figure 6. Net photosynthesis (measured at 25 C) of leaves of *Sorghum bicolor* seedlings 48 hours after exposure from 12 to 48 hours to chilling stress at 8 C in the light (600 $\mu$E m$^{-2}$sec$^{-1}$). SE of mean values are indicated on figure. (J. R. McWilliam, unpublished data.)**

and the leaves are maintained at a higher temperature. The difference in this respect between chilling-sensitive and chilling-resistant species is illustrated in Figure 7. The current hypothesis is that low temperature reduces the hydraulic conductivity of the root membranes of chilling-sensitive species; thus, in addition to reduced water uptake, the conductivity of the tonoplast of the guard cells is reduced to render them less sensitive to closure under the influence of reduced water potential. The combination of reduced water uptake and slow closure of stomates causes a reduction in leaf water potential leading to wilting and ultimately dehydration of tissue. My own experiments and those reported by Wilson (1976) indicate that dehydration and damage can be minimized by chilling plants in the dark, or if in the light, by maintaining the tops in a saturated atmosphere (Figure 8). These results suggest that water deficit and light are acting synergistically to cause dysfunction and injury to plant tissue exposed to chilling temperatures. Dehydration injury appears to be more pronounced in chilling-sensitive $C_3$ than in sensitive $C_4$ plants and may reflect a difference in the water relations of these two groups at low temperature (Crookston et al., 1974). Reduced leaf water

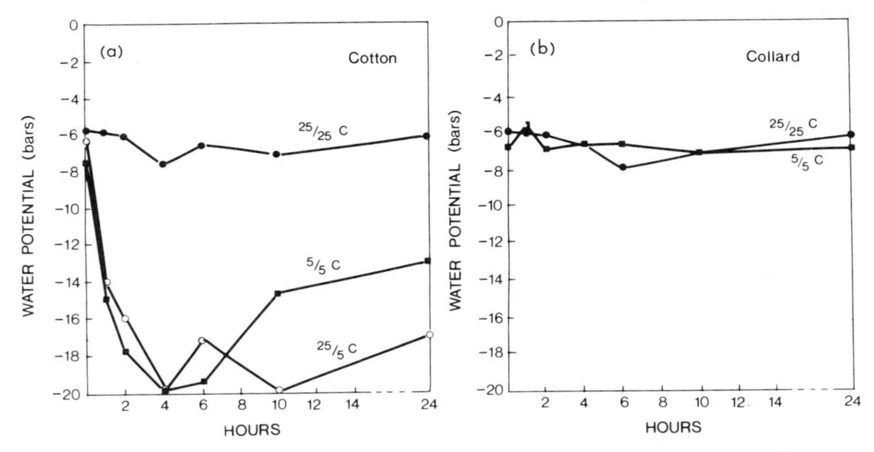


**Figure 7.** Leaf water potentials of cotton *(Gossypium hirsutum)* and collard *(Brassica oleracea)* during the first 24 hours after exposure to chilling (shoot/root) temperature of 5/5 C and 25/5 C by comparison with the nonchilled 25/25 C control. (J. R. McWilliam, unpublished data.)

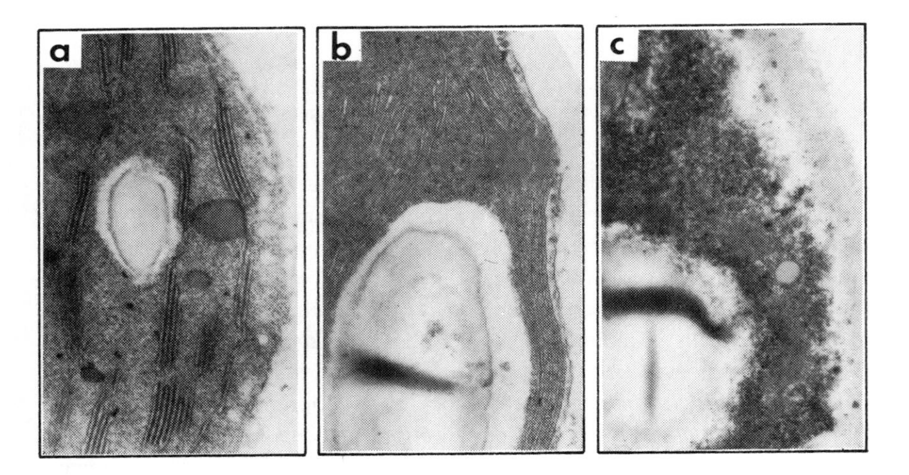


**Figure 8.** Ultrastructural changes in the chloroplasts from leaves of cotton seedlings harvested immediately after chilling at 5 C under the following conditions: (a) 72 hours, dark, 100 RH; (b) 72 hours, light (600 $\mu$E $m^{-2}sec^{-1}$), 100% RH; and (c) 24 hours, light, 70% RH, severely wilted. Micrographs are magnified 1600X. The gradual deterioration of the thylakoid membrane system and the degeneration of chloroplast structure which is apparent after prolonged chilling in the dark (a), is accelerated by exposure to light (b), and dramatically so in the presence of water stress (c).

potentials have been reported in sorghum after several days of chilling stress (Bagnall, 1979) but it appears to be a consequence rather than a causal factor in chilling injury.

Chilling induced water stress is observed in the field, particularly in the spring, with seedlings growing in cold soil and is undoubtedly a predisposing factor in the injury observed under these conditions.

*Vegetative Growth.* The depressing effect of low temperatures on the vegetative growth of chilling-sensitive species is a consequence of the various metabolic and physiological dysfunctions described in this review.

Young cotton seedlings show a marked reduction in the relative growth rate of roots, shoots, and leaves when returned to a favorable temperature (25 C) after chilling for 24 hours in the light at 5 C. The same exposure has little or no effect on the subsequent growth rate of a chilling resistant species such as collard *(Brassica oleracea)* (Table 1). The injury to roots under these conditions, also reported by Christiansen (1968), is undoubtedly a major factor influencing post-chilling recovery in the field.

Other consequences of chilling stress in crops such as rice, cotton and corn include reductions in leaf area, tillering, height, and dry weight and a delay in the time to flowering (Castleberry et al., 1978; Constable, 1976; Kaneda and Beachell, 1974; Patterson and Flint, 1979; Sato, 1972). These effects are ultimately reflected in a reduction in yield as illustrated in data for field beans (Table 2).

In all studies of the effect of chilling on vegetative growth of sensitive species, there is a strong interaction with irradiance. For example, the growth of sorghum is much less sensitive to a period of low temperature given during the night period than under conditions of high irradiance during the day (Bagnall, 1979). This adds support to my earlier claim that light accelerates and intensifies low temperature injury in chilling-sensitive tissue.

*Floral Sterility.* Sterility of spikelets following exposure to low temperature just prior to anthesis has been reported in rice and sorghum and may be a

**Table 1. Effect of chilling on the maturity (length of growing season) and yield of field beans *(Phaseolus vulgaris)* sown at three times during the spring in the Lockyer Valley, Qld., Australia (C. R. McMahon and J. R. McWilliam, unpublished data).**

| Minimal Soil Temperature At 15 cm | Length of Growing Season | Fresh Weight of Marketable Pods |
|---|---|---|
| *– days <15C –* | *– days –* | *– kg/ha –* |
| 30 | 76 | 5,330 |
| 18 | 73 | 5,280 |
| 0 | 59 | 15,640 |

**Table 2. Grain yields of tropical rice varieties exposed to water at 17 C during differend stages of floral maturity. Decrease in grain yield resulted from reduced fertility and fewer spikelets per panicle. (Adapted from IRRI Annual Report, 1978.)**

| Variety | Stage of Floral Maturity During Stress | | | |
|---|---|---|---|---|
| | Meiosis | Anthesis | Grain Filling | Control |
| | | *– grain yield (t/ha) –* | | |
| Josaengtongil | 3.6 | 4.8 | 5.2 | 6.3 |
| Suweon 264 | 2.8 | 3.9 | 4.5 | 4.7 |
| Len Kwong | 3.4 | 3.9 | 3.9 | 3.7 |
| Nong baek | 4.1 | 5.8 | 5.4 | 6.0 |
| Towada | 3.2 | 5.7 | 5.5 | 6.3 |
| Mean | 3.4 | 4.8 | 4.9 | 5.4 |
| (% Control) | (63%) | (89%) | (91%) | (100%) |

factor influencing seed yield in many other chilling-sensitive crops (Booking, 1976; Downs and Marshall, 1971; Satake et al., 1969). The most sensitive stage occurs during meiosis of pollen mother cells in both species; however, spikelet sterility, at least in rice, is the result of cold injury to developing pollen and ovules (Lin and Peterson, 1975). Sterility increases in direct proportion to the extent and duration of low temperature exposure and can be induced by low air or water temperature in the case of tropical rices which are extremely sensitive to chilling stress (Table 3).

Chilling temperatures are less frequent during the flowering period of most chilling-sensitive crops. However, in the case of rice which is widely grown in many temperate regions of the world, yields are often depressed if temperatures during the late summer flowering period fall much below 18 C (Lin and Peterson, 1975).

## MINIMIZING OR AVOIDING CHILLING INJURY

### Conditioning (Cold Hardening)

Many studies have demonstrated that certain chilling-sensitive species can be conditioned or hardened to reduce or avoid chilling injury by a gradual reduction of the growth temperature to about 12 C (Chen et al., 1980; St. John and Christiansen, 1976; Wheaton and Morris, 1967). Similarly exposure to water stress minimizes subsequent chilling injury associated with dehydration induced by low temperature in cotton leaves (Wilson, 1976). Bramlage et al. (1979) also have shown that chilling resistance during the germination of soybean seeds may be influenced by the hardening effect of cool temperature during the maturation of the seeds under field conditions.

Not all chilling-sensitive species, however, respond to cold hardening (McWilliam and Ferrar, 1974; Wilson, 1979). Many true tropical species

**Table 3. Effect of prior chilling in the light (24 hours at 5 C and 600 $\mu$E $m^{-1}sec^{-1}$) on the relative growth rates of young chilling-sensitive cotton *(Gossypium hirsutum)* and chilling-resistant collard *(Brassica oleracea)* plants grown at 25 C. (J. R. McWilliam, unpublished data.)**

| Species | Treatment | Relative Growth Rate | | |
|---|---|---|---|---|
| | | Roots | Shoots | Leaf Area |
| | | *– mg/mg/day –* | *– mg/mg/day –* | *– $cm^2/cm^2/day$ –* |
| Collard | Chill | 0.19 | 0.26 | 0.23 |
| | Control | 0.19 | 0.24 | 0.23 |
| | % Control | 100 | 108 | 100 |
| Cotton | Chill | - 0.03 | 0.04 | 0.02 |
| | Control | 0.14 | 0.13 | 0.12 |
| | % Control | 1 | 31 | 17 |

that never experience temperatures much below 15 C may lack the ability to acclimate.

Acclimation to chilling temperatures has been attributed to an increase in lipid unsaturation, especially linolenic acid (18:3) (St. John and Christiansen, 1976; Wilson and Crawford, 1974). This general correlation between low temperature and increased levels of unsaturated fatty acids has been observed in many species (Harris et al., 1978), and it has even been suggested that the unsaturated/saturated fatty acid ratio of seeds could be used to identify cotton genotypes with increased chilling resistance (Bartkowski et al., 1977). The existence of a causal relationship between lipid unsaturation and chilling resistance, however, has not been resolved.

Other symptoms of cold hardening include ultrastructural modifications of chloroplasts (Rogers et al., 1977), maintenance of nucleotide levels in tissue (Stewart and Ginn, 1971), and closure of stomates and increased hydraulic conductivity of roots (Wilson, 1976). Despite the differences in these apparent responses, the primary target in cold hardening, as in chilling injury, is probably the membranes.

For crops that do respond, cold hardening may be of significance in the autumn when the temperatures are declining, but it is probably of little value to plants in the early spring.

## Chemical Amelioration

No entirely successful chemical protection from chilling injury has been developed. Possible chemical approaches to modify the physical nature or composition of membranes have been discussed by Lyons and Breidenbach (1979). Respiratory inhibitors have been used by Amin (1969) to artificially induce chilling resistance in cotton. Antitranspirants, including abscisic acid (ABA), have been used directly to avoid dehydration chilling injury in cotton, cucumber and bean by closing stomates before exposure to low temperature (Christiansen and Ashworth, 1978; Rinkin et al., 1976; Wilson, 1976). ABA also has been used indirectly to avoid dehydration during chilling by treatment of roots to increase hydraulic conductivity (Markhart, 1979b; J. R. McWilliam, unpublished data).

At best these approaches merely delay the onset of chilling injury. To be successful in agriculture, a chemical cure must protect sensitive plants for an extended period, must not inhibit other metabolic processes, and must be relatively inexpensive to apply as a seed or seedling treatment on a field scale. Ultimately the most effective protection will be the development of genetic resistance to chilling stress.

## VARIATION AND SELECTION FOR CHILLING TOLERANCE

Variation in chilling sensitivity has been observed in a range of crop species including rice (Lin and Peterson, 1975), corn (Castlebery et al., 1978), cotton (Muramoto et al., 1971) and sorghum (W. Manokaran, 1979. Effect of chilling temperature on sorghum. M.Sc. Agric. Thesis, Univ. New England, Armidale, N.S.W., Australia.) and appears to be correlated with the climatic adaptation of the individual varieties. Even more striking patterns of low temperature variation have been described in cultivated species of the genus *Passiflora* (Patterson et al., 1976), in altitudinal races of *Zea mays,* and in the wild green-fruited tomato *(Lycopersicon hirsutum)* (Patterson et al., 1978; Vallejos, 1979).

Another example of the evolution of greater chilling tolerance is the widespread occurrence of $C_4$ tropical grasses, such as Johnson grass, paspalum, crab grass, barnyard grass, and couch grass, as persistent and serious weeds of crops in temperate environments. All these species have followed man's cultivation into higher temperate latitudes and now persist, some as perennials, in these more extreme thermal environments.

This evidence of adaptive variation for chilling sensitivity is supported by inheritance studies which indicate that variation for chilling sensitivity within species is largely additive and under polygenic control (Paul et al., 1979). Maternal inheritance has been reported for variation in response of seed hydration to chilling in cotton (Marani and Dag, 1962; Pinnell, 1949), which Christiansen and Lewis (1973) have suggested may be controlled by a cytoplasmic gene. These results also suggest that the genetic control of chilling-sensitivity may not be the same for all stages of crop development.

Increased chilling tolerance has been achieved by selection in cotton (Christiansen and Lewis, 1973), in corn (McConnell and Gardner, 1979; Mock and Bakeri, 1976), and in tropical varieties of rice (IRRI Ann. Rep. 1978). In these cases, selection pressure has been applied during the germination and emergence stage in controlled environments, because of the difficulty of screening sufficient numbers of large plants at controlled temperatures.

The transfer of genes for cold tolerance from wild species adapted to high altitudes in the topics is being attempted with sorghum (van Arkel, 1977) and has been successful in tomato by crossing the wild, chilling-tolerant *Lycopersicon hirsutum* to a commercial strain of *L. esculentum* and recovering chilling-tolerant progeny in the subsequent backcross populations (B. D. Patterson, personal communication).

Recently a number of rapid screening techniques have been proposed to assist in selecting for greater chilling tolerance (Paul et al., 1979). One of the most promising approaches is to select within single cell cultures following mutagenic treatment or somatic hybridization to increase the frequency of genes for chilling tolerance. Chilling-tolerant cell lines of *Nicotiana* and *Capsicum* have been isolated (Dix and Street, 1976), but problems associated with genetic stability and regeneration of whole plants from callus will have to be overcome before this technique can be applied routinely for incorporating new genes for chilling tolerance into sensitive crop species.

Another approach relies on chlorophyll fluorescence which can be monitored on intact leaves of plants during or after exposure to chilling stress (Melcarek and Brown, 1977). The fluorescence yield is reduced to a greater extent in chilling-sensitive plants and the technique can be adapted to measure detached leaf segments (Smillie, 1979). A similar approach using the rate of greening of small etiolated leaf segments in the light at a given chilling temperature has been developed by the author using visual or rapid photometric techniques to assess the difference in chlorophyll accumulation over a given time interval.

In both these cases the extent of damage to chloroplast membranes, which is among the earliest and most common symptoms of chilling stress, is measured indirectly by the assay. The extent to which the ability to maintain membrane integrity in chloroplasts is a reflection of general chilling tolerance requires further validation.

## CONCLUSIONS

Many plants of tropical and subtropical origin are injured and sometimes killed by chilling temperatures below about 15 C. The extent and severity of chilling injury is a function of the temperature and duration of the chilling stress and also the level of irradiation experienced during exposure. Other variables include the species and the condition and stage of plant development.

Low, nonfreezing temperatures cause a number of reversible changes in the physical state of the membranes (resulting in reduced permeability) and in the structure and function of enzyme proteins. If chilling is sustained, these primary events lead sequentially to a series of secondary lesions (including altered metabolic rates), biochemical and physiological dysfunctions, and ultimately to irreversible damage which is expressed in a variety of visible symptoms. In field crops, these symptoms include poor germination and establishment, wilting and chlorosis of young seedlings, reduced assimilation and growth, and for certain cereals, spikelet sterility and reduced grain yield.

These deleterious consequences of chilling can be reduced or modified in certain species by prior conditioning at low temperature or by the use of appropriate chemicals. The most permanent way of reducing chilling injury, however, will be to select for increased chilling-tolerance from within the range of variation present in most chilling-sensitive crop species.

## NOTES

J. R. McWilliam, Department of Agronomy and Soil Science, University of New England, Armidale, New South Wales 2351, Australia.

## LITERATURE CITED

Amin, J. V. 1969. Some aspects of respiration and respiration inhibitors in low temperature effects of the cotton plant. Physiol. Plant. 22:1184-1219.

Bartkowski, E. J., D. R. Buxton, F. R. H. Katterman, and H. W. Kircher. 1977. Dry seed fatty acid composition and seedling emergence of pima cotton at low soil temperatures. Agron. J. 69:37-40.

Bagnall, D. J. 1979. Low temperature response of three *Sorghum* species. p. 67-80. In J. M. Lyons, D. Graham, and J. K. Raison (eds.) Low temperature stress in crop plants. Academic Press, New York.

Bishop, D. G., J. R. Kenrick, J. H. Bayston, A. S. Macpherson, S. R. Johns, and R. I. Willing. 1979. The influence of fatty acid unsaturation on fluidity and molecular packing of chloroplast membrane lipids. p. 375-389. In J. M. Lyons, D. Graham, and J. K. Raison (eds.). Low temperature stress in crop plants. Academic Press, New York.

Brooking, I. R. 1976. Male sterility in *Sorghum bicolor* (L.) Moench induced by low night temperature. I. Timing of the stage of sensitivity. Aust. J. Plant. Physiol. 3:589-596.

Bramlage, W. J., A. C. Leopold, and J. E. Specht. 1979. Imbibitional chilling sensitivity among soybean cultivars. Crop Sci. 19:811-814.

Castleberry, R. M., J. A. Teeri, and J. F. Buriel. 1978. Vegetative growth responses of maize genotypes to simulated natural chilling events. Crop Sci. 18:633-637.

Chen, H. H., and P. Li. 1980. Characteristics of cold acclimation and deacclimation in tuber-bearing *Solanum* species. Plant Physiol. 65:1146-1148.

Christiansen, M. N. 1964. Influence of chilling upon subsequent growth and morphology of cotton seedlings. Crop Sci. 4:584-586.

Christiansen, M. N. 1967. Periods of sensitivity to chilling in germinating cotton. Plant Physiol. 42:431-433.

Christiansen, M. N. 1968. Induction and prevention of chilling injury to radicle tips of imbibing cottonseed. Plant Physiol. 43:743-746.

Christiansen, M. N., and E. N. Ashworth. 1978. Prevention of chilling injury to seedling cotton with anti-transpirants. Crop Sci. 18:907-908.

Christiansen, M. N., and C. F. Lewis. 1973. Reciprocal differences in tolerance to seed-hydration chilling in $F_1$ progeny of *Gossypium hirsutum* l. Crop Sci. 13:210-212.

Christiansen, M. N., and R. O. Thomas. 1969. Season-long effects of chilling treatments applied to germinating cottonseed. Crop Sci. 9:672-673.

Constable, G. A. 1976. Temperature effects on the early field development of cotton. Aust. J. Exp. Agric. Anim. Husbandry 16:905-910.

Creencia, R. P., and W. J. Bramlage. 1971. Reversibility of chilling injury to corn seedlings. Plant Physiol. 47:389-392.

Crookston, R. K., J. O'Toole, R. Lee, J. L. Ozbun, and D. H. Wallace. 1974. Photosynthetic depression in beans after exposure to cold for one night. Crop Sci. 14: 457-464.

Dix, P. J., and H. E. Street. 1976. Selection of plant cell lines with enhanced chilling resistance. Ann. Bot. 40:903-910.

Downes, R. W., and D. R. Marshall. 1971. Low temperature induced male sterility in *Sorghum bicolor.* Aust. J. Exp. Agric. Anim. Husbandry 11:352-356.

Downton, W. J. S., and J. S. Hawker. 1975. Evidence for lipid-enzyme interaction in starch synthesis in chilling-sensitive plants. Phytochemistry 14:1259-1263.

Duke, S. H., L. E. Schrader, and M. G. Miller. 1977. Low temperature effects on soybean *[Glycine max.* (L.) Merr. cv. Wells] mitochondrial respiration and several dehydrogenases during imbibition and germination. Plant Physiol. 60:716-722.

Forde, B. J., H. C. M. Whitehead, and J. A. Rowley. 1975. Effect of light intensity and temperature on photosynthetic rate, leaf starch content and ultrastructure of *Paspalum dilatatum.* Aust. J. Plant Physiol. 2:185-195.

Graham, D., D. G. Hockley, and B. D. Patterson. 1979. Temperature effects on phosphoenol pyruvate carboxylase from chilling-sensitive and chilling-resistant plants. p. 453-461. In J. M. Lyons, D. Graham, and J. K. Raison (eds.). Low temperature stress in crop plants. Academic Press, New York.

Harris, H. C., J. R. McWilliam, and W. Mason. 1978. Influence of temperature on oil content and composition of sunflower. Aust. J. Agric. Res. 29:1203-1212.

Hillard, J. H., and S. H. West. 1970. Starch accumulation associated with growth reduction at low temperatures in a tropical plant. Science 168:494-496.

Kaneda, C., and H. M. Beachell. 1974. Response of Indica-Japonica rice hybrids to low temperature. Sabro. J. 6:17-32.

Kramer, P. J. 1942. Species difference with respect to water absorption at low soil temperature. Amer. J. Bot. 29:828-832.

Leopold, A. C., and M. E. Musgrave. 1979. Respiratory changes with chilling injury of soybean. Plant Physiol. 64:702-705.

Levitt, J. 1972. Responses of plants to environmental stresses. Academic Press, New York.

Lin, S. S., and M. L. Peterson. 1975. Low temperature induced floret sterility in rice. Crop Sci. 15:657-660.

Lyons, J. M. 1973. Chilling injury in plants. Ann. Rev. Plant Physiol. 24:445-451.

Lyons, J. M., and R. W. Breidenbach. 1979. Strategies for altering chilling sensitivity as a limiting factor in crop production. p. 180-196. In H. Mussell and R. C. Staples (ed.) Stress physiology in crop plants. John Wiley & Sons, New York.

Lyons, J. M., J. K. Raison, and P. L. Steponkus. 1979a. The plant membrane in response to low temperature. p. 1-24. In J. M. Lyons, D. Graham, and J. K. Raison (eds.) Low temperature stress in crop plants. Academic Press, New York.

Lyons, J. M., D. Graham, and J. K. Raison. 1979. Epilogue. p. 543-548. In J. M. Lyons, D. Graham, J. K. Raison (eds.) Low temperature stress in crop plants. Academic Press, New York.

Marani, A., and J. Dag. 1962. Germination of seeds of cotton varieties at low temperature. Crop Sci. 2:267.

Markhart, A. H., E. L. Fiscus, A. W. Naylor, and P. J. Kramer. 1979a. Effect of temperature on water and ion transport in soybean and broccoli root systems. Plant Physiol. 64:83-87.

Markhart, A. H., E. L. Fiscus, A. W. Naylor, and P. J. Kramer. 1979b. Effect of abscisic acid on root hydraulic conductivity. Plant Physiol. 64:611-614.

McConnell, R. L., and C. O. Gardner. 1979. Selection for cold germination in two corn populations. Crop Sci. 19:765-768.

McWilliam, J. R., and P. J. Ferrar. 1974. Photosynthetic adaptation of higher plants to thermal stress. p. 467-476. In R. L. Bieleski, A. R. Ferguson, and M. M. Cresswell (eds.) Mechanisms of regulation of plant growth. Bulletin 12, Royal Soc. N.Z., Wellington.

McWilliam, J. R., W. Manokaran, and T. Kipnis. 1979. Adaptation to chilling stress in sorghum. p. 491-505. In J. M. Lyons, D. Graham, and J. K. Raison (eds.) Low temperature stress in crop plants. Academic Press, New York.

Melcarek, P. K., and G. A. Brown. 1977. Effects of chilling stress on prompt and delayed chlorophyll fluorescence from leaves. Plant Physiol. 60:822-825.

Millerd, A., and J. R. McWilliam. 1968. Studies on a maize mutant sensitive to low temperature. I. Influence of temperature and light on the production of chloroplast pigments. Plant Physiol. 43:1967-1972.

Mock, J. J., and A. A. Bakri. 1976. Recurrent selection for cold tolerance in maize. Crop Sci. 16:230-233.

Muramoto, H., J. D. Hesketh, and D. N. Baker. 1971. Cold tolerance in a hexaploid cotton. Crop Sci. 11:589-591.

Nishiyama, I. 1976. Effects of temperature on the vegetative growth of rice plants. p. 159-185. In Climate and rice. IRRI, Los Banos, Phillippines.

Obendorf, R. L., and P. R. Hobbs. 1970. Effect of seed moisture on temperature sensitivity during imbibition of soybean. Crop Sci. 10:563-566.

Patterson, B. D., T. Murata, and D. Graham. 1976. Electrolyte leakage induced by chilling in *Passiflora* species tolerant to different climates. Aust. J. Plant Physiol. 3: 435-442.

Patterson, B. D., R. Paull, and R. M. Smillie. 1978. Chilling resistance in *Lycopersicon hirsutum* Humb. & Bonpl., a wild tomato with a wide altitudinal distribution. Aust. J. Plant Physiol. 5:609-617.

Patterson, D. T., and E. P. Flint. 1979. Effect of chilling on cotton *(Gossypium hirsutum)*, velvetleaf *(Abutilon theophrasti)*, and spurred anoda *(Anoda cristata)*. Weed Sci. 27:473-479.

Paull, R. E., B. D. Patterson, and D. Graham. 1979. Chilling injury assays for plant breeding. p. 507-519. In J. M. Lyons, D. Graham, and J. K. Raison (eds.) Low temperature stress in crop plants. Academic Press, New York.

Pinnell, E. L. 1949. Genetic and environment factors affecting corn seed germination at low temperature. Agron. J. 41:562-568.

Raison, J. K. 1974. A biochemical explanation of low-temperature stress in tropical and sub-tropical plants. p. 487-497. In R. L. Bieleski, A. R. Ferguson, and M. M. Cresswell (eds.) Mechanisms of regulation of plant growth. Bulletin 12, Royal Soc. N. Z., Wellington.

Raison, J. K., J. M. Lyons, R. J. Mehlhorn, and A. D. Keith. 1971. Temperature-induced phase changes in mitochondrial membranes detected by spin labelling. J. Biol. Chem. 246:4036-4040.

Raison, J. K., and E. A. Chapman. 1976. Membrane phase changes in chilling-sensitive *Vigna radiata* and their significance to growth. Aust. J. Plant Physiol. 3:291-299.

Rikin, A., A. Blumenfeld, and A. E. Richmond. 1976. Chilling resistance as affected by stressing environments and abscisic acid. Bot. Gaz. 137:307-312.

Rogers, R. A., J. H. Dunn, and C. J. Nelson. 1977. Photosynthesis and cold hardening in *Zoysia* and Bermuda grass. Crop Sci. 17:727-732.

Satake, T., I. Nishiyama, N. Ito, and H. Hayase. 1969. Male sterility caused by cooling treatment at the meiotic stage in rice plants. II. The most sensitive stage to cooling and the fertilizing ability of pistils. Proc. Crop Sci. Soc. Japan 38:706-711.

Sato, K. 1972. Growth responses of rice plants to environmental conditions. 1. The effect of air temperatures on growth at the vegetative stage. Proc. Crop Sci. Japan 41: 388-393.

Simon, E. W. 1974. Phospholipids and plant membrane permeability. New Phytol. 73: 377-420.

Smillie, R. M. 1979. The useful chloroplast: a new approach for investigating chilling stress in plants. p. 187-202. In J. M. Lyons, D. Graham, and J. K. Raison (eds.) Low temperature stress in crop plants. Academic Press, New York.

Stewart, J. M., and G. Guinn. 1971. Chilling injury and nucleotide changes in young cotton plants. Plant Physiol. 48:166-170.

St. John, J. B. 1976. Manipulation of galactolipid fatty acid composition with substituted pyridazinones. Plant Physiol. 57:38-40.

St. John, J. B., and M. N. Christiansen. 1976. Inhibition of linolenic acid synthesis and modification of chilling resistance in cotton seedlings. Plant Physiol. 57:257-259.

St. John, J. B., and J. L. Hilton. 1976. Structure versus activity of substituted pyridazinones as related to mechanism of action. Weed Sci. 24:579-582.

Taylor, A. O., G. Halligan, and J. A. Rowley. 1975. Farris banding in panicoid grasses. Aust. J. Plant Physiol. 2:247-251.

Teeri, J. A., D. T. Patterson, R. S. Aberte, and R. M. Castleberry. 1977. Changes in the photosynthetic apparatus of maize in response to stimulated natural temperature fluctuations. Plant Physiol. 60:370-373.

Vallejos, C. E. 1979. Genetic diversity of plants for response to low temperature and its potential use in crop plants. p. 473-480. In J. M. Lyons, D. Graham, and J. K. Raison (eds.) Low temperature stress in crop plants. Academic Press, New York.

van Arkel, H. 1977. New forage crop introductions for the semi-arid highland areas of Kenya as a means to increase been production. Netherlands J. of Agric. Sci. 25: 135-150.

van Hasselt, P. R., and J. T. Strikwerda. 1976. Pigment degradation in discs of the thermophilic *Cucumis sativus* as affected by light, temperature, sugar application and inhibitors. Physiol. Plant. 37:253-257.

Wheaton, T. A., and L. L. Morris. 1967. Modification of chilling sensitivity by temperature conditioning. Proc. Amer. Hort. Sci. 91:529-533.

West, S. H. 1970. Biochemical mechanism of photosynthesis and growth depression in *Digitaria decumbens* when exposed to low temperature. p. 514-517. Proc. XIth Int. Grassl. Congr., Surfers Paradise, Australia.

Wilson, J. M. 1976. The mechanism of chill and drought hardening of *Phaseolus vulgaris* leaves. New Phytol. 76:257-270.

Wilson, J. 1979. Drought resistance as related to low temperature stress. p. 47-65. In J. M. Lyons, D. Graham, and J. K. Raison (eds.) Low temperature stress in crop plants. Academic Press, New York.

Wilson, J. M., and R. M. M. Crawford. 1974. The acclimatization of plants to chilling temperatures in relation to the fatty acid composition of leaf polar lipids. New Phytol. 73:805-820.

Wright, M. 1974. The effect of chilling on ethylene production, membrane permeability and water loss of leaves of *Phaseolus vulgaris*. Planta 120:63-69.

# 9
# INTERACTION OF CHILLING AND WATER STRESS

**J. M. Wilson**

Low temperature and water stress are perhaps the two most important factors limiting the growth and yield of semitropical and tropical crop plants such as *Gossypium hirsutum* and *Phaseolus vulgaris.* Estimates of the amount of damage to tropical crops caused by chilling temperatures of 0 to 10 C are difficult to find but the losses of cotton seedlings in the U.S.A. have been valued at $60 million per year. Chilling-injury is also of economic importance in the preservation of tropical fruits, as the storage life of fruits such as bananas cannot be prolonged by lowering the temperature into the chilling-injury range. My research has been focused mainly on the development of chilling-injury in the leaves. Water stress is more important in the development of injury to these organs than to tropical fruits.

The first symptoms of chilling-injury to the leaves of many agriculturally important species, such as *P. vulgaris,* are rapid leaf wilting and the development of sunken, light-green pits within 24 hours after the start of chilling. When the plants are rewarmed, the leaf margins and the light-green pits usually dry out and turn brown giving the leaf a very mottled and brittle appearance. Thus, chilling plants of *P. vulgaris* for 24 hours at 5 C and 85% RH results in rapid leaf wilting and, upon rewarming, approximately 50% leaf death. In other species, such as the extremely chill-sensitive ornamental *Episcia reptans* from tropical South America, a chilling treatment of only 2 hours at 5 C results in the loss of leaf turgor and the development of water-soaked patches on the leaf surface which become necrotic if chilling is prolonged or the plants are returned to the warmth. These observations led to investigations on whether the phase transitions in the membrane lipids of the plasmalemma and tonoplast, as described at low temperature by Lyons (1973), resulted in increased permeability of the leaf cells to water and electrolytes and, consequently, in rapid leaf dehydration.

## THE PREVENTION OF CHILLING-INJURY

### Chill-hardening

It is not possible to harden tropical and semitropical plants against chilling-injury as effectively as one can harden temperate species against frost-injury. For instance, chill-hardening of *P. vulgaris* for 4 days at 12 C and 85% RH will protect only against leaf wilting and injury at 5 C and 85% RH for approximately 9 days, and this protection cannot be significantly improved by increasing the hardening period. In contrast, hardening treatments of temperatures just above the freezing point of only short duration can protect temperate plants against frost throughout the winter.

### Drought-hardening

Injury to species such as *P. vulgaris* and *G. hirsutum* also can be prevented by drought-hardening the plants at 25 C and 40% RH by withholding water from the roots over a 4-day period so that the leaves wilt. Drought-hardening is as effective as chill-hardening in preventing chilling-injury (Wilson, 1976).

### Maintenance of a Saturated Atmosphere at 5 C

Chilling-injury to *P. vulgaris* and other species can be prevented for up to 9 days simply by maintaining a saturated (100% RH) atmosphere around the leaf at 5 C. This is most easily achieved by enclosing the plant inside a polythene bag before transfer from 25 C to 5 C. Chilling-injury to all species, however, cannot be prevented in this way. For instance, in the extremely chill-sensitive species *Episcia reptans,* maintaining a saturated atmosphere around the leaf does not delay significantly the onset of chilling-injury. Neither can this species be drought or chill-hardened. Therefore, chilling-injury to the less chill-sensitive, agriculturally important species is attributed primarily to water stress whereas chilling-injury in the more chill-sensitive species *(E. reptans)* is concluded primarily to be metabolic (Wilson, 1976), However, metabolic changes during chilling of *P. vulgaris* for 9 days at 5 C and 100% RH must lead eventually to cell death.

## THE CAUSE OF LEAF DEHYDRATION IN *P. VULGARIS* AT 5 C

### Importance of Membrane Lipids

The leakage of electrolytes (Figure 1) and the development of water soaked patches for leaves of chilled plants indicate an increase in membrane permeability at 5 C caused by a phase change in the membrane lipids of the plasmalemma and tonoplast and leading to a decrease in leaf turgor. Chill-hardening at 12 C increases the degree of unsaturation of the fatty acids, the temperature at which the phase transition or separation occurs may be lowered to below 5 C so that chilling-injury is prevented. Data of Wilson and Crawford (1974) agree with this postulate. During chill-hardening at 12 C an increase of 5 to 12% occurred in the degree of unsaturation of the fatty acids associated with the phospholipids of *P. vulgaris* and *G. hirsutum* (Table 1). These increases in unsaturation mainly were the result of an increase in the percentage of linoleic acid. No increase in the degree of unsaturation of the glycolipids was detected. In contrast, during ineffective hardening of *Episcia reptans* at 15 C, no increase in the degree of unsaturation of the phospholipids was detected. However, the increases in unsaturation during chill-hardening of *P. vulgaris* are not thought to be important in the prevention of chilling-injury since drought-hardening resulted in no increase in unsaturation and yet was as effective as chill-hardening (Table 1). In addition, the chill-hardening of *P. vulgaris* leaves at 12 C was not effective if the plants are maintained at 100% RH (Wilson, 1976). Although the degree of unsaturation of the phospholipids increased during this treatment, the lack of an increase in chill tolerance indicates that chill-hardening is not dependent on a highly

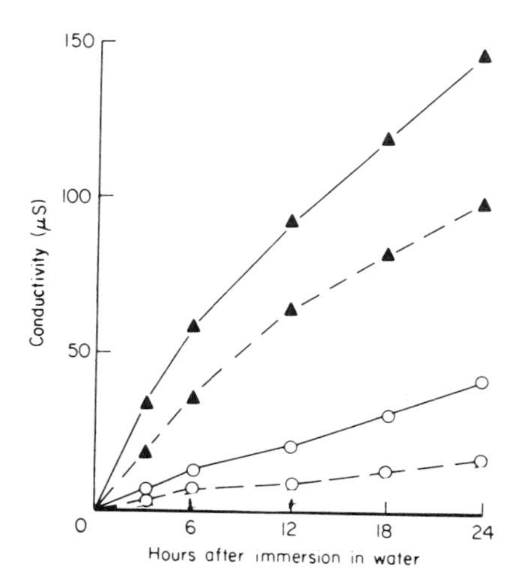


**Figure 1. Leakage of electrolytes from leaves of *Phaseolus vulgaris* chilled at 5 C, 85% RH for 24 hours (▲) in comparison to unchilled leaves (○), placed in water at either 25 C (—) or 5 C (- - -).**

**Table 1. Changes in percent fatty acid composition of phosphatidylcholine from leaves of *Phaseolus vulgaris* during chill-hardening at 12 C and 85% RH and drought-hardening at 25 C and 40% RH.**

| | Hardening Treatment | | |
|---|---|---|---|
| Fatty Acid[a] | Control | Chill-hardened at 12 C and 85% RH For 4 Days | Drought-hardened at 25 C and 85% RH For 4 Days |
| | – *% fatty acid composition of phosphatidylcholine* – | | |
| 14:0 | 3.8 | 2.1 | 5.3 |
| 16:0 | 20.4 | 12.8 | 24.3 |
| 16:1 | 0.9 | 1.0 | 0.7 |
| 16:2 | 0.9 | 1.1 | 0.4 |
| 18:0 | 6.5 | 4.3 | 6.2 |
| 18:1 | 4.0 | 3.5 | 2.0 |
| 18:2 | 27.5 | 40.0 | 24.1 |
| 18:3 | 36.0 | 35.2 | 37.0 |
| Total | 69.3 | 80.8 | 64.2 |

[a]The numbers shown are the ratios of the number of carbon atoms to the number of double bonds in the molecule.

unsaturated fatty acid composition. Furthermore, the phase-change hypothesis (Lyons, 1974; Raison et al., 1971) is unable to account for the prevention of chilling-injury to the leaves of *P. vulgaris* when the plant is enclosed inside a polythene bag before transfer to 5 C. Plants maintained in a saturated atmosphere at 5 C for 7 days do not wilt. If lipid phase transitions resulted in an increase in the permeability of the plasmalemma to water at 5 C, then the leaves would be expected to wilt even under conditions of 100% RH since the turgor pressure of the cell would facilitate the loss of water and electrolytes.

### Importance of Stomata and Root Permeability to Water

Rapid leaf wilting during the first 2 hours of chilling of *P. vulgaris* is caused by the "locking open" of the stomata at a time when the permeability of the roots to water is low. The locking open of the stomata at 5 C and 85% RH (Figure 2) is surprising as the leaf is wilted and, in most plants, the stomata close in the early stages of water stress before visible wilting occurs. The replacement of the water lost by evapotranspiration from the leaf is prevented by the low permeability of the roots to water at 5 C (Figure 3) resulting in rapid leaf dehydration and injury. Hence, the severity of chilling-injury depends on a synergistic effect between stomatal opening and reduced permeability of the roots to water at 5 C. When only the leaves of *P. vulgaris* were chilled at 5 C and 85% RH, there was no detectable fresh weight loss after 24 hours except for a slight wilt after 3 to 4 hours of chilling (Figure 4). Chilling only the leaves for 24 hours resulted in only 6% injury to the leaf

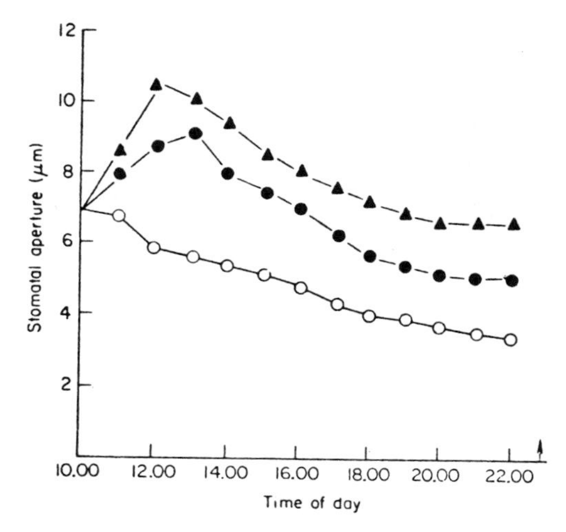


**Figure 2.** Changes in stomatal aperature on transferring entire plants of *Phaseolus vulgaris* directly from 25 C and 85% RH to 5 C and 85% RH (▲) and 12 C and 85% RH (●) compared to the controls maintained at 25 C and 85% RH (○). Arrow shows start of night period.

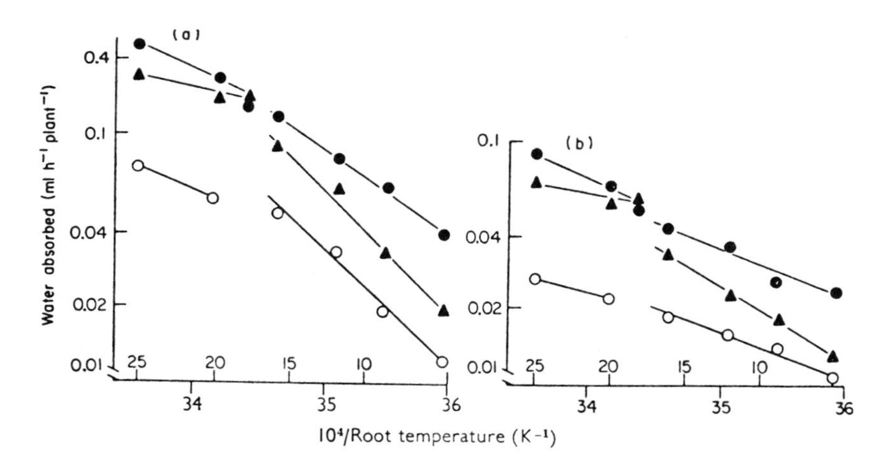


**Figure 3.** Arrhenius plots of the effect of root temperature on the rate of water absorption by *Phaseolus vulgaris* plants grown at 25 C and 85% RH (▲), chill-hardened at 12 C and 85% RH (●), and drought-hardened at 25 C and 40% RH (○). The rate of water uptake plus exudation under 50 cm Hg vacuum are shown in (a) and the rate of exudation alone is shown in (b). Each point represents the average value from at least five plants.

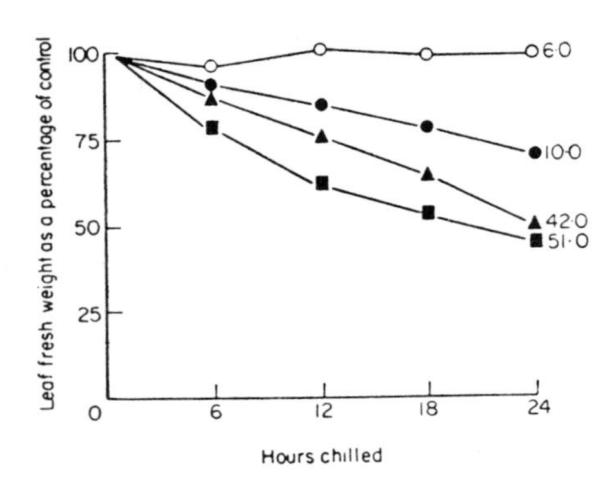


**Figure 4. Changes in leaf fresh weight on chilling the leaves alone (○), roots alone (●), or the whole plant (▲), of *Phaseolus vulgaris* for 24 hours at 5 C and 85% RH, and for whole plants (■) grown at 12 C and 100% RH for four days before being chilled at 5 C and 85% RH. The percentage of the leaf which became necrotic after 24 hours of chilling and two days of recovery at 25 C and 85% RH are given beside the points.**

after 2 days of recovery at 25 C and 85% RH (Figure 4). In the reverse experiment, when roots were chilled and the leaves held at 25 C and 85% RH, there was a 30% decrease in fresh weight after 24 hours but only 10% injury to the leaf after 2 days of recovery. Chilling the entire plant of *P. vulgaris* at 5 C and 85% RH produced a more rapid fresh weight loss than chilling either the roots or leaves alone. Chilling the whole plant for 24 hours resulted in a 50% decrease in fresh weight and 42% injury on return to 25 C (Figure 4). This reduction in the degree of dehydration and injury to the plant on cooling the roots alone is surprising. Greater injury might be expected when only the roots are cooled as the vapor pressure deficit across the leaf at 25 C and 85% RH is higher than at 5 C and 85% RH, and cooling the roots alone is thought to result in the locking open of the stomata (Tagawa, 1937). Probably more injury would have occurred to plants with roots alone chilled if the period of chilling was longer than 24 hours. It is not known if chilling the leaves alone results in the locking open of the stomata.

## MECHANISM OF CHILL-HARDENING AND DROUGHT-HARDENING

Chill-hardening at 12 C and 85% RH prevents leaf dehydration by conditioning the stomata so that they close on transfer to 5 C and 85% RH (Figure 5). Similarly, drought-hardening causes stomatal closure, and the stomata remain closed on transfer to 5 C and 85% RH. Although chill-hardening resulted in an increase in the permeability of the roots to water at low temperature, drought-hardening produced a large decrease in root permeability

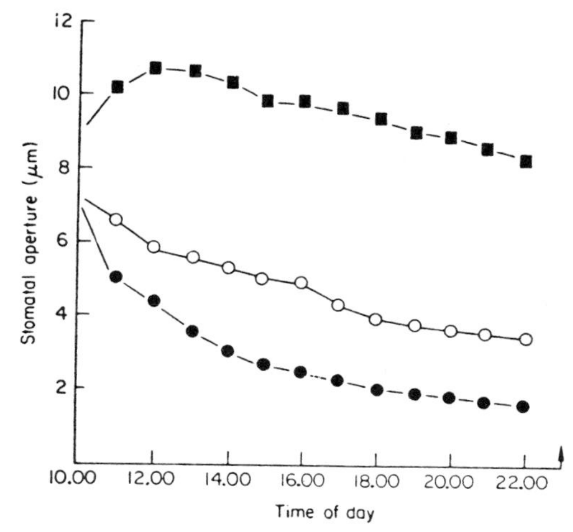


**Figure 5. Changes in stomatal aperture of *Phaseolus vulgaris* plants on chilling at 5 C and 85% RH after being hardened at 12 C and 85% RH (●) or ineffectively hardened at 12 C and 100% RH by enclosure in a polythene bag (■) compared to the controls maintained at 25 C and 85% RH (○). Arrow shows start of night period.**

(Figure 3); yet, drought-hardening was as effective as chill-hardening in preventing leaf injury. Therefore, the most important factor in the prevention of chill-injury to *P. vulgaris* during chill-hardening and drought-hardening is the closure of the stomata. This can be demonstrated by spraying the leaves of plants grown at 25 C and 85% RH with 100 μmol abscisic acid (ABA) which causes stomatal closure within 24 hours. On transfer to 5 C and 85% RH, the sprayed leaves do not wilt as the stomata remain closed. Injury is prevented for approximately 2 days by which time the effectiveness of the ABA has decreased. This decrease in the effectiveness of applied ABA is surprising as the plant might be expected to synthesize ABA during this 2-day period. Perhaps 5 C is too low a temperature for the synthesis of ABA in chill-sensitive plants.

During the chill-hardening at 12 C and 85% RH, the plant experiences a water stress as shown by the temporary wilting of the leaves. The water stress is caused by the locking open of the stomata (Figure 2) and the decrease in the permeability of the roots to water (Figure 3). However, at the intermediate temperature of 12 C the stress is not severe enough to result in damage, and the wilting vanishes after 12 hours. Similarly, during drought-hardening the water stress is imposed simply by withholding water from the roots under conditions of high evapotranspiration so that the leaves wilt. Plants maintained at 5 C or 12 C and 100% RH do not harden because they experience no water stress. Even though the stomata are open under these conditions, no water can be lost from the leaf, and the stomata remain fully open on transfer to 5 C and 85% RH (Figure 5). For plants held at 12 C and 100% RH for 4 days before being chilled for 24 hours at 5 C and 85% RH, a 55% decrease occurred in fresh weight during chilling and approximately 50% leaf injury occurred after 2 days of recovery at 25 C (Figure 4). Therefore, enclosure in a polythene bag is not a method which can be used to lower the hardening temperature below 12 C. The above experiments indicate that ABA is not synthesized in *P. vulgaris* leaves at 12 C and 100% RH, but it is not known if low temperature alone, in the absence of water stress, can induce ABA synthesis. The correlation between chill-hardening and drought-hardening has shown that an intermediate temperature of 12 C is not essential for hardening. Therefore, water stress and not low temperature per se is the primary factor inducing hardening against chilling-injury in *P. vulgaris* leaves.

## CAUSES OF CHILLING-INJURY TO *PHASEOLUS VULGARIS* LEAVES AT 5 C AND 100% RH

Although changes in root permeability to water and stomatal aperture are able to explain the rapid wilting, dehydration, and ultimately injury to the leaves during chilling at 5 C and 85% RH, an alternative explanation must be sought for the death of the leaves after 9 days at 5 C and 100% RH. The loss of turgor, bleaching, and necrosis of *P. vulgaris* leaves after 9 days at 5 C and 100% RH suggest that injury is probably due to a combination of factors which may develop from a phase transition event in the membrane lipids.

### Lipid and Fatty-acid Changes

Preliminary experiments on the changes in membrane lipid composition during prolonged chilling have revealed no significant decrease in the weights of phospholipids or glycolipids during the first 4 days at 5 C and 100% RH. This indicates that up to this time there is no increase in phospholipase activity. However, during this period there was a decrease in the percentage of linoleic acid associated with the phospholipids but no change in the fatty-acid composition of the glycolipids. Changes in the fatty-acid and lipid composition during prolonged chilling may lead to changes in membrane permeability and function and severely affect the reversibility of the phase change on return of the plants to the warmth.

### Photo-oxidation of Plant Pigments

The bleaching of some *P. vulgaris* leaves after 7 days of chilling at 100% RH indicates photo-oxidative degradation of the leaf pigments. Severe photo-oxidation of the leaf pigments and membrane lipids has been shown (van

Hasselt and Strikwerda, 1976) to occur in cucumber leaves after 2 to 3 days of chilling at 1 C. At temperatures higher than 1 C, this type of damage develops more slowly.

### Photosynthesis and Translocation

It is well known that photosynthesis is more sensitive to low temperature than respiration so that starvation of plant tissue may occur during prolonged chilling. Inhibition of translocation in chill-sensitive species at 5 C may lead to the starvation of nonphotosynthetic parts of the plant. The accumulation of starch in the chloroplast may further inhibit photosynthesis. Giaquinta and Geiger (1973) suggested that the cessation of translocation in chill-sensitive species at 5 C is caused by a phase change in the membrane lipids of the plasmalemma of the sieve tube which results in the collapse of the material lining the cell and the blockage of the sieve plate by the flow of cytoplasm, organelles, P-protein and membranes into it.

### ATP Supply

It has been suggested that impaired phosphorylation as a result of a phase change in the mitochondrial membrane lipids may lead to an insufficient level of ATP to maintain the metabolic integrity of the cytoplasm (Stewart and Guinn, 1969). However, experiments with *P. vulgaris* have shown that the ATP level increases, rather than falls, during chilling at 5 C and 100% RH (Wilson, 1978), probably because of a decrease in energy demand for growth at low temperature. Alternatively, the higher ATP level at 5 C may be related to the cold sensitivity of ATPase which is readily inactivated at low temperatures (Penefsky and Warner, 1965). Only when plants are chilled at 5 C and 85% RH, so that the leaves suffer rapid wilting and death, can one demonstrate a decline in ATP level during chilling (Wilson, 1978). Even in the extremely chill-sensitive species *E. reptans,* the level of ATP remains high during chilling at 5 C and 100% RH. The cause of chilling-injury to *E. reptans* is thought to be associated with both a three-fold increase in the rate of oxygen uptake at 5 C after only 2 hours of chilling in comparison to the control rate at 25 C (Wilson, 1978) and the appearance of unusual fibrils and filaments in the cytoplasm of the cells on rewarming (Murphy and Wilson, 1980). Although we have an idea of some of the metabolic differences which make some plants chill sensitive and others chill resistant, our greater understanding of the changes in the water relations of tropical plants at chilling temperatures should be put to practical use.

## AGRICULTURAL IMPLICATIONS

The yield of *P. vulgaris,* which has a higher water requirement than many other vegetable, cereal and forage crops (Ludlow and Wilson, 1972; Moldau, 1973) is reduced by water stress at various growth stages (Salter and Goode, 1967). The sensitivity of *P. vulgaris* and other tropical plants to water stress at 5 C can be attributed to the locking open of the stomata and to the greater reduction in the permeability of the roots to water on transfer to low temperature than occurs in many chill resistant species (Kramer, 1942). However, it is considered that the responses reported for *P. vulgaris* in the present investigation on transfer from 25 C and 85% RH to 5 C and 85% RH are extreme and that the events described will not happen as rapidly in the field. In these experiments the plants were grown under growth-cabinet conditions of constant 25 C, high humidity (85% RH), and adequate water supply so that the plants experienced very little water stress and had stomata that were nearly fully open at the start of chilling (Figure 2). Field-grown plants would experience greater water stress during early stages of growth because of the far greater fluctuations in temperature and relative humidity and, therefore, would be expected to be more resistant to leaf dehydration at 5 C. Also, in the experiments reported above, chilling was started by transferring the plants directly from a cabinet at 25 C to a cabinet already equilibrated at 5 C so that cooling was very rapid. The cooling of the roots was also rapid as the plants were grown in small, 8-cm diameter plastic pots. It has been shown by Böhning and Lusanandana (1952) that rapid cooling leads to a greater reduction in root permeability than slow cooling to the same temperature. Therefore, the rapid fall to 5 C of outside air temperature around the crop will result only in leaf dehydration if the soil also cools rapidly. Protection against this type of damage may be afforded by agricultural improvements to the soil, such as improved drainage which would tend to increase soil temperature.

The rate of rewarming is also very important in determining the extent of dehydration and injury of the leaves of *P. vulgaris.* Slow rewarming under high relative humidity and low light intensity causes less injury by reducing the rate of evapotranspiration and, since the soil temperature has more time to increase, by facilitating the absorption of water. The severity of chilling-injury to the crop, therefore, will be less if the chilling period is followed by a cool, cloudy day with high relative humidity and low wind speed. More research is needed to determine the most favorable rewarming conditions and to relate these to environmental changes experienced in the field.

The extent of injury to field-grown plants can be reduced if a period of darkness is experienced prior to chilling. Darkness results in stomatal closure and the stomata remain closed in the dark so that the leaves remain turgid during chilling (Wilson, unpublished data). As the incidence of low temperatures in the field is more frequent at night, the extent of chilling-injury will tend to be reduced. Hence, Crookston et al. (1974) were unable to detect any change in the water potential of *P. vulgaris* leaves when whole plants were chilled overnight, but they were able to detect a decrease in water potential from -2 to -6.8 bars on return to the light at 18 C. This decrease in water potential could not be attributed to an increase in the rate of evapotranspiration related to stomatal opening as the stomatal resistance remained high for many hours after return to the warmth. The reason for this type of water deficit is not known. These deficits only appear after exposure to light and even occur in plants held at 24 C overnight (Pasternak and Wilson, 1972). Ehrler et al. (1966) also have reported that the sudden exposure of cotton plants to light results in the development of temporary water deficits. Apparently, these deficits can be exaggerated by incubating the plants at very low or very high night temperatures. As the leaf resistance to evapotranspiration on return to the warmth was high, this water deficit cannot be attributed to above normal rates of evapotranspiration and probably is caused by resistance to water movement in the roots and shoots. Localized resistance to water movement in the leaf may account for the very scattered distribution of necrotic patches over the leaf surface in response to chilling. Alternatively, the necrotic patches may be areas where the stomata were in the fully open condition at the start of the chilling treatment, as it is well known that small groups of stomata undergo opening and closing cycles over the entire leaf at frequent intervals during the day. This is thought to be in response to short-term fluctuations in the water status of localized areas of the leaf (Shaner and Lyon, 1978).

Although field-grown plants may be less susceptible to leaf dehydration at chilling temperatures than ones grown under growth-cabinet conditions, glasshouse-grown plants transferred to the outdoors may be very susceptible as they are grown at relatively high temperatures and plentiful water supply. Even chill-resistant species such as cabbage can be susceptible to this type of leaf dehydration if they are transferred directly from glasshouse to the field with no intermediate hardening in a cold frame. Chilling-injury to cotton seedlings can be prevented by spraying with ABA or by spraying with a film anti-transpirant. However, in the case of chill-resistant brussel sprout plants treated with ABA before transplanting, McKee (1978) was unable to show an improvement in their water balance and growth.

## FUTURE RESEARCH

### Stomatal Regulation at Low Temperatures

The cause of the locking open response of the stomata of *P. vulgaris* at 5 C is not known. It is possible that the rapid reduction in the permeability of the roots to water at 5 C may lead to loss of turgor of the subsidiary cells and thereby prevent stomatal closure. Alternatively, the movement of $K^+$ ions out of the guard cells may be prevented by phase transitions in the membrane lipids of the plasmalemma which inactivate $K^+$-$H^+$ ATPase pumps so that the guard cells remain turgid. At present, we are investigating in a variety of species the differences in stomatal response that may result from passive effects as a consequence of changing leaf water relations and turgor potentials during cooling or from active responses of changed $K^+$ ion uptake and organic acid accumulation.

The application of ABA to plants before chilling results in a very significant improvement in the water balance at 5 C and 85% RH. ABA has most of the properties of an ideal anti-transpirant (Mansfield, 1976), although applications to the leaves may reduce the rate of shoot and root growth. Studies with ABA analogues (Orton and Mansfield, 1974) have not indicated a more desirable compound that ABA itself. However, considering the large number of possible structural variations of the ABA molecule, very few so far have been analyzed. Although a cheap and ideal anti-transpirant may be found, the usefulness of this compound in relation to chilling-injury would be dependent on our ability to predict the occurrence of chilling temperatures since, to induce stomata close before the temperature drops, applications must be made well in advance of the incidence of chilling. However, it should be noted that application of ABA is not always beneficial. Davies et al. (1978) have reported that the application of ABA to broad bean seedlings, which show partial stomatal closure in response to water stress, may reduce subsequent stomatal closure as water stress develops and, thus, have an adverse effect on plant turgor. ABA does appear to have beneficial effects on plant metabolism in prolonging the life of cotton leaves chilled at 4 C under conditions of 100% RH (Rikin et al., 1978). It can be speculated that, in addition to improving the water balance at chilling temperatures, ABA makes membranes, proteins, and cytoplasm more stable at low temperatures in some way.

It is well known that enzyme levels, chlorophyll synthesis, and cell growth may be reduced at bulk tissue water potentials higher than those which normally cause stomatal closure. The severity of these water deficits may have been increased in many crop varieties as an unforeseen result of extensive breeding programs (Davies et al., 1978). Maximization of yield is a primary consideration for plant breeders, and unless selection programs are carried out in environments dominated by water deficits, it is possible that some physiological responses to water stress will be bred out of modern cultivars. For example, the lack of a $CO_2$ response by the stomata of several crop plants may be an accidental result of breeding programs (Pallas, 1965).

## Root Permeability to Water

The greater decrease in the permeability to water of the roots of chill-sensitive plants on transfer from 25 C to 5 C than in chill-resistant species can be attributed to a greater increase in viscosity of cytoplasm, a decrease in membrane permeability, or both. The phase-change theory suggests that any lipid phase change in the root cell membranes of the endodermal cells should lead to an increase in permeability to water because of a decrease in membrane thickness and changes in the arrangement of the polar head groups (Träuble and Haynes, 1971). Therefore, the greater resistance to water uptake in chill-sensitive plants at 5 C than in chill-resistant species appears to be the result of a much greater increase in protoplasmic viscosity at low temperatures. Protoplasmic changes during chilling have not been studied extensively, but Patterson et al. (1979) and Murphy and Wilson (1980) have reported that among the earliest observable responses to chilling of the shoot hair cells of several chill-sensitive species are the disappearance of linear protoplasmic strands in the cytoplasm and the appearance of very many small cytoplasic vesicles which coincide with the cessation of cytoplasmic streaming. If such changes occur in endodermal cells, it can be expected that these events will have a profound effect on cytoplasmic viscosity and root permeability.

Chill-hardening of *P. vulgaris* plants before chilling increases the permeability of the roots to water at chilling temperatures (Figure 3). Whether this improved root permeability is achieved by modifications to the membranes, the cytoplasm, or both is not known. However, it has been suggested that increases in the degree of unsaturation of the root phospholipids during hardening may increase root permeability to water. Alternatively, hardening may reduce cytoplasmic viscosity at low temperatures perhaps by limiting the amount of vesiculation of the cytoplasm of endodermal cells. Although we need to know more about the mechanisms controlling root permeability to water, it is clear that breeding programs for selection of more chill-resistant varieties should be directed toward selection of those strains with high root permeability to water at low temperature. The selection of varieties with high ABA levels also may help to preserve a better water balance by enhancement of the hydraulic conductivity of the root (Glinka and Reinhold, 1971) to increase root permeability to water, as well as by stomatal regulation.

**Depression of Photosynthetic Capacity on Return to the Warmth**

Chilling *P. vulgaris* for as little as one night can result in severe reductions in photosynthesis the following day. This reduction in photosynthetic capacity is caused by increased stomatal resistance attributable to temporary water stress on return on of the plants to the light and warmth. The water stress had no effect on the activities of ribulose diphosphate carboxylase and malate dehydrogenase (Crookston et al., 1974). Following more prolonged durations of chilling during the important early stages of plant establishment, depressed photosynthesis can persist for 2 to 3 days after return to the warmth and may retard growth. Plant breeders may be able to select for plants that have stomata which open quicker on return to the warmth and, therefore, are able to grow more rapidly in marginal areas.

## CONCLUSIONS

Many tropical and semitropical crop plants are very sensitive to water stress at low temperatures. Improvements in the drought resistance of plants are likely to increase their chill-resistance. More fundamental research is needed to explain the locking open behavior of the stomata of tropical plants, the changes in the permeability of the roots to water, and the effects of ABA on membrane permeability and cell metabolism at 5 C.

### NOTES

John M. Wilson, School of Plant Biology, University College of North Wales, Bangor, United Kingdom.

### LITERATURE CITED

Böhning, R. H., and B. Lusanandana. 1952. A comparative study of gradual and abrupt changes in root temperature on water absorption. Plant Physiol. 27:475-488.

Crookston, R. K., J. O'Toole, R. Lee, J. L. Ozbun, and D. H. Wallace. 1974. Photosynthetic depression in beans after exposure to cold for one night. Crop Sci. 14:457-464.

Davies, W. J., T. A. Mansfield, and P. J. Orton. 1978. Strategies employed by plants to conserve water: can we improve on them? p. 45-54. In Opportunities for chemical plant growth regulation, British Crop Protection Council Monograph 21.

Ehrler, W. L., C. H. m. van Bavel, and R. S. Nakayama. 1966. Transpiration, water absorption, and internal water balance of cotton plants as affected by light and changes in saturation deficit. Plant Physiol. 41:71-74.

Giaquinta, R. T., and D. R. Geiger. 1973. Mechanism of inhibition of translocation by localized chilling. Plant Physiol. 51:372-377.

Glinka, Z., and L. Reinhold. 1971. Abscisic acid raises the permeability of plant cells to water. Plant Physiol. 48:103-105.

Kramer, P. J. 1942. Species differences with respect to water absorption at low soil temperatures. Amer. J. Bot. 29:828-832.

Ludlow, M. M., and G. L. Wilson. 1972. Photosynthesis of tropical pasture plants. IV. Basis and consequences of differences between grasses and legumes. Aust. J. Biol. Sci. 25:1133-1145.

Lyons, J. M. 1973. Chilling injury in plants. Ann. Rev. Plant Physiol. 23:445-451.

Mansfield, T. A. 1976. Chemical control of stomatal movements. Phil. Trans. Roy. Soc. Lond. Bull. 273:541-550.

McKee, J. M. T. 1978. The effect of abscisic acid on the water balance and growth of vegetable transplants during reestablishment. p. 63-68. In Opportunities for chemical plant growth regulation, British Crop Protection Council Monograph 21.

Moldau, H. 1973. Effects of various water regimes on stomatal and mesophyll conductances of bean leaves. Photosynthetica 7:1-7.

Murphy, C., and J. M. Wilson. 1981. Ultrastructural features of chilling injury in *Episcia reptans*. Plant, Cell Envir. 4:261-265.

Orton, P. J., and T. A. Mansfield. 1974. The activity of abscisic acid analogues as inhibitors of stomatal opening. Planta 121:263-272.

Pallas, J. E. 1965. Transpiration and stomatal opening with changes in carbon dioxide content of the air. Science 147:171-173.

Patterson, B. D., D. Graham, and R. Paull. 1979. Adaptation to chilling: survival, germination, respiration and protoplasmic dynamics. p. 25-35. In J. M. Lyons, D. Graham, and J. K. Raison (eds.) Low temperature stress in crop plants: the role of the membrane. Academic Press, Inc., New York.

Penefsky, H. S., and R. C. Warner. 1965. Partial resolution of the enzymes catalyzing oxidative phosphorylation. IV. Studies on the inactivation of mitochondrial adenosine triphosphatase. J. Biol. Chem. 240:4694-4702.

Raison, J. K., J. M. Lyons, R. J. Mehlhorn, and A. D. Keith. 1971. Temperature-induced phase changes in mitochondrial membranes detected by spin labelling. J. Biol. Chem. 246:4036-4040.

Rikin, A., D. Atsmon, and C. Gitler. 1978. Chilling injury in cotton (*Gossypium hirsutum* L.): prevention by abscisic acid. Plant Cell Physiol. 20:1537-1546.

Salter, P. J., and J. E. Goode. 1967. Crop responses to water at different stages of growth. Commonwealth Agricultural Bureau, Farnham Royal, Bucks, England.

Shaner, D. L., and J. L. Lyon. 1979. Stomatal cycling in *Phaseolus vulgaris* L. in response to glyphosate. Plant Sci. Letters 15:83-87.

Stewart, J. M., and G. Guinn. 1969. Chilling injury and changes in adenosine triphosphatase of cotton seedlings. Plant Physiol. 44:605-608.

Tagawa, T. 1937. The influence of temperature of the culture water on the water sbsorption by the root and on the stomatal aperature. Hokkaido Imp. Univ. Journ. Faculty Agr. 39:271-296.

Träuble, H., and D. H. Haynes. 1971. The volume change in lipid bilayer lamellae at the cyrstalline-liquid crystalline phase transition. Chem. Phys. Lipids 7:324-344.

van Hasselt, P. R., and J. T. Strikwerda. 1976. Pigment degradation in discs of the thermophilic *Cucumis sativus* as affected by light, temperature, sugar application and inhibitors. Physiol. Plant. 37:253-257.

Wilson, J. M., and R. M. M. Crawford. 1974. The acclimatization of plants to chilling temperatures in relation to the fatty acid composition of leaf polar lipids. New Phytol. 73:805-820.

Wilson, J. M. 1976. The mechanism of chill and drought hardening of *Phaseolus vulgaris* leaves. New Phytol. 76:257-270.

Wilson, J. M. 1978. Leaf respiration and ATP levels at chilling temperatures. New Phytol. 80:325-334.

## SECTION IV

# REDUCTION OF INJURY BY MANAGEMENT

*This section is introduced by a paper on the estimation of risk, followed by groups of papers on crop management. Barfield, Palmer, and Haan discuss the items to be considered in estimating the economic risks and probability of return on capital invested in irrigation facilities. Among these factors are an evapotranspiration model, a crop simulation model, the frequency of drought, and the designed capacity of the irrigation system. When irrigation capacity is limited, there is increased necessity for accurate crop simulation models to predict the occurence of critical periods in crop development for irrigation. An example is the need to predict accurately the time of tasseling of corn. Better models of phenological development also are needed for other crops.*

*Cassel discusses problems limiting the amount of water available to roots that can be stored in the soil and the geographical distribution of problem soils in the eastern half of the United States. Among the problems are soils with shallow profiles, soils with deep, sandy profiles, and soils with physical and chemical barriers to root penetration within the profile. Physical barriers, or pan layers, can be corrected by management practices such as subsoiling. He suggests that we need varieties with roots that can penetrate impermeable layers in the soil and are tolerant of aluminum. In addition to more work on tillage systems, we also need to learn more about measurement of available soil water. Most of the participants agreed that the permanent wilting point is at about -15 bars, which is well correlated with the actual wilting of sunflowers.*

*Sneed and Patterson state that in the humid Southeast supplemental irrigation produces significant increases in yield of the important crops. The increasing costs of land, equipment, and fertilizer make supplemental irrigation more profitable. Also, irrigation makes it possible to use varieties with a high yield potential regardless of their behavior under water stress. An important consideration in cost of supplemental irrigation in humid, temperate regions is the fact that rarely are more than a few centimeters of water needed to carry the crop through a stress period. This requires much less water and energy than in more arid regions where irrigation is necessary during much or all of the growing season. One problem with irrigation in humid regions is injury from saturated soil if irrigation is followed by a heavy rain. This emphasizes the need for better prediction of rainfall. Better surface drainage also is needed to handle excess rainfall. Perhaps the major problem is to decide when to irrigate. Kriedemann and Barr also state that a major problem in irrigation is to learn to apply water more efficiently. They think there will be increased combination of fertilization, especially nitrogen, with irrigation. There still is need to consider the best methods of irrigation: flooding, trickle, sprinkler, or subsurface, depending on water quality, crop and soil characteristics.*

*Curry and Eshel discuss simulation modeling as a potential tool in stress-avoidance research and management. They begin the discussion with a description of SOYMOD/OARDC which should be ready for use in crop management within a few years. Use of such models can reduce research time in selecting the most likely combinations of planting date, stand density, canopy architecture, and irrigation scheduling to minimize effects of stress on crop yields. Models also can be used to study the feasibility of other strategies of stress avoidance before testing them in field situations. Research required to enhance the usefulness of simulation modeling as a tool in stress studies includes: improved understanding of the physiological basis of stress effects on plant growth and the timing of phenological events, improved understanding of the soil and root environment complex involved in plant response to changes in soil moisture, and construction of models of tillage and management to be coupled with the plant model.*

# 10

# SIMULATING THE ECONOMICS OF SUPPLEMENTAL IRRIGATION FOR CORN

**B. J. Barfield, W. L. Palmer, and C. T. Haan**

Using simulation models with site specific parameters to evaluate the economics of supplemental irrigation can help determine if supplemental irrigation is economically feasible in semi-humid regions. Since crop growth occurs without irrigation in these regions, the question to be answered is whether the return from increased yields from supplemental irrigation offsets the extra expenditures necessary. In many instances, a farmer may lose money due to an investment in irrigation. Surface water is the main source of water supply for irrigation in many semi-humid regions, therefore, sizing of a reservoir is necessary. Due to limited information on the stochastic nature of irrigation water requirements, sizing of reservoirs for irrigation water supply in semi-humid regions has historically been based on rules of thumb. These estimates tend to be conservative resulting in frequent oversizing and excess cost. In a site specific evaluation of the economics of supplemental irrigation when using surface water supply, consideration should be given to optimum sizing of the water supply reservoir. An analysis of the optimum reservoir size, as well as the economics of supplemental irrigation, must consider variations in soil moisture holding capacity, hydrologic characteristics of the watershed supplying the runoff, climate conditions, agronomic practices, and irrigation management practices.

In the analysis given in this paper, simulation models are used to determine grain yields, availability of irrigation water, and the capital outlay involved in irrigation. Climatic data, agronomic practices, and irrigation managements practices are inputs to these models. Daily water demand for irrigation, water flow into a reservoir and a mass balance is used to determine the reservoir size which will supply water at all times for the study period. This reservoir volume is incrementally reduced and thus limits the availability of irrigation water, which results in reduced irrigated yields. For each reservoir size, the grain yields and irrigation expenses are calculated. A present worth analysis is used in evaluating the economics with the increased income from grain yield, additional expenses from irrigation and reservoir construction cost as the factors being considered. A family of curves is generated at different risk levels which indicates the amount of capital needed for investment in the irrigation systems as a function of reservoir size. These curves can be used as a guide in deciding if irrigation is economically feasible.

## SYSTEM DESCRIPTION AND BASIC ASSUMPTIONS

In the analysis conducted for this project, corn growth calculations were generated by the Duncan SIMAIZ Model (Duncan, 1975). This model generates daily growth calculations and ultimately grain yield as a function of temperature, solar radiation, rainfall, and irrigation water applied. In addition to growth calculations, SIMAIZ also calculates daily evapotranspiration and subtracts it from the current plant available water to give a measure of the daily soil water deficit; thus, the model can be used to generate the daily irrigation water requirements. Plant available water (PAW) is the amount of water in the rotting depth of the soil which is available for plant growth. The potential PAW is that which exists between field capacity and wilting point (15 bars). A listing of the inputs required is shown in Table 1. The outputs used from the SIMAIZ model are irrigated and non-irrigated grain yields and irrigation water demand.

Daily water yield into a water supply reservoir is generated by the Haan Water Yield Model (Haan, 1972; Jarboe and Haan, 1974). Haan's Water Yield Model calculates the daily water budget of a watershed by distributing daily rainfall among surface runoff, deep seepage, and evapotranspiration. Site specific parameters for the watershed are used in the model.

By using the SIMAIZ and the Haan Water Model, it is possible to predict, on a daily basis, the irrigation water demands and the water flow into a water supply reservoir. If these daily inflows and outflows are combined with other reservoir water losses and gains, as described by Palmer, et al. (1981), then the water supply available for irrigation can be calculated on a daily basis. If the water supply is less than that required for irrigation, crop moisture stress and, depending upon the size of the deficit and weather subsequent to the date of the irrigation demand, a yield reduction may occur. Therefore, for a minimum reservoir size over a long period of time, there can be periods when the water supply is less than the irrigation demand which will result in re-

**Table 1. Input data for SIMAIZ. The values listed are those which are most likely to change for location and variety of corn. A more complete listing of the inputs can be found in Palmer et al. (1981).**

| Variable | Description |
|---|---|
| XSTRES | Table relating soil moisture deficit and plant stress. |
| EVAPK | Table relating evapotranspiration and soil water deficit. |
| FRULCH | Factor modifying surface evaporation and used to simulate no-till. |
| H2OPRO | Water in rooting profile at first climatic information, expressed in inches. |
| H2OCAP | Available water held in soil to rooting depth, expressed in inches. |
| U | Amount of water that must evaporate from bare soil before it ceases to act as a free water surface, expressed in mm. |
| ALPHA | Constant dependent upon the hydraulic properties of the soil. |
| STYLE | Maturity classification of the variety: Early = 1, Intermediate = 2, Late or full season = 3. |
| CLASSM | Fine adjustment of maturity classification in days later (–) or earlier (+) than standard. |
| CORNMX | Approximate weight of a whole plant grown under ideal conditions at a low plant population. |
| EARMAX | Maximum weight of the grain on a full ear. |
| EARNMX | Maximum expected ears per stalk at low low populations. |
| PCORRF | Factor for relative photosynthetic rate of the variety. |
| WTKERN | Weight of a maximum sized kernel of the variety, expressed in g. |
| CLIMAR (I,J) | An array consisting of daily rainfall, maximum and minimum temperature, and solar radiation. |

duced yields. The amount of yield decline can be determined on a probabilistic basis by using the simulation models over a time period of suitable length. In the analysis conducted in this report, daily climatic values for 25 years are used. Initially, a reservoir size is determined so that during the 25 years the reservoir will always have water when needed. Once this size is determined, the size is incrementally reduced and the data for the 25 years are used to calculate the probability of grain yields for each reservoir size. An example is shown in Fig. 1.

In addition to the analysis mentioned above, the irrigation costs are also determined for each year and reservoir size to allow an economic analysis to be made. In this analysis, the increased income resulting from the yield improvement attributable to irrigation is used as the benefit to be evaluated

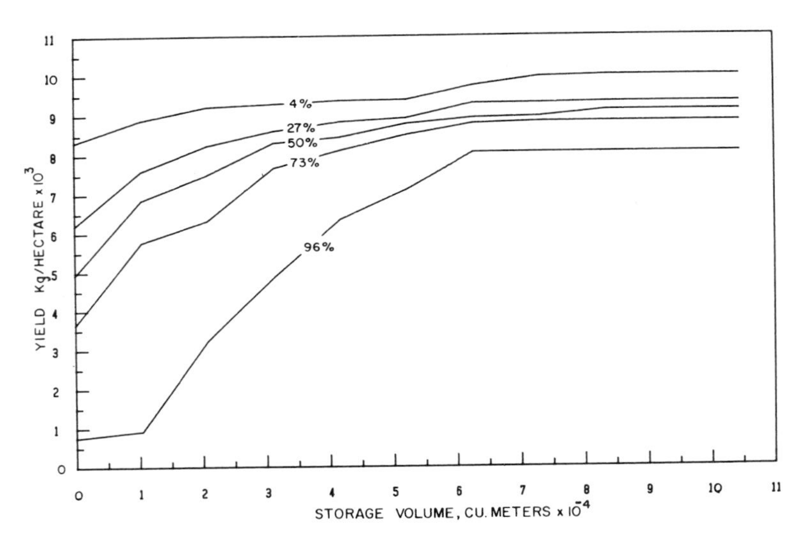


**Figure 1. Probability curve of yield versus storage volume.**

against the cost for irrigation. An analysis of the economic feasibility of supplemental irrigation is made, assuming that growing corn is economical. In semi-humid regions the total rainfall amount during the growing season is generally in excess of the evaporative demand. Any irrigation water would be used to supplement rainfalls during periods of drought. There is not the need for changes in agronomic practices as there is in more arid regions when converting from dryland to irrigated farming where irrigation water represents a significant portion of the plant water use. Because of the uncertainty of projecting new farming trends, it is assumed in the analysis that crop production practices (except for the application of water) will not change over the life of the system.

The reader is referred to Palmer et al. (1981) for details of the reservoir sizing routine in this model. The data inputs for each component are shown in Tables 1 through 4. From an analysis of these inputs one can see that the model has the ability to conform to individual site conditions, thus enabling the analysis to be site specific. The remainder of this report will be directed towards the economic sections of the model.

## ECONOMICS MODEL

In an economic analysis, all cash flows are evaluated at some reference time. This simulation model uses a present worth analysis (Taylor, 1975) for irrigation economics. In a present worth analysis, all cash flows over the life of the system are converted into an equivalent present value by using appro-

**Table 2. Input data for Haan's Water Yield Model.**

| Variable | Description |
|---|---|
| VARB1 | Infiltration rate. |
| VARB2 | Maximum possible seepage rate. |
| VARB3 | Maximum soil moisture capacity for watershed. |
| VARB4 | Constant defining the fraction of seepage that becomes runoff. |
| RML | Volume of water which is less readily available for evapotranspiration. |
| RMV | Volume of water which is readily available for evapotranspiration. |
| HD | Distribution of hourly rainfall within a day. (SCS Type II or Type I curve.) |
| PR | Distribution of rainfall within an hour. (See Haan, 1972). |
| PET | Potential monthly evapotranspiration. (Thornthwaite's values used.) |
| R | Daily rainfall. |

**Table 3. Input data for Water Budget Routine.**

| Variable | Description |
|---|---|
| PACRE | Pond area of reservoir site at incremental elevations. |
| WIDTH | Width of the centerline at the proposed dam site for same elevation as PACRE. |
| AREARO | Area of watershed contributing to runoff. |
| H | Initial dam height assumed. |
| HH | Interval between elevation for PACRE and WIDTH. |
| HFIX | Maximum height of dam not to be exceeded. |
| | *– Data necessary for calculations of seepage through dam –* |
| ZZ | Distance from riser outlet to top of dam. |
| W | Width of the top centerline of dam. |
| ANG | Angle formed by downstream and upstream faces and the ground surface. |
| CK | Hydraulic conductivity of the material comprising the least permeable section of the dam. |

**Table 4. Input data for Farm Irrigation Scheme.**

| Variable | Description |
|---|---|
| AREAIR | Area of corn to be irrigated. |
| POPPLT | Population of corn. |
| H2ODEF | Amount of soil water deficit which signals irrigation. |
| H2OIRR | Amount of irrigation water to be applied. |
| H2OLIM | Minimum amount of irrigation water to be applied during one application. |
| SACIRR | Amounts of water to be stored in reservoir until tasseling occurs. |

priate factors to account for interest. In the economic analysis, the items considered are: (1) reservoir construction, (2) yearly maintenance, (3) pumping system operation cost, and (4) additional income from increased grain yields.

To calculate the cost of the dam, it is necessary to know the volume of fill. By knowing the reservoir site typography, a relation between volume of storage water and dam size is made. The length of the dam can also be measured from the topographic map, and the total volume of the dam can then be calculated allowing a means of determining reservoir construction cost in relation to reservoir volume. Reservoir cost is calculated from:

$$\text{Reservoir Cost} = [(\text{Volume of Dam}) \times (\text{Fill Price})] + \text{DAMCST} \qquad [\text{I}]$$

Fill price is an input which may vary for different locations. The value of $2.00/m^3 used in this analysis, was recommended by the Soil Conservation Service in Kentucky as reasonable for small earth dams. DAMCST is used when additional expenses, such as sealing problems and special out-flow structures, are encountered. Yearly maintenance cost of the reservoir is figured as a constant percentage of the reservoir cost.

Pumping costs are figured from the volume of water pumped and dynamic head. In the examples, electricity is used as the power source. Other fuels may be used if a conversion factor or appropriate equations are substituted. The equation used for pumping cost is:

$$\text{Pumping Cost} = (\text{CKWH} \times \text{TDH} \times \text{H2OAD})/(11.7 \times \text{EFFP} \times \text{EFFM}) \qquad [\text{II}]$$

where CKWH is the cost per kilowatt hour, TDH is the total dynamic head which varies with site, H2OAD is the total volume of water applied during each season, EFFP is efficiency of the pump, and EFFM is efficiency of the motor.

System set-up cost is based on the number of times irrigation was performed during each season. A total cost of $8.00 for labor was assumed for each time irrigation was performed. This will vary with different type systems and is an input to the model.

The value of the increased yield is calculated from:

$$\text{Value of Increased Yield} = (\text{Difference in yield due to irrigation}) \times \text{GRPR} \qquad [\text{III}]$$

where GRPR is the price of grain. A fixed price of grain of $2.50/bu ($9.84/quintal) was used in this analysis. By using a fixed value for GRPR, no account is made for the stochastic nature of grain prices. It would be possible to account for price variation if an acceptable prediction model could be developed. In the absence of such a model, one can input a best estimate of the projected long-term value for GRPR or, if so desired, evaluate the economics under a family of values for GRPR.

The present worth using these cash flows would be a value which must exceed the cost of the irrigation hydraulics system (i.e., pipes, pumps, sprinklers) to realize a profit from irrigation. The equation used is:

PW = [(Value of increased yield - Maintenance Cost - Pumping Cost - Operating Cost) x USPWF] - Reservoir Construction Cost [IV]

where USPWF is the Uniform Series Present Worth Factor.

A summary of the equations used to compute the costs is given in Table 5. To make comparisons between investment decisions, comparisons must be made on the present worth of any future costs and benefits, including system set-up costs, pumping costs, value from selling the corn, etc. By comparing present worth values, recognition is made of the fact that interest causes money invested immediately to have a different cost to the project than money invested at subsequent periods over the life of the project. These differences are attributable to the facts that interest must be paid on monies borrowed for a project, and capital available for a project could be invested to produce interest income.

The reservoir has initial and yearly maintenance costs. The initial cost is based on the volume of earth to be moved. This volume depends upon dam

**Table 5. Summary of equations used for calculating costs and returns in the model and values used in example computations. English units, given in parentheses, were used in the computations.**

RESERVOIR COST:

Initial Cost = [(Volume of dam) x (Fill price)] + DAMCST [I]

Volume of dam is calculated for each reservoir size.

Fill price is used to determine the cost of constructing the reservoir based on the volume of the dam – $2.00/m^3$ ($1.50/cu yd)

DAMCST is used when additional expenses are encountered for sealing and other related expenses.

Maintenance Cost = a constant yearly cost.

PUMPING COST = (CKWH x TDH x H2OAD)/(11.7 x EFFP x EFFM) [II]

| | | |
|---|---|---|
| CKWH | = | cost per kilowatt hour – $0.08/kW·hr |
| TDH | = | total dynamic heat – 47 m (100 ft) |
| H2OAD | = | total volume of water used – determined in model |
| EFFP | = | efficiency of pump – 0.65 |
| EFFM | = | efficiency of motor – 0.90 |

SET-UP COSTS = $8.00 for labor each time irrigation is performed.

VALUE OF INCREASED YIELDS = (Difference in yield due to irrigation) x GRPR

GRPR = value used for price of grain – $9.84/quintal ($2.50/bu) [III]

height, which in turn depends upon the storage volume desired. The yearly maintenance costs are figured as a percentage of the initial costs and the present worth computed from the Uniform Series Present Worth Factor (Taylor, 1975) using the average life of the irrigation system as a time base. The uniform series present worth gives the present worth (PW) of a series of *n* uniform annual investments (UI) at the interest rate *i*, or

$$PW = UI\ ([1-(1+i)^{-n}]\ /i) \qquad [V]$$

The return on investment could be figured two ways. One method would be to select an actual irrigation system (e.g., center pivot, side roll, hand moved, etc.), determine its initial cost and the operational costs, and perform an economic analysis. A separate analysis would need to be conducted for each type system. Another option is to calculate all costs except the investment in the system itself (e.g., pipe and sprinklers), and to subtract those costs from the anticipated sales of the increased yields due to irrigation. The result would be a prediction of the amount of money available for an investment in an irrigation system, making the analysis less system specific. Thus, for each reservoir size, the average return on an investment would be calculated over the period of simulation. To make the later analysis, the following steps need to be taken:

(1) Calculate the yield increases due to irrigation on a yearly basis for the expected life of the system (a 10-year period was used in this analysis).
(2) Calculate the present worth of the increased yield minus the increased annual cost due to irrigation (not including equipment amortization). Sum over the life of the system.
(3) Subtract the cost of the reservoir (construction plus present worth of the maintenance costs) from the value obtained in (2). This will be the capital available for investment in an irrigation system (CAIS).

An example is shown in Fig. 2 for an irrigation system with a 10-year life. The characteristics of the watershed and crops analyzed are given in Table 6 and investment costs in Table 5.

To use the analysis system described above, cost and return decisions must be based only on the long-term average values for return on investment without information on the level of risk involved on an annual basis or over the life of the system. To develop information on the level of risk involved, one would need to simulate a large number of values (30 to 40) for each point on the curve shown in Fig. 2 and do a return period analysis on the results. Such a simulation would require either adequate weather records for 300 to 400 years or the use of a technique to simulate daily maximum and minimum temperatures, wind speeds, relative humidities, and incoming solar radiation. Neither of these methods is presently feasible. The computational

**Table 6. Farm and watershed characteristics for example calculations. English units, given in parentheses, were used in the computations.**

| Inputs | Characteristic or Value |
|---|---|
| Farm inputs: | |
| Area of corn | 40 ha (100 acres) |
| Plant available water | 7.6, 11.4, and 20.3 cm (3.0, 4.5, and 8.0 in.) |
| Corn variety | Pioneer 3369A |
| Plant density | 74,071 plants/ha (30,000 plant/acre) |
| Maturity class | Medium maturity |
| Water inputs: | |
| Watershed area | 121 ha (300 acres) |
| Maximum infiltration rate | 0.21 cm/hr (0.53 in./hr) |
| Maximum seepage rate | 0.025 cm/day (0.063 in./day) |
| Fraction of seepage to runoff | 0.23 |
| Total soil moisture capacity | 2.1 cm (5.33 in.) |
| Irrigation management scheme: | Irrigate when 1/3 of the plant available water is depleted. |

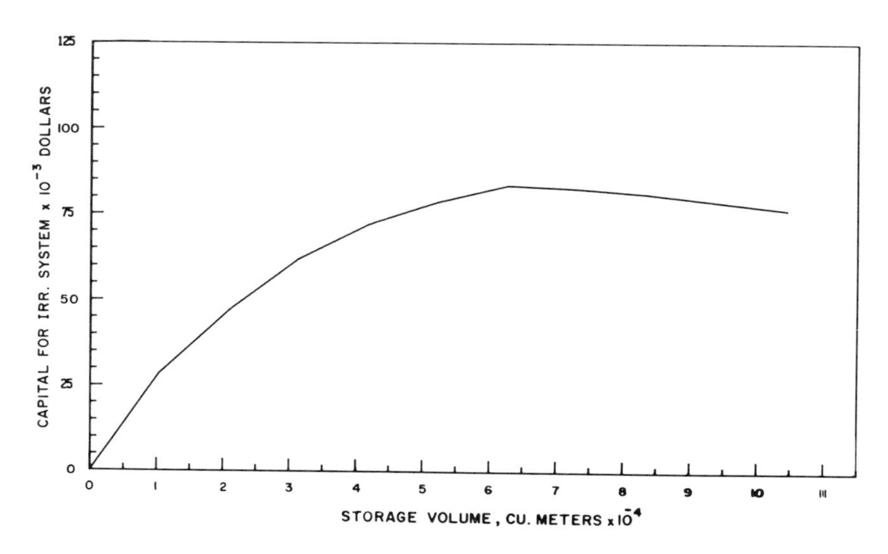


**Figure 2. Average capital versus reservoir size.**

cost of such an analysis, if techniques were available, would be excessive even with state-of-the-art computers.

An alternative approach would be to use the Uniform Series Present Worth Factors, as outlined below:

(1) On a yearly basis, calculate the yield increases, irrigation set-up times, and volume of water pumped for each reservoir size evaluated.
(2) Calculate the present worth of the yield increase minus the annual cost of irrigation using the Uniform Series Present Worth Factor for each year and reservoir size.
(3) Calculate the capital available for investment in an irrigation system as the amount figured in (2) above minus the cost of the reservoir.
(4) Rank capital available for investment in an irrigation system for each reservoir size and calculate the associated probability of occurance.

An example of the output using this method outlined is shown in Fig. 3 for the example farm and watershed in Tables 5 and 6. For a reservoir of 55,000 $m^3$ constructed for this site, a return on investment could be expected in 50% of the years (point A) if $80,000 or less were invested in the irrigation system itself (pipes, sprinklers, and pumps). If a system could be purchased for $40,000 or less (point B), a return on investment could be expected in 73% of the years.

This type of analysis could be used to offer farmers additional informa-

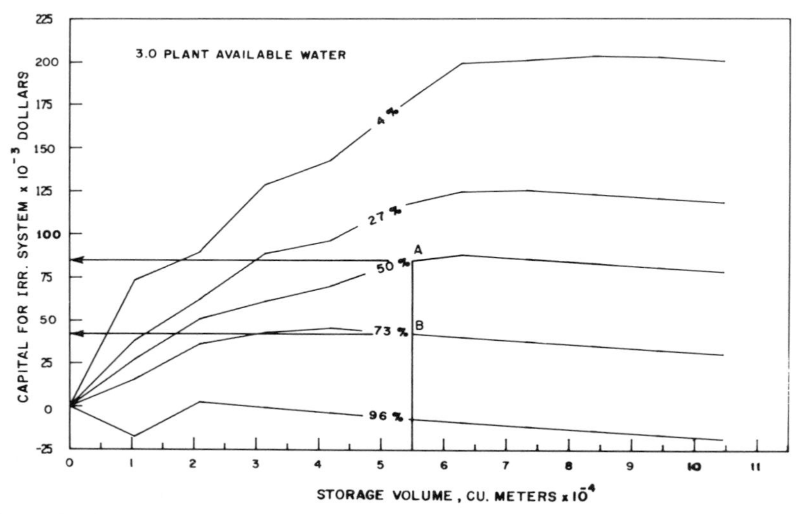


**Figure 3. Capital available for investment as a function of reservoir size for various probability levels for example farm near Lexington, Kentucky, based on annual values.**

tion to use in deciding whether to invest in an irrigation system; however, it essentially tells the investor the probability of recovering enough return on the investment in any given year to meet the payment on the system at a given interest rate. Information is still needed on the level of risk over the life of the system.

An approach to the analysis of risk over the life of the system can be developed using the Central Limit Theorem. The probabilities of the present worth values for a given reservoir size in Fig. 3 were calculated from a return period analysis on *n* years of records (25 in this case). These present worth values have a mean $\mu_{CAIS_i}$ (estimated by $CAIS_i$) and a standard deviation of $\sigma_{CAIS_i}$ estimated by $S_{CAIS_i}$) where the subscript *i* on CAIS refers to the estimation made based on annual values for a given reservoir size. Based on the Central Limit Theorem, it can be shown that the Present Worth based on an average over *n* years of return will be normally distributed with a mean of

$$\mu_{CAIS_n} = \mu_{CAIS_i} \quad [VI]$$

and standard deviation of

$$\sigma_{CAIS_n} = \sigma_{CAIS_i} / \sqrt{n} \quad [VII]$$

Using this fact and the standard normal curves, the risk (probability) levels can be developed for values of capital available for investment in an irrigation system based on a *n*-year life ($CAIS_n$).

The values in Fig. 3 are transformed this way and plotted in Fig. 4. A comparison of Figs. 3 and 4 can be made for any specific reservoir size. We will use the example of 55,000 $m^3$. The 50% probability level is $80,000 on other Figs. as expected from equation [VI]. Thus, one could expect the increased return on an investment of $80,000 in an irrigation system to be greater than the annual payment in 50% of the years and could be 50% sure of a return on investment of $80,000 amortized over a life of 10 years. However, from the 73% curve in Fig. 3, one could expect the increased return on an investment of $40,000 in an irrigation system to more than meet the annual payment in 73% of the years. However, one can be 75% sure of a return amortized over the 10-year life of an irrigation system with an investment of $70,000 (Fig. 4). The difference in the two values is due to the fact that the annual return on investment much more than meets the annual payment in some of the years.

A comparison based on a given potential investment is a more appropriate way of using Figs. 3 and 4. For example, if one considered investing in a system (pumps and pipe) that costs $50,000 and construction of a 55,000 $m^3$ reservoir, the chance of making a return on the investment over a ten-year system life would be better than 96% (Fig. 4). Approximately 70% of

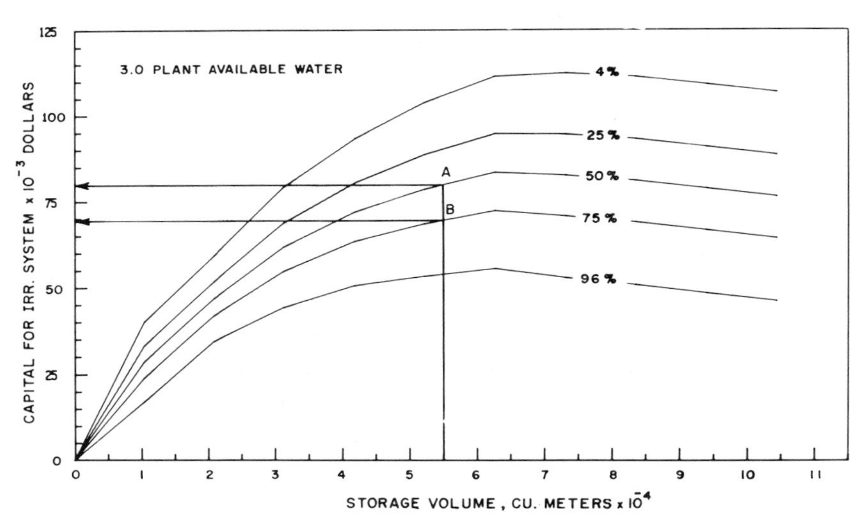


**Figure 4. Capital available for investment as a function of reservoir size for various probability levels for example farm near Lexington, Kentucky, based on averages over a ten-year life of the system.**

of the years would have an increased return on investment equal to, or greater than, the annual payment required for the money invested (Fig. 3). Therefore, the risk of obtaining a return on investment over the life of a system can be determined using Fig. 4, and an estimate can be made of the probability of obtaining a return in any one year using Fig. 3. The use of Fig. 3 would be important only when cash flow is a problem and the investor is not willing to accept a high risk of no return on investment in any one year.

## SENSITIVITY OF PREDICTIONS TO SELECTED INPUTS

### Sensitivity to Reservoir Volume

The effects of reservoir volume on average capital available for investment in an irrigation system ($CAIS_n$) are shown in Figure 4. Apparently, volumes of more than 60,000 $m^3$ have little effect on $CAIS_n$, although increased reservoir volumes above 60,000 $m^3$ apparently had little effect on corn yield (Fig. 1). Since the increased volume results in an increased dam cost, this seems somewhat surprising at first glance. For the site being evaluated, however, the geometry is such that a very small increase in dam height above that required for 60,000 $m^3$ results in a large increase in reservoir volume, and $CAIS_n$ is relatively insensitive to increases in reservoir sizes greater than 60,000 $m^3$ over the range of sizes evaluated.

**Effects of Plant Available Water**

The relationship between reservoir size and $CAIS_n$ at various levels of plant available water (PAW) is shown in Fig. 5 for climatic conditions near Lexington, Kentucky. As one would expect, both the size of reservoir needed and $CAIS_n$ are affected by changes in PAW. With increasing PAW, expensive above-ground storage of water is substituted for storage in the root zone. One can conclude from this analysis that PAW is an important parameter in determining $CAIS_n$.

**Effects of Management Decisions**

One management option would be to reserve enough water during the period prior to tasseling for one complete irrigation, since stress at pollination causes a significant yield decrease. A simulation using this management option (Plan 2 in Fig. 6) showed that it is not cost effective when compared with not reserving water (Plan 1) at 7.6 cm of PAW.

Another management option would be to replace only part of the depleted PAW on each irrigation in view of the fact that rainfall is probable. This option was evaluated by comparing 100% replacement of PAW with 67% replacement of PAW. In the 67% replacement of PAW, 67% of the soil moisture deficit is replaced when evapotranspiration has depleted one-third of the PAW. During dry periods, the 67% replacement of PAW would require

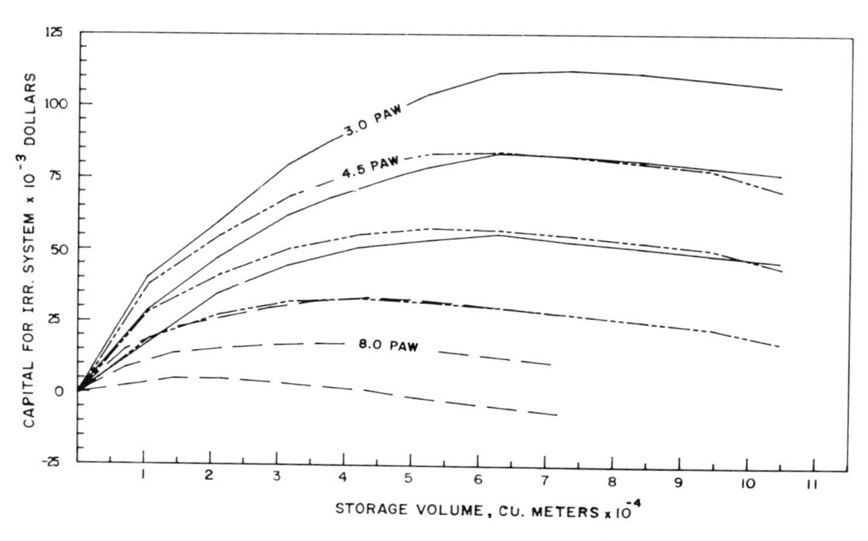


**Figure 5. Comparison of the effects of plant available water (PAW) on capital available for investment in an irrigation system.**

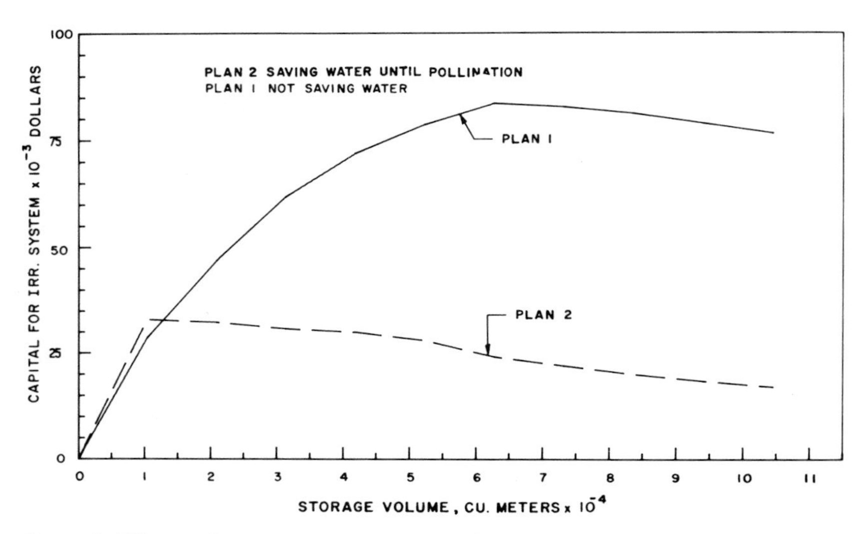


**Figure 6. Effects of saving water until time of pollination compared with not saving water.**

more frequent irrigation; however, long periods without rainfall are rare in humid areas. The results of this simulation shown in Fig. 7, indicate that frequent light irrigations are as cost effective as filling up the soil profile. Since nitrogen losses would be expected to be higher when a full soil profile is subjected to a heavy precipitation event, the light irrigation may be advantageous.

## RECENT WORK

Model refinements have been made subsequent to the presentation of the material included in this paper. Procedures have been programmed to include the effects of inflation and to consider the economics of supplemental irrigation when using a groundwater supply. Information on these procedures as well as a detailed model description are given in Palmer et al. (1981, 1982a, 1982b).

Procedures similar to those described in this report have been utilized to evaluate the economics of supplemental irrigation of selected horticultural crops. A description of those analysis procedures is given in Sands et al. (1982).

## SUMMARY

A procedure is discussed which uses the Duncan SIMAIZ Model, the Haan Water Yield Model for Water Supply, and other models to predict the economics of supplemental irrigation. Allowance is made in the procedures for variations in management decisions. The procedures are described and

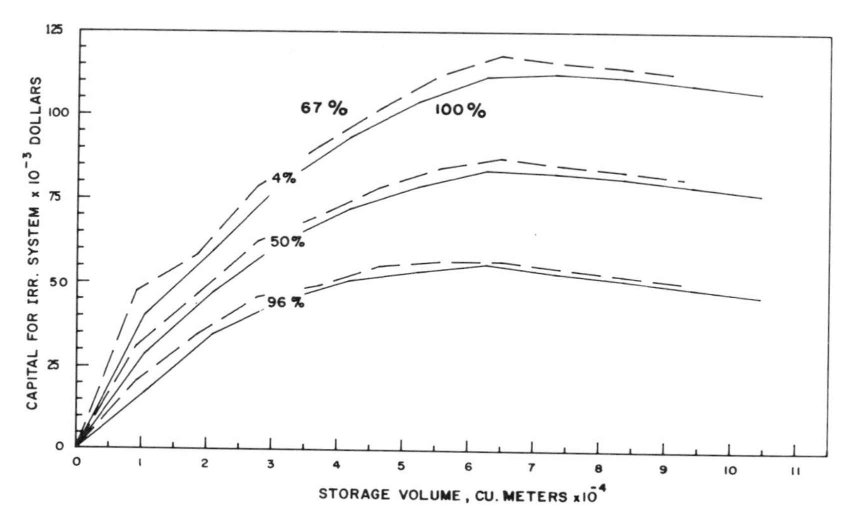


**Figure 7. Effects of replacing 100% of the moisture deficit versus replacing 67% of the moisture deficit.**

example computations presented. The conclusion can be made that a site specific analysis is required for accuracy.

## NOTES

W. L. Palmer, Wright McLaughlin Engineers, Denver, Colorado; B. J. Barfield, Department of Agricultural Engineering, University of Kentucky, Lexington, Kentucky 40546; C. T. Haan, Agricultural Engineering Department, Oklahoma State University, Stillwater, Oklahoma 74074.

## LITERATURE CITED

Duncan, W. G. 1975. SIMAIZ: A model simulating growth and yield in corn. p. 32-48. In D. N. Baker, R. G. Creech, and F. G. Maxwell (eds). An application of system methods to crop production. Mississippi Agric. For. Exp. Stn., Mississippi State, Mississippi.

Haan, C. T. 1972. A water yield model for small watersheds. Water Resour. Res. 8: 58-69.

Jarboe, J. E., and C. T. Haan. 1974. Calibrating a water yield model for small, ungaged watersheds. Water Resour. Res. 10: 256-262.

Palmer, W. L., B. J. Barfield, M. E. Bitzer, and C. T. Haan. 1981. Simulating the water and economic feasibility of irrigation of corn in the Midwest. Research Report No. 125, University of Kentucky Water Resources Research Institute, Lexington, Kentucky.

Palmer, W. L., B. J. Barfield, and C. T. Haan. 1982a. Sizing farm reservoirs for supplemental irrigation. Part I: Modeling reservoir size-yield relationships. Trans. ASAE 25: 372-376.

Palmer, W. L. B. J. Barfield, and C. T. Haan. 1982b. Sizing farm reservoirs for supplemental irrigation. Part II: Economic analysis. Trans. ASAE 25: 377-380.

Sands, G. R., I. D. Moore, and C. R. Roberts. 1982. Supplemental irrigation of horticultural crops in humid regions. Water Resour. Bull. 18.

Taylor, G. A. 1975. Managerial and engineering economy. 2nd Ed. D. Van Nostrand Company, New York.

11

# EFFECTS OF SOIL CHARACTERISTICS AND TILLAGE PRACTICES ON WATER STORAGE AND ITS AVAILABILITY TO PLANT ROOTS

D. K. Cassel

Factors in the soil environment which affect plant growth and yield are soil water potential, soil temperature, mechanical impedance (the resistance of the soil to root penetration), aeration, and the nature of the chemical environment (Shaw, 1952). Of these four factors, soil water potential, a measure of the energy level of water held in the soil, is the one which most often limits growth. If roots of an initially turgid plant do not take up water from the soil as rapidly as water is transpired from the leaves, the plant enters a period of water stress. Depending upon the stage of plant development and the duration and intensity of the stress period, reductions may result in growth and harvestable yield.

The amount of available water retained by soils is highly variable. Available water is defined as that water held by the soil between *in situ* field capacity and the permanent wilting percentage. *In situ* field capacity is the amount of water retained after drainage from a previously wetted soil becomes negligibly small. Permanent wilting percentage, the soil water content at which the plants wilt and fail to recover, is commonly estimated by the water content at -15 bars soil water potential. All water retained by soil between *in situ* field capacity and the permanent wilting percentage is considered to be available to plants although the roots absorb water at increasingly slower rates as the soil in the root zone becomes dryer. Moreover, the rate of water uptake by plant roots is also dependent upon root density and climatic conditions.

Many kinds of soil situations occur that can lead to plant water stress under natural rainfall conditions and, in some cases, even under irrigated conditions. Some soils have inherently low available water holding capacities. Examples of such soils are deep, coarse-textured soils and soils underlain by bedrock or gravel at shallow depths. These soils retain only enough available water to supply actively transpiring plants for several days. Another group of soils has greater available water holding capacities than those listed above, but the presence of some physical or chemical barrier(s), which impairs or prevents roots from growing deeper into the soil, prevents effective utilization of the available water stored in and below the barrier. Physical barriers can be natural (e.g. fragipans, duripans, clay pans) or man-induced (plowsoles, tillage-induced pans, and surface compaction). In humid, temperate climates, chemical barriers usually consist of soil horizons or zones with low pH and which usually have excessive amounts of aluminum or manganese.

The purposes of this chapter are: (1) to identify and discuss the effects of several soil physical and chemical barriers upon water uptake and root growth of agricultural crops; (2) to indicate the approximate areal extent in the humid, temperate region of the United States of soils affected by these barriers; and (3) to illustrate how tillage or other management practices may be used to amend or remove some of these barriers.

## FACTORS AFFECTING AVAILABLE WATER

The amount of available water retained by a soil is dependent upon the pore size distribution which is determined by soil texture, soil structure, and soil depth. Each of these three factors will be discussed briefly.

### Soil Texture

Soil texture is defined as the relative proportions of sand, silt, and clay in a soil. In general, soil texture varies with depth in the soil profile. For example, soil texture may gradually change from sandy loam in the A horizon to clay loam in the B1t horizon to loam in the B2 horizon. In addition, changes in soil texture also occur in the lateral or horizontal direction. Soils which are predominantly composed of sand-sized particles (0.05 to 2 mm effective diameter) have large pores and in the absence of a high water table, are excessively drained. Such soils retain only small amounts of available water and are said to be droughty. On the other hand, soils consisting primarily of particles smaller than 0.05 mm diameter have smaller, but many times more numerous pores, and retain greater amounts of available water. Most soils, of course, are composed of mixtures of both coarse (sand) and fine (silt and clay) particles; thus, soils have a wide range in pore sizes and in the amount of available water stored. Figure 1, adapted from Brady (1974), shows, in general, how the amounts of total, available, and unavailable water

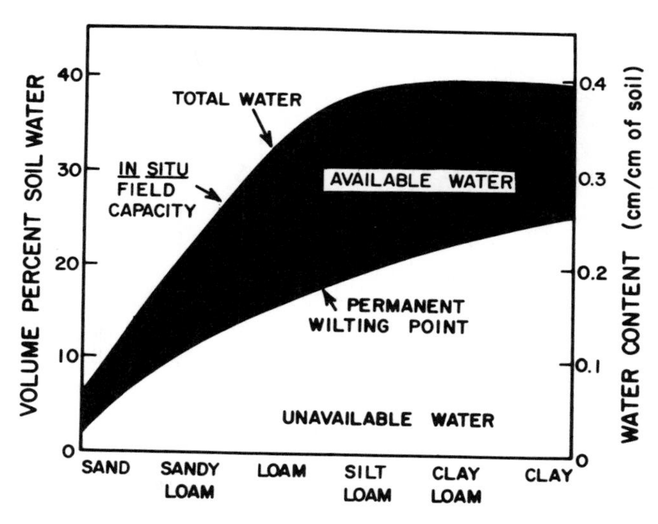


**Figure 1. General relationships between total, available, and unavailable water holding capacities as a function of texture (adapted from Brady, 1974).**

retained by soils vary with soil texture. The amount of water at *in situ* field capacity is related to pore size and more water is retained as texture becomes finer. Unavailable water is more closely related to surface area of the soil particles than to pore size. For this reason, the amount of unavailable water retained by a soil increases as soil texture becomes finer. The net result is that the amount of available water increases as texture becomes finer until it reaches a maximum for silt loam-textured soils.

The importance of soil texture upon the water holding capacities of field soils is demonstrated below. Figure 2A shows a schematic of the coarse-textured Lakeland soil profile [Typic Quartzipsamment] sand content of all horizons is 90% or greater. Even though no physical or chemical barriers to root growth may exist in this soil, the high sand content and associated large pore sizes result in a low available water holding capacity and limit the usefulness of this soil for nonirrigated agriculture. At *in situ* field capacity, a 150-cm deep profile of Lakeland sand retains 8 to 12 cm of available water. The areal distribution of deep (>150 cm) coarse-textured soils similar to Lakeland in the humid, temperate region of the U.S. is shown in Figure 3A.

One management practice developed to increase the available water storage capacity of deep, sandy soils is the installation of asphalt barriers (Erickson et al., 1968; Saxena et al., 1968). Continuous asphalt barriers installed at a depth of approximately 60 cm increase the quantity of stored available water by slowing the rate of drainage. The installation of asphalt

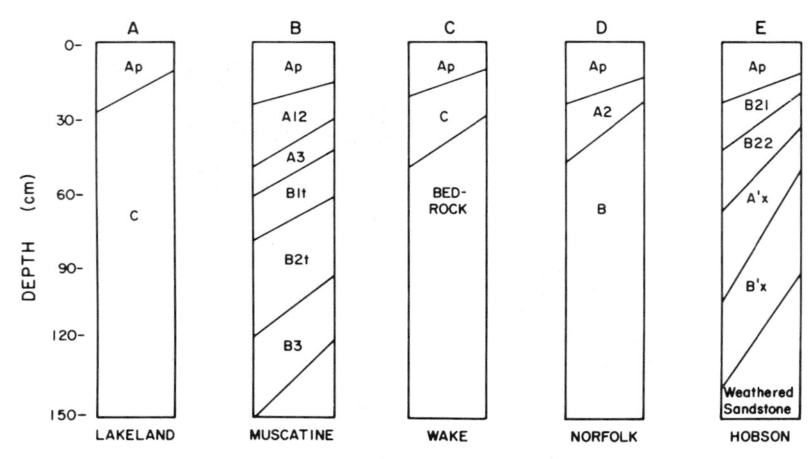


**Figure 2. Schematic diagrams of: (A) deep, coarse-textured soil (Lakeland); (B) soil with no physical or chemical barriers to root growth (Muscatine); (C) shallow soil underlain by bedrock (Wake); (D) coarse-textured surface soil subject to tillage pan formation (Norfolk); (E) soil with fragipan (Hobson).**

barriers in deep, sandy soils, however, is not economical except for high valued horticultural crops.

### Soil Structure

Soil structure refers to the arrangement of individual soil particles and is intimately related to soil texture and clay mineralogy. The structural units may be strong and resist destruction when external pressure is applied to the soil. Many soils, however, have weak structure and are quite susceptible to compaction. Soils composed of particles having approximately the same diameter are not easily compacted whereas soils composed of particles having a broad range in diameters are more easily compact (Bodman and Constantin, 1965; Coughlan et al., 1978). For the latter case, the smaller soil particles fit into the voids among the larger particles allowing a more densely packed configuration. In addition to particle size distribution, the soil water content (Chancellor, 1976), the shapes of the individual particles (Cruse et al., 1980), and the compactive effort also affect the density to which a given soil is compacted (Hegarty and Royle, 1978).

Soil structure influences root growth and activity. Root elongation occurs as the direct result of elongation of meristematic cells behind the root tip. As root elongation occurs, the root tip follows the path of least mechanical resistance. In fact, Drew (1979) proposes that root growth can be explained by the roots extending preferentially into zones of low mechanical impedance (and therefore higher water content) rather than growing in strict

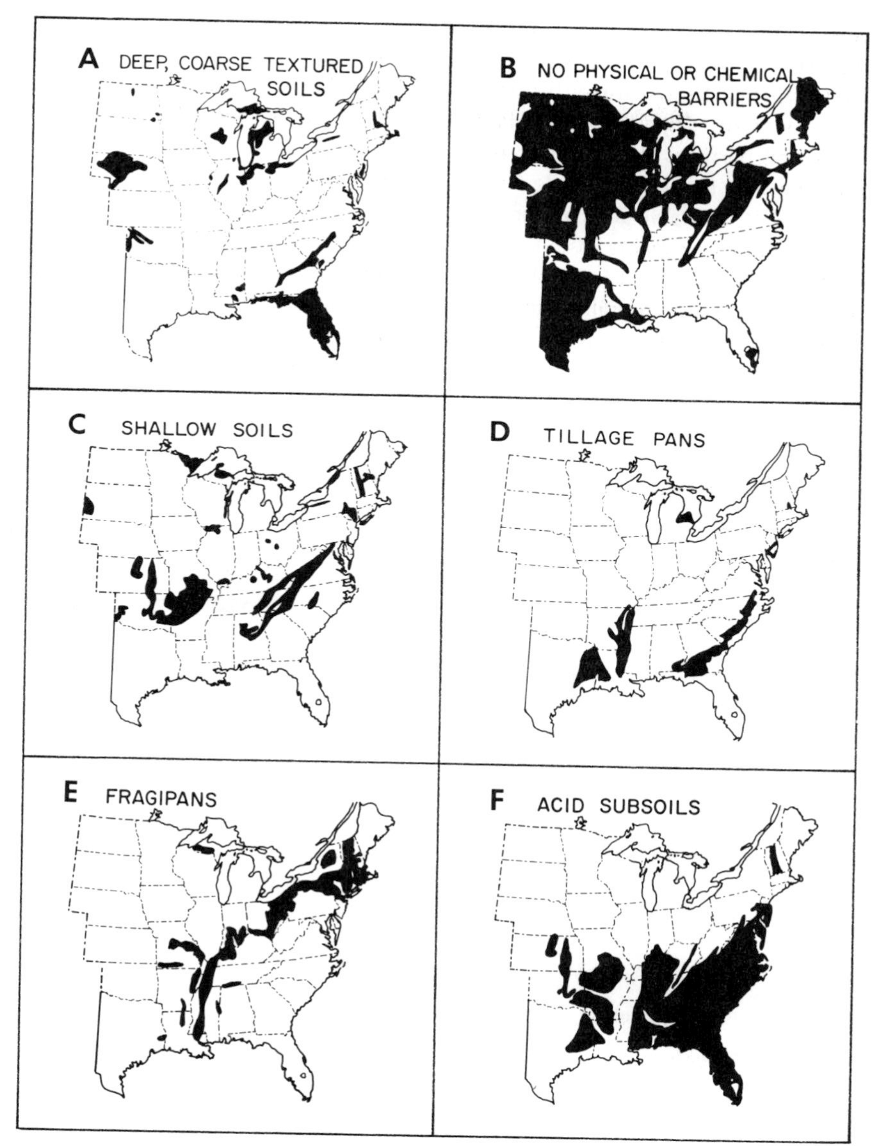


**Figure 3.** Areal distribution in the United States of: (A) deep, coarse-textured soils; (B) soils with no physical or chemical barrier to root growth; (C) shallow soils; (D) soils with tillage pans; (E) fragipan soils; (F) soils with low pH or Al toxicity. These maps were prepared from information supplied by the State Soil Scientists (Soil Conservation Service) from most states in the humid, temperate region.

accordance to hydrotropic response. Because plant roots cannot reduce the diameter of their apical meristems and pass through rigid, narrow pores, the roots must: (1) grow through pores which have diameters as large as or larger than the root, (2) grow along planes of fracture of soil peds, (3) physically move soil particles thus creating larger pores, or (4) fail to penetrate deeper into the soil (Wiersum, 1957; Goss and Drew, 1972). If the structure of a deep soil is favorable for root growth, a plant with the genetic potential for deep rooting is capable of extracting water and nutrients from considerable depth in the absence of chemical barriers. On the other hand, restriction of root development in soils with poor structure, which is often accompanied by high mechanical constraint to root elongation, may result in water stress even though adequate soil water is present in the soil below the region of mechanical restraint (Lutz, 1952).

### Soil Depth

Soil depth is the physical thickness of the soil solum. The Lakeland profile in Figure 2A is a deep profile although the amount of available water it retains is low as a result of its coarse texture. The Muscatine profile [Aquic Hapludoll] in Figure 2B is another deep soil, and because of its finer texture, it retains 25 to 30 cm of available water in a 150-cm deep profile. The areal distribution of deep soils is shown in Figure 3B. The Wake soil [Lithic Udipsamment] shown in Figure 2C is representative of soils that are shallow ($\leqslant$50 cm) to bedrock. Other physical barriers which give rise to shallow soils are gravel deposits or impermeable clay layers. The reduced physical thickness of these soil profiles reduces the available water storage capacity for a given soil texture. Hence, plants growing on shallow soils are more susceptible to water stress compared to those growing on deeper soils. The areal distribution of shallow soils in the humid, temperate region of the United States is shown in Figure 3C. Many of these soils remain forested. Crop production on shallow soils without irrigation is a high risk venture, even in humid regions.

## SOIL PHYSICAL BARRIERS

### Soils With No Physical Barriers

The Muscatine soil (Figure 2B), which was identified earlier as a representative of deep soils, is in fact a representative of a more important group of soils, i.e. deep soils with no physical or chemical barriers. That portion of the humid, temperate region of the United States occupied primarily by deep soils with no physical or chemical barriers is shown in Figure 3C. The mapping units on a map of this scale are, of course, not pure; for this reason, the smaller areas of soils that do have physical or chemical barriers cannot be delineated.

Soils which do not have physical barriers are, however, susceptible to surface compaction. Surface compaction refers to compaction of soil in the upper portion of the soil profile in response to the weight of wheeled vehicles and farm machinery as they ride upon the soil surface. The degree of compaction is especially dependent upon the ground contact pressure, the soil water content at the time of compaction, and the number of machinery trips, i.e. the number of times the soil is subjected to contact pressure. The first step in the compaction process is the breaking of bonds of the aggregating agents which hold soil particles together into structural units. Soil particles are then reoriented into a configuration having a higher bulk density. Virtually all soils are susceptible to compaction although those soils with high percentages of sand tend to be more susceptible. Lindstrom and Voorhees (1980) caution against excess wheel traffic in fields because the slower infiltration rate on compacted soils reduces the amount of water stored for plant use. In addition, the additional runoff from compacted soils greatly increases the erosion hazard. In humid, temperate regions, where much of the rainfall for the growing season occurs during intense thunderstorms of short duration, the amount of water stored is greater for noncompacted soils than for compacted soils.

Compaction increases soil bulk density which, in turn, increases the resistance the soil offers to root penetration. Mechanical impedance is defined as the resistance of soil to penetration of a penetrometer, i.e. a pointed rod. Root penetration decreases as mechanical impedance increases. For many studies, bulk density is commonly measured to assess the degree of compaction and is highly correlated with mechanical impedance. Increasing bulk density for a soil increases the resistance of the soil to root penetration by reducing the size and number of larger pores. In general, the upper limits of bulk density at which roots fail to penetrate in relatively wet soils range from 1.75 $g/cm^3$ for coarser-textured soils to 1.4 to 1.6 $g/cm^3$ for the finer-textured ones. No one critical bulk density value exists which prevents root penetration for all soils. Figure 4, adapted from Taylor and Gardner (1963), shows that cotton [*Gossypium hirsutum* L.] roots fail to penetrate the soil as the soil becomes both denser and dryer. Water acts as a lubricant to make it easier for the thrust developed by the roots to move the soil particles to create larger pores.

Raghaven and McKyes (1978) studied the effects of tractor traffic upon corn production. They related soil water content and root distribution in Ste. Rosalie clay to compaction intensity. The soil was subject to 1, 5, 10, and 15 wheeled-vehicle passes between each corn row, either before or after planting, at contact pressures of 0.32, 0.42, or 0.63 $kg/cm^2$. Measured soil water contents at various times during the growing season showed that the soil remained significantly wetter as the compaction treatment became more intense. The wetter soil resulted from reduced water uptake which was a

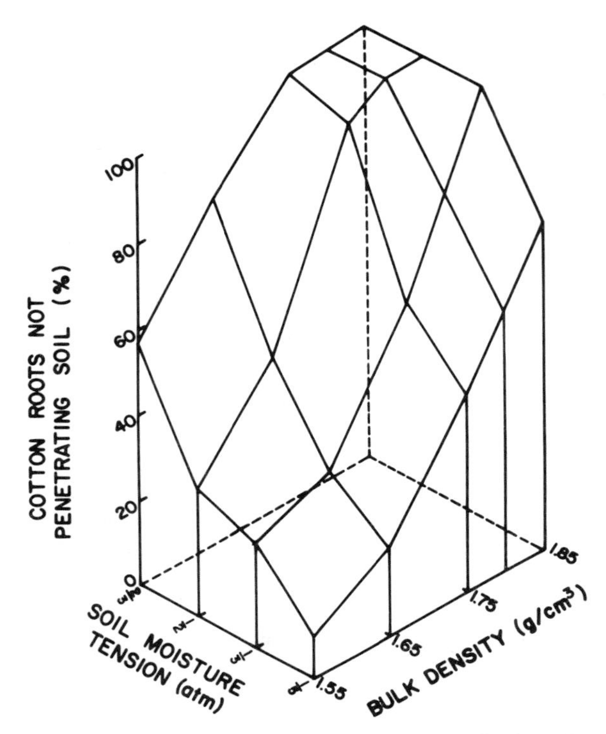


**Figure 4. Percent cotton roots failing to penetrate Amarillo fine sand loam as a function of bulk density and soil moisture tension (after Taylor and Gardner, 1963).**

consequence of reduced root proliferation. Figure 5 shows the rooting zone in the Ste. Rosalie soil subjected to compaction before seeding for 1, 5, 10, and 15 tractor passes at contact pressures of 0.32 and 0.63 kg/cm$^2$. Corn yield was significantly reduced as a direct result of poor root penetration in the compacted soil which was a function of the number of tractor passes and the contact pressure.

**Tillage-Induced Pans**

Large acreages of soils in the Atlantic and Gulf Coastal Plains and in the Mississippi River Basin are susceptible to tillage pan formation (Figure 3D). Tillage pans are compacted zones formed in response to pressure exerted by various tillage implements or by wheeled or tracked vehicles pulling them through the field. Moldboard plowing is one tillage operation which causes pan development in some soils. The pan is caused not by the action of the plowshear sliding across and orienting soil particles, but instead by compaction of soil immediately below the tractor wheel as it rides on top of moist

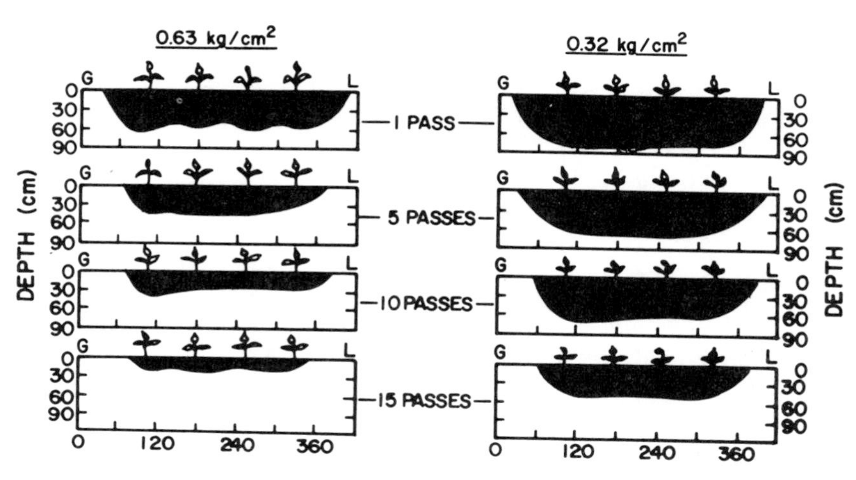


**Figure 5. Root distribution of Ste. Rosalie clay subjected to compaction at contact pressures of 0.32 and 0.63 kg/cm$^2$ for 1, 5, 10, and 15 passes prior to seeding (from Raghaven and McKyes, 1978).**

soil in the furrow bottom. After several years of moldboard plowing, the plowpan may be continuous throughout an entire field. The soils most susceptible to pan formation appear to be those which are coarse textured immediately below the depth of plowing.

The Atlantic Coastal Plain has extensive acreages of sandy-surfaced soils which are subject to tillage pan formation. Many soils in this region have sandy loam to loamy sand Ap and A2 horizons, and clay loam B horizons. The norfolk series [Typic Paleudult], shown in Figure 2D, is typical of such soils. The areal extent of soils subject to severe tillage pan formation is shown in Figure 3D. When many of these soils are plowed in early spring or late winter, the soil is too wet because of a perched water table above the B horizon. Tillage in this wet condition promotes compaction and high mechanical impedance.

*In situ* field capacity, permanent wilting percentage, and available water for the 150-cm deep Norfolk soil profile shown in Figure 2D are presented in Table 1. In the absence of rooting barriers a 36-day supply of water is retained when the soil is at *in situ* field capacity. However, a tillage-induced pan typically occurs at 15 to 25 cm, the depth of moldboard plowing. This pan in sandy material, which has a bulk density of 1.75 to 2.0 g/cm$^3$ and very few macropores, is extremely effective in reducing the depth of rooting and the rate of root extension (Campbell et al., 1974; Taylor and Bruce, 1968; Reicosky et al., 1977). By preventing or severely restricting rooting below 25 cm, the pan limits available water to a 6-day supply. Many reports

**Table 1. Bulk density, *in situ* field capacity, 15-bar water content, available water, and cumulative days water storage in a 150-cm deep Norfolk soil profile.**

| Horizon | Depth | $D_b$ | *In Situ* Field Capacity | 15-Bar Water Content | Avail. Water | Cumm. Avail. Water | Cumm. Water Supply[a] | Cumm. Water Supply With Pan |
|---|---|---|---|---|---|---|---|---|
| | *–cm–* | *g/cm³* | | *– cm –* | | | *–days–* | *–days–* |
| Ap | 0-25 | 1.62 | 4.2 | 1.1 | 3.1 | 3.1 | 6.2 | 6.2 |
| A21 | 25-35 | 1.75 | 1.8 | 0.5 | 1.3 | 4.4 | 8.8 | – |
| A22 | 35-45 | 1.70 | 1.6 | 0.6 | 1.0 | 5.4 | 10.8 | – |
| B1 | 45-75 | 1.66 | 8.7 | 4.5 | 4.2 | 9.6 | 19.2 | – |
| B2 | 75-150 | 1.51 | 19.9 | 11.3 | 8.6 | 18.2 | 36.4 | – |

[a]An E.T. rate of 0.5 cm/day is assumed.

substantiate reduced crop yields during years when several dry periods of 1 to 2 weeks duration occur during a critical growth stage.

The effects upon soybean [*Glycine max* (L.) Merr.] rooting and grain yield of a pan induced by the moldboard plow were reported by Kamprath et al. (1979). In this study rooting and grain yields for a Wagram loamy sand [Arenic Paleudult] with a pan resulting from conventional tillage were compared to those for the same soil subjected to two different deep tillage practices before planting. The deep tillage practices were imposed to "break up" or modify the physical properties of the pan. Conventional tillage, i.e. the tillage method used by farmers in the area, consisted of moldboard plowing to a 25-cm depth followed by three diskings prior to planting. The second treatment employed a chiselplow, an implement with shanks spaced 30 cm apart and extending 27 cm below the surface. The third tillage operation was in-row subsoiling to a depth of 45 cm. The soil was mounded, or "bedded," over the subsoil slit in the same tillage operation; soybeans were planted in the beds.

Mechanical impedance of the Wagram soil as a function of soil depth and position (distance from the row) two weeks after planting is shown for all three tillage treatments in Figure 6. Mechanical impedance was measured with a cone penetrometer (Cassel et al., 1978) when the soil water content was at *in situ* field capacity. For the conventionally tilled soil, mechanical impedance attained a value of 20 $kg/cm^2$ at the 17- to 20-cm depth. The hardness intensified to greater than 30 $kg/cm^2$ in the 20- to 30-cm zone. Taylor et al. (1966) have shown that root elongation problems become important for cotton at mechanical impedance values of 20 $kg/cm^2$. This value is also used to indicate when mechanical impedance becomes limiting for other crops. Chiselplowing reduced mechanical impedance of the pan to values $<$20 $kg/cm^2$ in the upper 30 cm. Likewise, the subsoil-bedded operation

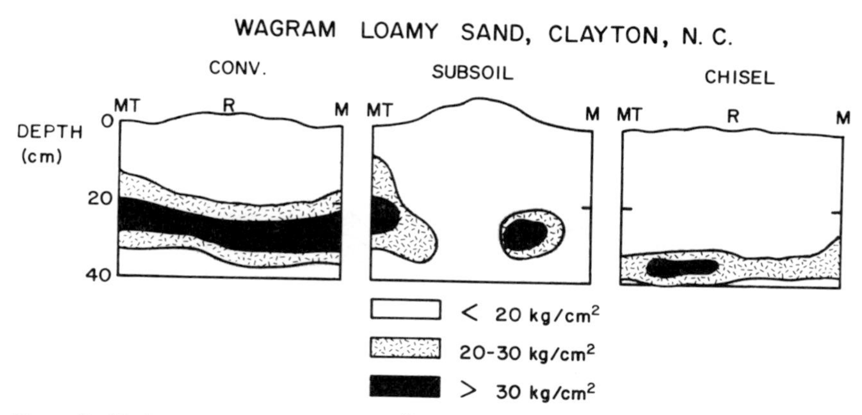


**Figure 6. Mechanical impedance (kg/cm$^2$) at *in situ* field capacity of Wagram loamy sand for conventional, chisel, and subsoil plus bedding tillage practices.**

was effective in reducing mechanical impedance of the tillage pan. Effectiveness of the subsoiler in reducing mechanical impedance decreased with distance from the row whereas the chiselplow uniformly reduced mechanical impedance at all distances from the row.

The effect of the three tillage operations upon water uptake during the growing season by soybean roots from the Wagram soil is shown in Figure 7. An extended dry period began during the first part of July and ended the second week in September. In Figure 7, soil moisture tension in millibars, measured by a network of small tensiometers, is shown as a function of depth and position. On 1 July, soil moisture tension profiles were identical regardless of the kind of tillage operation. By 16 July, the 800 millibar soil moisture tension isopleth was deeper for both the chisel and subsoil-bedded treatments than for the conventionally tilled soil. Higher soil moisture tension indicates dryer soil profiles; thus, roots extracted more water from below the 20- to 30-cm depth for the chiselplowed and subsoil-bedded treatments than for the conventionally tilled soil where the tillage pan remained intact. As the growing season progressed, the 800 millibar soil moisture tension isopleth moved downward in the profile more rapidly for the chisel and subsoil-bedded treatments than for the conventionally tilled treatment. On 7 September, the last day of the drought, the 800 millibar isopleth was deeper than 90 cm for the subsoil-bedded treatment and was approximately at the 90-cm depth for the chiselplowed treatment, but was at only the 35- to 50-cm depth for the conventionally tilled soil. These results demonstrate that in dry years water uptake by soybeans from soils with tillage pans can be increased by using the appropriate tillage practices.

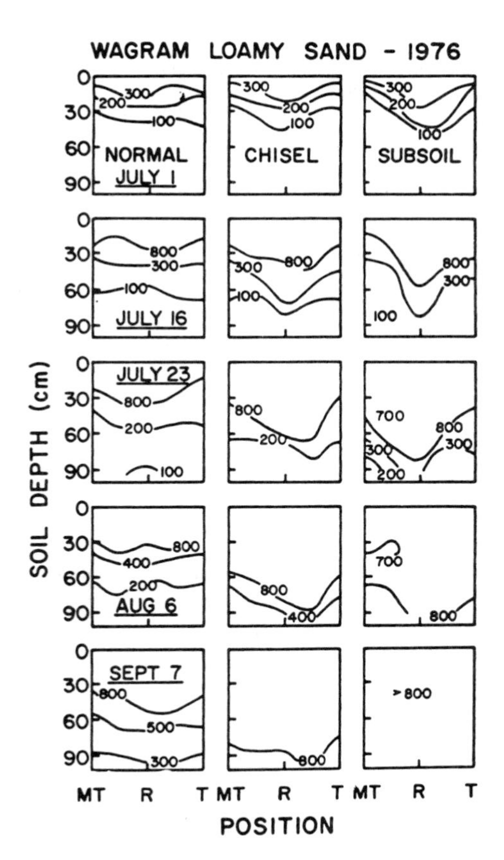


**Figure 7. Soil moisture tension (millibars) profiles for five dates for soybeans on Wagram loamy sand subjected to conventional, chisel, and subsoil tilllage practices.**

Root data were also obtained in the same study. Dry weight and relative distribution of secondary soybean roots at full bloom are presented in Table 2 as a function of depth. Only 4% of the roots penetrated below 30 cm for the conventional treatment compared to 14 and 27% for the chiselplow and subsoil-bedded operations, respectively. These latter two treatments, which reduced mechanical impedance of the tillage pan, promoted deeper root penetration with subsequent greater utilization of subsoil moisture. Soybean grain yield for both deep tillage treatments was 29% greater than the yield of 2,755 kg/ha yield for the conventional treatment.

The depth to tillage pan is an important parameter. The pan usually occurs at the 15- to 25-cm depth, slightly above or at the Ap - B horizon boundary. A 6-day supply of available water is retained in the Ap (Table 1).

**Table 2. Soybean secondary root dry weight and relative distribution with depth in a Wagram loamy sand as influenced by tillage treatments, 1976 (from Kamprath et al., 1979).**

| Depth[a] | Conventional | Chisel Plow | Subsoil |
|---|---|---|---|
| *–cm–* | | *–dry weight, mg/1,000 cm$^3$–* | |
| 0 to 10 | 334 | 323 | 326 |
| 10 to 20 | 219 | 276 | 198 |
| 20 to 30 | 64 | 237 | 101 |
| 30 to 45 | 14 | 44 | 65 |
| 45 to 60 | 10 | 48 | 74 |
| 60 to 75 | 6 | 59 | 87 |
| *–cm–* | | *–% distribution–* | |
| 0 to 10 | 52 | 33 | 38 |
| 10 to 20 | 34 | 29 | 23 |
| 20 to 30 | 10 | 24 | 12 |
| >30 | 4 | 14 | 27 |

[a] $L.S.D._{0.05}$ for comparison of 2 treatments at a depth = 36.

If the pan occurs at a shallower depth, even less available water is held. As a consequence, the plant will enter a period of water stress at an earlier date for the soil with the shallower pan. The effect on seed cotton yield of depth to tillage pan (Lowry et al., 1970) is shown in Figure 8. At mechanical impedance values greater than 20 kg/cm$^2$, a reduction in pan depth decreased seed cotton yield.

**Fragipans**

A fragipan is a natural subsurface horizon with high bulk density relative to the soil above. It seemingly is cemented when dry, but has moderate to weak brittleness when moist (Rich, 1979). Fragipans restrict air and water movement, have high soil strength, and are usually extremely acid. They are present in noncalcareous cultivated and virgin soils. A profile of Hobson silt loam [Typic Fragidalf], a soil containing a fragipan found in Missouri is shown in Figure 2E. The extremely pronounced physical and chemical properties of fragipans form a barrier to root growth. Consequently, if the fragipans are shallow, crop yields are reduced by water shortages in years with limited rainfall and by damage to root systems by perched water tables above the fragipan during years of excessive rainfall. If fragipans are deep, detrimental effects upon root growth may not occur.

Areas where soils with fragipans are common in the humid, temperate region of the United States are shown in Figure 2E. The effects of a fragipan upon rooting, water use, and crop yield were reported for Hobson silt loam by Bradford and Blanchar (1977). The top of the fragipan occurred at a

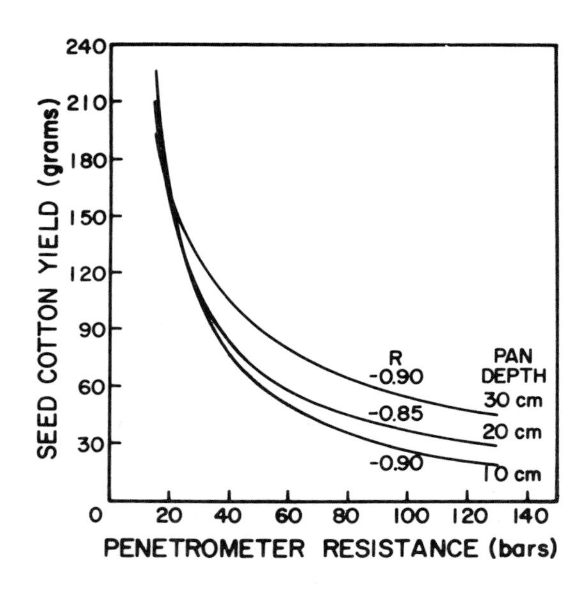


**Figure 8. Effect of penetrometer resistance and pan depth upon seed cotton yield (from Lowry et al., 1970).**

depth of 53 cm, and its thickness varied from 30 to 80 cm. The depth and thickness of fragipans preclude the use of subsoilers to alter their physical characteristics. To modify the fragipan horizons, Bradford and Blanchar (1977) used a trencher to uniformly mix the soil profile to a depth of 150 cm and incorporated various combinations of lime, fertilizer, and saw dust during the mixing process. Profile modification of the Hobson soil increased saturated hydraulic conductivity and total pore space, and reduced mechanical impedance at pan depth from its original values of 150 to 200 kg/cm$^2$ to the low value of 3 to 9 kg/cm$^2$ after mixing. Grain yield of sorghum [*Sorghum bicolor* (L.) Moench] increased from 1841 kg/ha with the fragipan present to 4322 kg/ha when the fragipan was mixed throughout the 150-cm profile. The researchers attributed the yield increase to increased root penetration and to greater infiltration and storage of water below the 53-cm depth.

The brief discussion above cites only several examples of pans, how they affect water utilization, and how they can be effectively managed. Additional research to establish quantitatively the effects of pans on root growth, water utilization, and crop yield must be continued. Information concerning the effects of deep tillage and profile modification on soils with various types of pans has been summarized by Unger (1979).

## CHEMICAL BARRIERS

Chemical barriers to root growth in soils in humid, temperate regions are associated with zones or horizons with low pH, or excessive soil acidity. High concentrations of Al, and sometimes Mn, usually are found in acid soils. Isolated soils in humid, temperate regions have chemical barriers due to high concentrations of Zn, Cu, or possibly some other toxic metal. Pearson (1966) distinguished between two kinds of plant behavior in response to unfavorable chemical environments: (1) those conditions that directly affect the part of the root system exposed to the unfavorable environment, and (2) those conditions which indirectly affect root development by limiting overall plant growth. We will be concerned only with plant behavior in response to low pH and Al-toxicity.

### Soil Acidity and Exchangeable Aluminum

Excessive subsoil acidity is defined for this treatise as soil pH values less than 5.0. Such values may affect root growth either through the effects of H ion activity or the effects of Al ion on root tissues (Pearson, 1966). However, because Al ion activity increases below pH 5.0, it has been difficult to separate the individual effects of excessive H and Al ions on root growth in field soils. The preponderance of evidence, however, indicates that Al-toxicity is more important than H-toxicity in soils with low pH values (Foy, 1974).

Exchangeable Al is often the predominant cation in acid soils, especially the leached acid soils in the southeastern U.S. (Kamprath, 1970). Reserve acidity in soils exists not as exchangeable H, but as Al ions, hydroxy-Al ions [$Al(OH)^{++}$ or $Al(OH_2)^{+}$], and charged hydroxy-Al complexes. Virtually no $Al^{+++}$ exists in soils above pH 5.5 but its concentration rises rapidly below pH 5.0 Foy (1974) presents an excellent discussion on several proposed mechanisms of $Al^{+++}$ interference with root growth.

The deleterious effect of soluble Al upon plant roots is well documented (Doss and Lund, 1975; Adams et al., 1967; Pearson, 1969; Phillips and Kamprath, 1973; Foy, 1974). A specific "critical" concentration level of soluble Al, above which deleterious effects to root growth occur and below which no deleterious effects occur, has not been established for all soils (Phillips and Kamprath, 1973; Adams and Lund, 1966). Rather, the critical level varies with individual soils. A better indicator of Al toxicity in soils appears to be the percent aluminum saturation, i.e. the ratio of the milliequivalents of Al ions retained by the cation exchange complex to the total cation exchange capacity (Kamprath, 1970) or to aluminum activity (Adams and Lund, 1966). Roots growing in soils with toxic levels of Al become dark brown as opposed to the white color of healthy roots. The tap root thickens and becomes distorted and many short, thick laterals develop.

Soils having zones or horizons with pH <5.0 are common in the humid, temperate region of the United States. The map in Figure 3F shows the areal extent of soils with subsurface barriers stemming from pH <5.0 or Al toxicity. (Considerable error may be present in this map because many of the soil scientists felt that inadequate data was available. Yet, I have attempted to summarize the information provided.) Subsoil pH generally decreases from west to east in this region. Reports of Al toxicity come primarily from the southeastern states where subsoil pH values decrease with profile depth as opposed to that of many soils in the midwestern region of the U.S. where pH values increase with depth.

The work of Doss and Lund (1975) illustrates the effect of subsoil pH (and of Al toxicity) upon water utilization by cotton. Cotton was grown on Lucedale fine sandy loam [Rhodic Paleudult] with pH ranging from 4.4 to 6.2 in the 15- to 30-cm depth. At pH 4.4, few roots extended below 15 cm and the taproot either died or turned 90 degrees and grew horizontally when it reached the low pH layer. When the subsurface pH was 5.0, the taproot extended as deep as 60 cm and 50% of the total roots by both weight and by number permeated the 15- to 60-cm depth. Conversely, only 14% of the total weight and 18% of the total root number were found at this depth when pH was <5.0. The amount of water extracted from the soil below 15 cm reflects the measured rooting distributions (Figure 9). Extraction of subsurface available water 17 days after the soil profile was recharged by rainfall

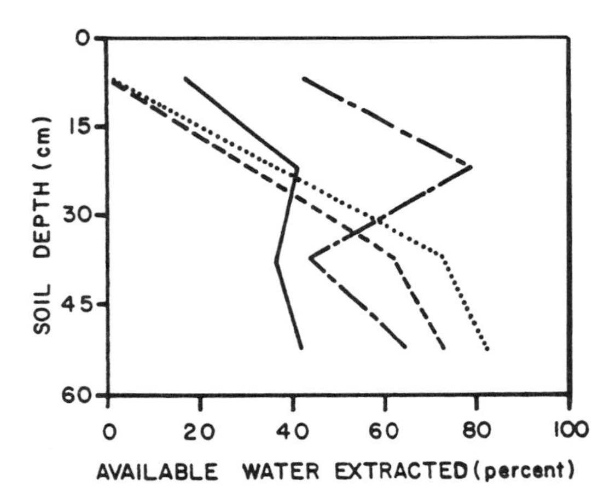


**Figure 9. Effect of subsoil pH on percent of available water extracted by unirrigated cotton on Lucedale fine sandy loam (from Doss and Lund, 1975). —— = pH 4.4 to 4.6; —-— = pH 4.7 to 4.9; ------ = pH 5.0 to 5.4; and · · · · = pH 5.5 to 6.2.**

increased from about 40% for pH 4.4 to 4.6 to nearly 80% when soil pH's were 5.5 to 6.2. For this particular soil, a subsoil pH of 5.0 is the "critical point" for affecting cotton root growth, shoot growth, and yield (Figure 10).

Subsurface horizons with low pH presently create a nearly unsurmountable management problem because no economically feasible management practice for uniform, deep incorporation of lime has been developed to increase pH below the normal depth of moldboard plowing. There is one approach, however, that merits further study. Ritchey et al. (1980) have attempted to reduce toxic levels of Al in subsoils of Oxisols in Brazil by applying excess amounts of $CaCO_3$, $CaSO_4$, or $CaCl_2$ to the surface soil and allowing rainfall to leach the Ca ions into the subsoil to reduce the percent Al saturation of the exchange complex. Using $CaSO_4$, they were able to demonstrate a decrease in the Al saturation percentage of the subsoil from 64 to 24%, thus allowing corn to be produced on their soils.

An alternative to chemically correcting subsoil acidity is to utilize Al tolerant crop species. A wide range in Al tolerance of plants exists (Foy, 1974). Presently available alfalfa and cotton cultivars have little tolerance to soluble aluminum whereas corn and soybean cultivars have a wider range of tolerance. Kamprath (1970) reported good growth of a tolerant corn cultivar at Al saturation percentages up to 44% for several Coastal Plain soils.

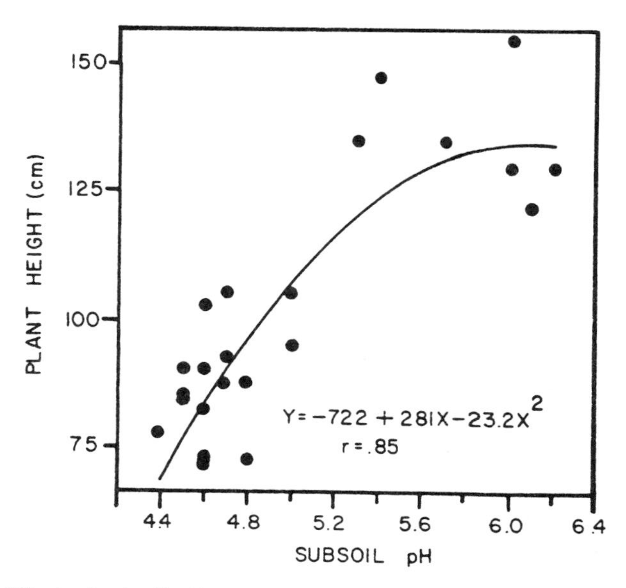


**Figure 10. Effect of subsoil pH on yield of unirrigated cotton on Lucedale fine sandy loam (from Doss and Lund, 1975).**

## SUMMARY

We have examined the areal distribution of several soil physical and chemical characteristics which adversely affect plant growth. The areal extent of the deep sandy soils, for example, is not as widespread as that for soils with tillage pans or low pH's. Both physical and chemical barriers exist for some soils. This is especially true for many of the soils in the southeastern U.S. which have coarse textured, sandy loam to loamy sand surface and subsurface horizons. Thomas and Cassel (1979) estimated the available water capacity for such soils by assuming that the depth of rooting was controlled by the shallowest root-restrictive barrier. For some sites, the depth of the physical barrier (tillage-induced pan) and the chemical barrier (Al toxicity) are identical. The reason for this phenomenon is that lime is incorporated into the Ap horizon by moldboard plowing. The plowing operation, in turn, forms a pan at the base of the Ap horizon. Such soils may be subsoiled or chiselplowed to alleviate the tillage pan but the chemical barrier will still persist. However, many soils with tillage pans in the southeastern Coastal Plain will permit deeper rooting in response to deep tillage even though some degree of soil acidity exists in subsurface layers.

It must be reemphasized that the soil maps in Figure 3 are regional in nature. A given field or part of a field within any one of the six mapping units may not have the same soil characteristics as the mapping unit even though the majority of soils in the area do have the soil properties defined by the mapping unit. Hence, the soil barrier problem is one that is site specific within a given field or part of a field. Soil barriers existing in a particular field can be identified only by *in situ* examination of the soil.

## NOTES

D. K. Cassel, Department of Soil Science, North Carolina State University, Raleigh, North Carolina 27650.

This paper was approved for publication as paper number 6706 of the journal series of the North Carolina Agricultural Research Service, Raleigh, North Carolina 27650.

## LITERATURE CITED

Adams, F., R. W. Pearson, and B. D. Doss. 1967. Relative effects of acid subsoils on cotton yield in field experiments and on cotton roots in growth chamber experiments. Agron. J. 59:453-456.

Adams, F., and Z. F. Lund. 1966. Effect of chemical activity of soil solution aluminum on cotton root penetration of acid subsoils. Soil Sci. 101:193-198.

Allmaras, R. R., W. W. Nelson, and W. B. Voorhees. 1975a. Soybean and corn rooting in southwestern Minnesota: I. Water-uptake sink. Soil Sci. Soc. Amer. Proc. 39: 764-771.

Allmaras, R. R., W. W. Nelson, and W. B. Voorhees. 1975b. Soybean and corn rooting in southwestern Minnesota: II. Root distributions and related water inflow. Soil Sci. Soc. Amer. Proc. 39:771-777.

Bodman, G. B., and G. K. Constantin. 1965. Influence of particle size distribution on soil compaction. Hilgardia 36:567-591.

Bradford, J. M., and R. W. Blanchar. 1977. Profile modification of a Fragiudalf to increase crop production. Soil Sci. Soc. Amer. J. 41:127-131.

Brady, N. C. 1974. The nature and properties of soils. Eighth Ed. McMillan, New York.

Campbell, R. B., D. C. Reicosky, and C. W. Doty. 1974. Physical properties and tillage of Paleudults in the southeastern Coastal Plains. J. Soil Water Conserv. 29: 220-224.

Cassel, D. K., H. D. Bowen, and L. A. Nelson. 1978. An evaluation of mechanical impedance for three tillage treatments on Norfolk sandy loam. Soil Sci. Soc. Amer. J. 42:116-120.

Chancellor, W. J. 1976. Compaction of soil by agricultural equipment. Univ. Calif. Ext. Bull. 1881.

Coughlan, K. H., R. J. Loch, and W. E. Fox. 1978. Binary packing theory and the physical properties of aggregates. Aust. J. Soil Res. 16:283-289.

Cruse, R. M., D. K. Cassel, and F. G. Averette. 1980. Effect of particle surface roughness on densification of coarse textured soils. Soil Sci. Soc. Amer. J. 44:692-697.

Doss, B. D., W. T. Damas, and Z. F. Lund. 1979. Depth of lime incorporation for correction of subsoil acidity. Agron. J. 71:541-544.

Doss, B. D., and Z. F. Lund. 1975. Subsoil pH effects on growth and yield of cotton. Agron. J. 67:193-196.

Drew, M. C. 1979. Root development and activities. p. 573-606. In R. A. Perry and D. W. Goodall (eds.) Arid-land ecosystems: structure, functioning and management. Cambridge Univ. Press, Great Britain.

Drew, M. G., and M. J. Goss. 1973. Effect of soil physical factors on root growth. Chemistry and Industry 21:679-684.

Erickson, A. E., C. M. Hansen, and A. J. M. Smucker. 1968. The influence of subsurface asphalt barriers on the water properties and the productivity of sand oils. 9th Internat. Congr. Soil Sci. Trans. 1:331-336.

Fehrenbacher, J. B., and H. J. Snider. 1954. Corn root penetration in Muscatine, Elliot and Cisne soil. Soil Sci. 77:281-291.

Foy, C. D. 1974. Effects of aluminum on plant growth. p. 601-642. In E. W. Carson (ed.) The plant root and its environment. Univ. Press of Virginia, Charlottesville.

Goss, M. J., and M. C. Drew. 1972. Effect of mechanical impedance on growth of seedlings. p. 35-42. Agric. Res. Council Letcomb Lab. Annual Report for 1971. Wantage, Oxfordshire, England.

Hegarty, T. W., and S. M. Royle. 1978. Combined effects of moisture content prior to compaction, compactive effort and rainfall quantity on soil crust strength. J. Soil Sci. 29:167-173.

Kamprath, E. J. 1970. Exchangeable aluminum as a criterion for liming leached mineral soils. Soil Sci. Soc. Amer. Proc. 34:252-254.

Kamprath, E. J., D. K. Cassel, H. D. Gross, and D. W. Dibb. 1979. Tillage effects on biomass production and moisture utilization by soybeans on Coastal Plain soils. Agron. J. 71:1001-1005.

Lindstrom, M. J., and W. B. Voorhees. 1980. Planting wheel traffic effects on interrow runoff and infiltration. Soil Sci. Soc. Amer. J. 44:84-88.

Linscott, D. L., R. L. Fox, and R. C. Kipps. 1962. Corn root distribution and moisture extraction in relation to nitrogen fertilization and soil properties. Agron. J. 54: 185-189.

Lowry, F. E., H. M. Taylor, and M. G. Huck. 1970. Growth rate and yield of cotton as influenced by depth and bulk density of soil pans. Soil Sci. Soc. Amer. Proc. 34: 306-309.

Lutz, J. F. 1952. Mechanical impedance and plant growth. p. 43-71. In B. T. Shaw (ed.) Agronomy. 2. Soil physical conditions and plant growth. Academic Press, New York.

Merrill, S. D., and S. L. Rawlins. 1979. Observations of root growth through ports covered with polyethylene sheeting as compared with other methods. Soil Sci. 172: 351-357.

Pearson, R. W. 1966. Soil environment and root development. p. 95-126. In W. H. Pierre, D. Kirkham, J. Pesek, and R. Shaw (eds.) Plant environment and efficient water use. American Society of Agronomy, Madison Wisconsin.

Phillips, J. A., and E. J. Kamprath. 1973. Soil fertility problems associated with land forming in the Coastal Plain. J. Soil Water Conser. 28:69-73.

Raghaven, G. S. V., and E. McKyes. 1978. Effect of Vehicular traffic on soil-moisture content in corn (Maize) plots. J. Agric. Eng. Res. 23:429-439.

Reicosky, D. C., D. K. Cassel, R. L. Blevin, W. R. Gill, and G. C. Naderman. 1977. Conservation tillage in the southeast. J. Soil Water Conserv. 32:13-19.

Rich, C. I. 1979. Glossary of Soil Science Terms. Soil Science Society of America, Madison, Wisconsin.

Ritchey, K., D. M. G. Souza, E. Lobata, and O. Correa. 1980. Calcium leaching to increase rooting depth in a Brazilian savannah Oxisol. Agron. J. 72:40-44.

Saxena, G. K., L. C. Hammond, and H. W. Lundy. 1968. Effect of asphalt moisture barrier on soil salinity and yield of vegetables under variable irrigation and fertilization. Soil and Crop Sci. Soc. Fla. Proc. 28:310-318.

Shaw, B. T. (ed.). 1952. Soil physical conditions and plant growth. Agronomy 2. Academic Press. New York.

Sivakumar, M. V. K., H. M. Taylor, and R. H. Shaw. 1977. Top and root relations of field-grown soybeans. Agron. J. 69:470-473.

Taylor, H. M. 1980. Soybean growth and yield as affected by row spacing and by seasonal water supply. Soil Sci. Soc. Amer. J. 72:543-547.

Taylor, H. M., and R. R. Bruce. 1968. Effect of soil strength on root growth and crop yield in the southern U.S. 9th Internat. Congr. Soil Sci. Trans. 11:803-811.

Taylor, H. M., and H. R. Gardner. 1963. Penetration of cotton seedling taproots as influenced by bulk density, moisture content, and strength of soil. Soil Sci. 153-156.

Taylor, H. M., and B. Klepper. 1973. Rooting density and water extraction patterns for corn (*Zea mays* L.) Agron. J. 65:965-968.

Taylor, H. M., and B. Klepper. 1974. Water relations of cotton: I. Root growth and water use as related to top growth and soil water content. Agron. J. 66:584-588.

Thomas, D. J., and D. K. Cassel. 1979. Land forming Atlantic Coastal Plain soils: Crop yield relationships to soil physical and chemical properties. J. Soil Water Conserv. 34:20-24.

Unger, P. W. 1979. Effects of deep tillage and profile modification on soil properties, root growth, and crop yields in the United States and Canada. Geoderma 22:275-295.

Wiersum, L. K. 1957. The relationship of the size and structural rigidity of pores to their penetration by roots. Plant Soil 9:75-85.

## 12

# THE FUTURE ROLE OF IRRIGATION IN A HUMID CLIMATE

**Ronald E. Sneed and Robert P. Patterson**

From the beginning of recorded history and probably before, man has depended upon irrigation for survival. We find in Genesis 2:10 "And a river went out of Eden to water the garden and from thence it was parted and became into four heads." Hammurabi, the great conqueror and architect of the Babylonian Empire, counted his irrigation works among his more important deeds. The pages of Persian, Turkish, Indian, Chinese and Roman history are dotted with accounts of the importance of irrigation. The Inca Indians of Peru had a high culture based on irrigation. Evidence uncovered in recent years indicates that the Hohokam Indians of southwestern Arizona were irrigating as early as 100 BC. In the early 1800's Franciscan priests operated large irrigated farms in Mexico and California. It was the Mormons, however, who settled in the Great Salt Basin of Utah about 1847, who introduced irrigation on a large scale in the United States. The 1860 U.S. census listed 752 irrigation enterprises supplying water to 162,906 hectares of land.

Irrigation has continued to grow since that time. In 1979, the irrigated acreage in the U.S. was about 25 million hectares (Morey, 1979), an increase of more than 25% in the past decade. The major irrigated area is in the arid West. During the past decade, however, the irrigated area in the humid East increased 77% compared to 17% for the 17 western states and 23% for the nation. In the 1955 Yearbook of Agriculture, the following statement was made: "The greatest potential for irrigation in the United States lies not in reclaiming the deserts, but in correcting the seasonal deficiencies of moisture in the sections where we consider the rainfall to be adequate." This statement is truer today than in 1955.

With this background, let us define some parameters. What is a humid climate, or when is rainfall considered to be adequate? For discussion purposes, a humid area is defined as an area that receives an average annual rainfall of 102 cm or more, somewhat uniformly distributed during the year, but with short moderate to severe drought periods that might adversely affect yields and quality of certain crops. Even with this high annual rainfall, droughts do occur. Fieldhouse and Palmer (1965) showed that for the period 1929 to 1963 in Maryland and Delaware, approximately 48% of the months had drought ranging from incipient to extreme. Data from Van Bavel and Verlinden (1956), Van Bavel and Lillard (1957) and Van Bavel et al. (1957) indicate that the minimum number of drought days in North Carolina, South Carolina, and Virginia for the period from April through September for soils with a 4.6-cm moisture storage capacity in five out of ten years will vary from less than 20 cm in the mountains of North Carolina to more than 60 cm in the Piedmont of South Carolina. For a probability of 10% and a 4.6-cm moisture storage capacity, the minimum number of drought days will increase by approximately 30. For soils with a 2.3-cm storage capacity and a 50% probability of drought, the minimum number of drought days will increase by approximately 30. Sopher et al. (1973) reported that in the Coastal Plain of North Carolina, 40 to 60% of the variation in corn yield was associated with drought.

## YIELD INCREASES FROM SUPPLEMENTAL IRRIGATION IN HUMID REGIONS

Except for the very intensively managed crops such as greenhouse production, crops such as rice and cranberries that are flooded for weed control, some irrigation for water-table control on grasses and sugar cane and isolated cases of vegetable crop irrigation, irrigation in humid areas really began following World War II with the introduction of lightweight aluminum tubing. For about 15 years there was slow but steady growth in irrigation in humid areas. Crops that were irrigated include tobacco, tree fruits, vegetables, small fruits, nursery crops, and turf (primarily golf courses). Studies conducted by Clark et al. (1956), Hawks et al. (1968), and Lynn et al. (1958) show that in an average year irrigation will increase tobacco yields by 280 kg/ha and quality by 10%. The data are less definitive on tree fruits, but Unrath and Sneed (1974), and Unrath (1972a; 1972b) indicated that yields, total red color, and size of Red Delicious apples could be increased, and harvest period, cork spot, and bitter pit could be reduced with irrigation. Feldstein and Childers (1957; 1965), Morris et al. (1962), Reynolds and Rogers (1958), and Ballinger et al. (1963) reported variable results on peach irrigation. In general fruit size and total number of fruits were increased, but the response depended upon rainfall distribution during final swell.

The area of citrus irrigated in Florida has greatly increased since 1962. Today, more than half of the citrus grown in that state probably is irrigated. Pecans constitute a major crop in several of the southeastern states, including Georgia, South Carolina, Mississippi, and Arkansas. There has been considerable interest in recent years in use of trickle irrigation for pecan irrigation, and some research work is presently being conducted in several states.

There have been very limited studies on irrigation of small fruits, and most of it was conducted in the fifties when present varieties were not being produced. Generally, strawberries are sprinkler irrigated for frost and freeze protection and to provide soil moisture. The irrigated area of blueberries and grapes is very small.

A number of irrigation studies have been conducted on vegetable crops. Reynolds and Rogers (1958) showed that yields of fresh market cucumbers increased by 14,236 kg/ha with irrigation, the percent of malformed fruit decreased, and the fruit had a darker green, more desirable color. Pickling cucumber yields were increased 1989 to 3226 kg/ha depending upon whether it was a spring or fall crop. Cucumbers show a maximum response to irrigation from pollination through harvest.

Vittum et al. (1963) reported that in the northeastern United States greenbean yields increased in 18 of 30 trials. Reynolds and Rogers (1958) reported greenbean yield increases of 82% with irrigation. In addition to yield increases, irrigation increased the size of pods, decreased the percent of severely crooked pods, lowered the seed content, and tended to reduce the fiber content of the beans. The most critical water stress period is during and immediately following bloom. Carreker and Cobb (1963) reported yield increases of 655 to 1517 dozen ears of sweet corn per hectare. Vittum et al. (1963) indicated that in the northeastern United States yield increases of sweet corn were significant in 5 of 13 trials. Irrigation increased the average weight per ear, gross weight and percent of usable corn cut from the ear, but slightly delayed maturity.

Vittum et al. (1963), in studies conducted with tomatoes from 1950 to 1960, reported average yield increases of 17,412 kg/ha. Significant yield increases were reported in 13 of 31 crop years. Carreker and Cobb (1963) reported yield increases of tomatoes in Georgia of 11,200 kg/ha on a clay soil compared to an increase of 2240 kg/ha on a silt loam soil. Benefits of sprinkler irrigation during the early part of the season included larger plants, more uniform stands, increase in fruit set, control of blossom-end rot, extension of the season, and during the late part of the season provided frost and freeze protection.

Significant yield increases for Irish potatoes grown in the northeastern

U.S. were obtained in 34 of 43 crop years from 1949 to 1961 (Vittum et al., 1963). An average of three 2.54-cm irrigations were required per year. Yield increases of 2475 to 11,075 kg/ha of sweet potatoes have been reported. Hernandez et al. (1965) found that drought periods between 40 to 70 days after transplanting in Louisiana caused the greatest reduction in sweet potato yield. Carreker and Cobb (1963) found that, if drought occurred during a two- to three-week period in September, irrigation could double yield of sweet potatoes in Georgia.

Hagan et al. (1967), in studies on other vegetables such as artichokes, asparagus, beets, cabbage, carrots, cauliflower, celery, greens, and onions, found that moisture stress reduced yields and caused malformed, poor quality, often unmarketable products.

In the late sixties interest began to develop in irrigation of field crops such as corn, peanuts, cotton, and soybeans. Research studies have been conducted on a number of these crops. Carreker and Cobb (1963), Snell (1968), Sneed and Martin (1969), and Wesley (1979), in studies conducted in Georgia, South Carolina and North Carolina, reported yield increases of corn from 1443 to 4704 kg/ha. Factors which affect profitability of corn irrigation, in addition to soil moisture, are soil type, fertility level, plant population, and variety. The most critical period of moisture demand is from just prior to tassel emergence to early dent or dough stage.

White (1961) and Spooner (1961) in Arkansas, Privette et al. (1980) in South Carolina, Grisson et al. (1952) in Mississippi, and Whitt and Van Bavel (1955) reported yield increases of 314 to 815 kg/ha for irrigated soybeans. Irrigation increased seed size and percent germination and slightly reduced protein and oil content. Maximum yield increases occurred when irrigation was begun at the full bloom stage of growth.

In studies conducted on sandy loam and loamy sand soils in North Carolina for several years, both mid-season and late-season corn varieties reponded significantly to irrigation when the water was supplied at the tasseling/silking period for that variety (R.P. Patterson, unpublished data). Percent nitrogen in both the ear leaf and grain were enhanced as a result of soil moisture adjustment by sprinkler irrigation. Moreover, efficiency of utilization of additional fertilizer nitrogen (80 kg N/ha following an initial application of 90 kg N/ha) was much higher for the irrigated corn. Significant increases of grain yield generally were obtained when total water potential of the ear leaf was not allowed to fall below about –9 bars (10 a.m. measurement), in contrast to stressed plants whose potential had fallen to –12 bars. Nitrate reductase activity of the ear leaf was quite sensitive to degree of leaf water deficit, suggesting that this measurement may be a useful plant indicator of when the corn crop will respond to soil moisture adjustment by irrigation.

The results of our own experiments under field conditions (R. P. Patterson, A. B. B. Tambi, and O. B. Yusof, unpublished data) show that the most critical period of moisture supply for soybeans of Groups V, VI, and VII in eastern North Carolina is from the late-flowering and early-pod-formation stages through the mid-pod-fill stage. Varieties representative of each of these maturity groups produce significantly higher yields when the moisture level in the Ap soil horizon is not allowed to fall below about 50% of field capacity, and the upper canopy leaves are no more stressed than −12 bars (11 a.m. measurement), in contrast to grain yields from plots where the plants were stressed to −18 bars at a soil moisture content of about 25% of field capacity. A water stress during the flowering period results in flower abortion. When suitable growing conditions are resumed before pod formation has been completed, however, the varieties studied appeared to be able to compensate for the stressed period by producing additional flowers which do not abort unless another stress episode occurs. Water stress which occurs after the mid-pod-fill period will reduce yield, but not to the extent that yield is suppressed when the drought occurs during pod formation. Of several varieties studied, 'Ransom' (Group VII) appears to be especially resilient, in that the flowering period is extended over a period of 5 to 6 weeks should soil moisture adjustment occur following a drought during the initial flowering period. Thus, should a prolonged drought occur in a region with soils of limited water storage capacity, as is the case for virtually all the soils in the soybean-growing region of North Carolina , a well-timed irrigation at the critical growth stage can provide much needed yield stability for the soybean grower.

Data on irrigation of peanuts is somewhat limited. Sneed and Martin (1969) reported yield increases of 672 kg/ha with irrigation. Some more recent unpublished work in North Carolina indicates that average yield increases of 784 kg/ha. The major response of peanuts to irrigation occurs during the fruit enlargement stage. Irrigation may reduce the percent of Sound Mature Kernels and Extra Large Kernels, but generally will improve the total grade resulting in a higher price per kilogram.

Studies conducted by Scarsbrook et al. (1961) in Alabama, Carreker and Cobb (1963) in Georgia, White (1961) in Alabama, Wilson and Snell (1968) in South Carolina, and Sneed and Martin (1969) in North Carolina indicate yield increases on irrigated cotton of 34 to 616 kilograms of lint cotton per hectare. Irrigated cotton had longer staple length, more uniform and better matured fibers, and lower lint percentage and breaking strength. Maximum demand for moisture by the cotton plant begins during the early flowering stage and continues until most of the bolls are mature.

Pasture and forage crops are produced in all areas in the humid zone, but the irrigated acreage is small. Woodhouse et al. (1958) and Van Horn et al. (1956) reported that irrigation of pasture for dairy cattle increased the carrying capacity per hectare, increased the length of the growing season, maintained a better stand of grass, and resulted in increased production of milk by the cows. Kincaid et al. (1960) reported that irrigation of pasture for beef cattle increased carrying capacity of the pasture, weight gain per animal, and carcass grade in three out of six years.

## IRRIGATION EQUIPMENT

What does all the data that have been collected on these many crops actually demonstrate? They indicate that crops in humid regions do respond to irrigation depending upon the time of the moisture stress in relation to the growth stage of the plant and the severity of the drought. But will the return from irrigation justify the purchase and operation of equipment, and is there equipment available that can be used?

Since 1975, an irrigation revolution has been taking place in the southeastern United States. It began in Georgia, spread to the surrounding states, and then moved north. The irrigated acreage of corn and peanuts, and to some limited extent of soybeans and cotton, has greatly expanded. Three factors have fueled this revolution. Severe droughts of longer-than-average duration have occurred during the last half of the decade. Cost of production has risen to the point that many growers are not producing profitable yields without irrigation. Equipment has been introduced and perfected to the point that large acreages of crops can be irrigated within minimal labor input.

In examining the development of irrigation equipment over the last three decades, observations will be confined mainly to sprinkler irrigation with an occasional reference to surface or drip irrigation. In the fifties, the major equipment used was hand-move, portable aluminum pipe systems with small sprinklers. These systems were limited to irrigation of small acreages. In the late fifties and early sixties, gun sprinklers were introduced. These reduced somewhat the moving of pipe, but large amounts of labor still were required and there was a limit to the acreage that could be irrigated. In the mid-fifties, side-roll wheel-move equipment was introduced. Aluminum pipe was used as the axle for wheels and a small engine was used to move the lateral line across the fields rather than using hand labor. The side-roll wheel-move system is limited to low-growing crops on flat, rectangular or square fields. Several versions of this machine have appeared. The latest version is an automatic side-roll with a computer controlling the automatic functions of the machine including starting the engine drive unit, moving a predetermined distance, stopping, sprinkling for a predetermined time, shutting off, draining, and repeating the cycle up to five times with the system moving a maximum distance of 50 m. The feeder hose must then be moved manually to another riser along the main pipeline and the sequence repeated. The system still has the limitations of being low clearance and best suited to flat, rectangular or square fields. The initial cost of the machine, based on cost per hectare, is reasonable. The side-roll wheel-move system is a medium pressure system that uses mainly rotary impact sprinklers.

During the same time frame, some growers were beginning to use solid-set aluminum pipe systems and permanent systems to reduce the labor required to irrigate. While these systems require a minimum of labor, they have a high initial cost per hectare. They are best suited to high value annual and perennial crops such as tomatoes, strawberries, nursery crops, and certain tree crops. For environmental modification, they are the recommended systems. These systems are medium pressure and use rotary impact or gear drive sprinklers.

About 25 years ago a farmer introduced the first center-pivot irrigation system. The growth of these systems was slow initially, but today they constitute the largest number of mechanical-move irrigation systems in use. The center pivot consists of an aluminum or steel lateral line supported by "A" frame trusses on wheels. Sprinklers are located along the length of the lateral line. The machine moves in a circle, irrigating a circular pattern. The early machines were propelled by water motors, water pistons, or hydraulic motors. The majority of the machines sold today are electrically driven. The early center-pivot used medium to high pressure rotary impact sprinklers. The trend today is toward medium to low pressure systems to reduce the cost of operation. The low pressure systems use spray nozzles, wabblers, low-pressure rotary impact and gear-drive sprinklers, and polyethylene tubes that drag on the ground applying water as trickle or drip irrigation. Some of the low pressure systems mount the nozzles above the lateral pipe, others are mounted on drops below the lateral pipe, and still others are mounted on booms fore and aft of the lateral pipe. Most of the nozzling of center pivots is done with computers.

Another new development in center pivot systems is corner attachments to irrigate corners or odd-shaped fields. This is accomplished in a variety of ways, including placement of something on the center pivot, irrigation of the corners with solid set or permanent systems, or using surface irrigation systems.

One manufacturer uses an extended boom concept. The boom remains underneath the main lateral line until the system approaches a corner. As the system moves toward the corner area, the boom extends itself into the corner. Among the limitations to this concept are: the boom weight, strength requirements, and counter balancing weight requirements. Water is fed from the main pipeline to the boom by a hose. Sprinklers are turned on in sequence as the boom extends into the corner.

Several manufacturers use an extended arm which trails behind the system until the system goes into a corner. As the system approaches a corner, the extended arm gradually moves in line with the main lateral line.

As the extended arm moves further into the corner, more sprinklers are turned on in sequence as the arm is extended to maximum length. As the system moves away from the corner, the arm is pulled gradually behind the system and the sprinklers are turned off in sequence.

One manufacturer uses an arm which trails next to the main pipe until the system is in line with the corner. The main system then stops, and the arm moves into the corner. When the arm has moved 90°, it is beyond the wetted area of the main lateral. The sprinklers are then turned off on the main pipe and the sprinklers on the arm are turned on. The sprinklers on the arm stay on and the arm rotates 180° one way and 180° back. During the time the arm is retracted the last 90° over the previously wetted area, the sprinklers on the main lateral lines are again turned on. When the arm returns in line with the main lateral line, the main lateral line moves as a typical center-pivot.

Corners of some center pivot irrigated fields are irrigated with aluminum pipe solid-set, hand-move systems or gated pipe surface systems. Water is supplied to systems by permanent line or by aluminum lines from the center-pivot. One manufacturer uses a control device and a drop line that automatically stops in each corner, connects to the solid-set system, shuts off the sprinklers on the main lateral line, and delivers water to the solid-set. After a set time, the sequence is reversed and the system moves on and irrigates as a typical center-pivot.

Another method of irrigating corners is to use large gun sprinklers on the end of the main lateral line. On some pivots, the sprinklers on the main lateral line are shut down and all the water goes to the end gun. The end gun is only operated in the corners.

An offshoot of the center-pivot is the linear-move system which moves in a straight line instead of a circle and irrigates a rectangular area. The lateral move is a low-pressure system designed to be used on flat, rectangular fields up to 780 m in length. Water is supplied at the center or end of the lateral line by pumping from a ditch or by a flexible rubber supply hose connected to a pressurized main pipeline with outlets. A four-wheel "boss" tower contains the pump, power unit, fuel, and guidance components that power the unit and guide it across the field. Guide posts, a guide wire, or a small wheel traveling in a ditch keep the unit moving in a straight line across the field.

There are two other automatic irrigators on the market. The Robot-Rain is a four-wheel carriage with a gun sprinkler mounted on a steel riser. The carriage is steered by feeler wheels that keep the machine traveling along the mainline which is laid on top of the ground. The supply line is laid out in a closed loop around the field with outlet valves spaced so that the sprinkler can irrigate the whole field evenly as it moves from valve to valve along the main pipeline. An electro-mechanical drive board controls the automatic operation of the sprinkler. The cycle begins with the sprinkler carriage moving along the mainline until a mainline outlet valve is located. The sprinkler column lowers, clamps over the outlet valve and turns on the water. The sprinkler operates for a set period of time, then the sprinkler riser pipe is withdrawn, shutting off the water. The carriage automatically moves along the main pipeline to the next riser valve where the cycle is repeated. This unit is battery driven, with the irrigation water being used to operate a hydraylic-motor driver generator. The batteries supply power to operate electric motors on the sprinkler carriage.

The Noble Automatic Linear Irrigation System is similar to other linear move systems, but it moves along a pressurized mainline changing outlet valves automatically as the Robot Rain. The lateral length is variable up to 78 m in length and is center fed from the water supply. This is a low pressure system using spray nozzles. The central carriage has a diesel engine coupled to a generator which supplies power to motors on the individual towers, guidance equipment, and the automatic valve connector equipment.

There are two types of self-propelled gun travelers. Cable-tow travelers were introduced in the United States in the late sixties. The machine consists of a two-, three-, or four-wheel trailer on which is mounted a single gun sprinkler, a cable reel, and a power supply that can be a water piston, water motor, turbine, or internal combustion engine. A hose reel may be mounted on the trailer or may be a separate machine. Water is supplied to the machine by a rubber or polyvinyl chloride plastic coated synthetic textile fiber hose. Water is supplied to the hose with underground or above-ground pipe with outlet valves at convenient spacings. The machine follows a steel cable through the field and the hose is pulled by the machine. Several sizes of machines are available. Hose size varies from 6.3 to 12.7-cm inner diameter and in lengths from 100 to 400 m. Hose size will determine the effective sprinkler capacity. The machine is moved from lane to lane by a tractor or other prime mover. One machine is equipped with its own power supply for moving from lane to lane.

Hose-reel or hose-pull travelers were introduced in Europe in the late sixties and in the United States in the late seventies. The machine consists of a large diameter hose reel on a two- or four-wheel trailer, a thick wall polyethylene plastic hose that supplies water to the sprinkler and pulls the sprinkler cart through the field, a sprinkler cart on which a gun sprinkler is mounted, and a power supply on the trailer that turns the reel. The power supply is a water drive turbine, bellows or an auxililary internal combustion engine. Several different sizes of machines are available with hoses from 6.3 to 11.4-cm inner diameter and in lengths from 200 to 400 m. Hose size will determine the effective sprinkler capacity. The sprinkler cart is pulled into position by a tractor or other prime mover and the hose reel is moved from lane to lane by a tractor.

One European manufacturer has a self-propelled traveler that uses a low pressure rubber hose to supply water to the traveler. The machine is a high clearance, four-wheel trailer on which is mounted a pump, power unit, and gun sprinkler. The machine is driven across the field laying out the hose. Once the hose is laid out, the power unit and pump on the traveler are started and the machine follows the hose, picking up the hose and winding it on the hose reel as it travels and applies water. The hose can be laid in almost any configuration and the machine will follow it.

The traveler units are high pressure systems. They are adapted to rolling terrain, but operate best in flat, rectangular fields. They are being used on slopes up to 30% and in fields with varying length rows. They require more labor than the center-pivots or linear moves, but less labor than hand-move systems.

All of the mechanical-move systems are reasonably priced, with an initial cost per hectare of $750 to $1500. Each of the systems has limitations. However, there is equipment available that allows growers to irrigate large areas with minimum labor and with a reasonable cost per hectare. The mechanical-move systems are best suited to the medium-to-large operations. There are travelers adapted to farms as small as 10 hectares.

Another type of irrigation is trickle or drip. It is most adapted to high cash value annual crops and perennials. It is a low pressure system that requires very clean water for satisfactory operation. These systems use a minimum amount of water to supply crop needs.

Surface irrigation is used in several of the southeastern states mainly to irrigate rice, soybeans, and cotton. Some subirrigation using water-table control is being used on vegetable crops, forage crops, and sugar cane and small acreages of other crops. Both of these are low-pressure systems. With the introduction of laser-controlled land-forming equipment, it is much easier to prepare fields for surface irrigation. Subirrigation is limited to the coastal flatwood areas that have a high natural water table and soils underlain with a sand layer to provide rapid lateral transmission of water.

The technology is available to allow most growers in humid areas to irrigate. Why do more growers not irrigate, what are the problems, how will they determine when to irrigate, and what is the future? The major factors in determining whether or not to irrigate are if water is the limiting factor in production and if the returns from irrigation justify the cost of owning and operating the equipment. For many growers, these factors have been satisfied. A subsequent question arises: Are adequate water supplies available or can they be obtained at a reasonable cost? Most water supplies in humid areas are developed by the grower at his expense with no assistance from federal or state agencies. In some areas, water availability will be a limiting factor.

Some growers may decide that to gamble on a 5- to 10-year payback period may be too risky. They may attempt to produce crops that are more drought tolerant, risk crop failure, or simply cease farming.

Once a decision is made to irrigate, the grower is faced with alternatives. Does he continue with his present crop mix? He may be forced to continue because of the need for crop rotations to control disease, insect, and weed problems. What additional inputs does he add? Is his level of management sufficient? When does he irrigate and how does he determine when to irrigate to obtain maximum economic response? This brings us to irrigation scheduling. In the more arid areas many growers are now scheduling irrigations based on soil moisture data obtained with tensiometers, gypsum blocks, or other devices, on rainfall and temperature data, and on soil type and crop. Such scheduling is just being implemented in the humid areas. Some of the computer programs developed in the arid areas, such as Ag-Net in Nebraska, are not totally transferrable to humid areas. There is a regional program in the Southeastern United States headed by Dr. Jerry Lambert, Clemson University, directed toward developing an acceptable computer program. Several consultants are working with growers in humid areas on irrigation scheduling. The are using mainly soil moisture readings to schedule irrigation. This practice will continue to grow. A number of growers, however, still are using the "soil-feel" method, simple rainfall-evapotranspiration bookkeeping methods, and crop appearance to schedule irrigation. There are reasonably good data on the critical moisture period for many crops, but more refined data are needed. More data are needed on depth of root penetration over time, especially in many of the soils of the humid areas that have chemical and physical barriers to root penetration. More data need to be obtained on varietal reponse to irrigation, spacing of plants, optimum pH and fertility levels, etc.

The question asked more often than any other is: "When do I irrigate to get the maximum economical response?" There are other questions such as: "Why can growers in south Georgia produce more than 200 bushels/acre (12,544 kg/ha) of corn on deep sand and I cannot produce but 175 bushels (10,976 kg/ha) on much better soil?"

## CONCLUSIONS

The future of irrigation in humid areas is bright. Manufacturers have provided equipment and they will improve the equipment to allow crops to be irrigated at reasonable cost. In 1980 dollars, it is costing from $175 to $325 per hectare to own and operate the mechanical-move equipment that is presently available. Many growers, even for field crops such as corn, peanuts, cotton, and soybeans, can realize a net return in excess of the cost of irrigation. The arid west is experiencing water shortages. Food must be produced to meet worldwide demand. Irrigation costs in the humid areas are less than in the arid areas. How fast growers adopt irrigation will depend upon the price of the crops they produce, water availability, and weather patterns. Many growers, however, are finding that even in favorable years increased yields result from one or two well-timed irrigations. New varieties that have the potential for higher yields will be helpful. Better weather forecasting will allow growers to make more effective use of irrigation.

There are some problems in expanding irrigation in humid regions. The degree of sophistication of irrigation scheduling must improve. New, more responsive varieties are needed. Low prices of commodities produced or excessive energy costs can be very detrimental. Finally, more research is needed on when to irrigate.

## NOTES

Robert P. Patterson, Department of Crop Science, North Carolina State University, Raleigh, North Carolina 27650; Ronald E. Sneed, Department of Biological and Agricultural Engineering, North Carolina State University, Raleigh, North Carolina 27650.

The use of trade names in this publication does not imply endorsement by the North Carolina Agricultural Experiment Station of the products named, nor criticism of similar products not mentioned.

## LITERATURE CITED

Ballinger, W.E., A. H. Hunter, F. E. Correll, and G. A. Cummings. 1963. Interrelationships of irrigation, nitrogen, fertilization pruning of Redhaven and Elberta peaches in the Sandhills of North Carolina. Amer. Soc. Hort. Sci. Proc. 83:248-258.

Carreker, J. R., and C. Cobb, Jr. 1963. Irrigation in the Piedmont. Georgia Agric. Exp. Stn. Tech. Bull. 29.

Clark, F., J. M. Myers, H. C. Harris, and R. W. Beldsoe. 1956. Yield and quality of flue-cured tobacco as affected by fertilization and irrigation. Florida Agric. Exp. Stn. Bull. 572.

Feldstein, J., and N. F. Childers. 1957. Effect of irrigation on fruit size and yield of peaches in Pennsylvania. Amer. Soc. Hort. Sci. Proc. 69:125-130.

Fieldhouse, D. J., and W. C. Palmer. 1965. The climate of the Northeast, meteorological and agricultural drought. Delaware Agric. Exp. Stn. Bull. 353.

Hagan, R. M., H. R. Haise, and T. W. Edminster. 1967. Irrigation of agricultural lands. Amer. Soc. of Agron. Monograph No. 11.

Hawks, S. N., Jr., W. K. Collins, H. Ross, R. E. Sneed, and J. G. Allgood. 1968. Field irrigation of tobacco. North Carolina Agric. Ext. Serv. Circ. 491.

Hernandez, T. P., T. P. Hernandez, J. C. Miller, and L. G. Jones. 1965. The value of sweet potato production in Louisiana. Louisiana Agric. Exp. Stn. Bull. 607.

Kincaid, C. M., R. C. Carter, J. S. Copenhaven, F. S. McClaugherty, J. H. Lillard, J. N. Jones, J. E. Moody, and R. E. Blaser. 1960. Supplemental irrigation of permanent pastures for beef cattle. Virginia Agric. Exp. Stn. Bull. 513.

Lynn, H. P., F. H. Hedden, and J. M. Lewis. 1958. Tobacco irrigation in South Carolina. South Carolina Agric. Ext. Serv. Cir. 438.

Morey, R. W. 1979. Irrigation survey, 1979. Irrigation Journal. 29(6):58A-58H.

Morris, J. R., A. A. Kaplan, and E. H. Arrington. 1962. Responses of Elberta peaches to interactive effects of irrigation, pruning and thinning. Amer. Soc. Hort. Sci. Proc. 80:177-189.

Privette, C. V., J. H. Palmer, J. P. Zublena, C. N. Nolan, J. W. Jordan, and D. B. Smith. 1980. Irrigating corn and soybeans in South Carolina. South Carolina Agric. Ext. Serv. Cir. 598.

Reynolds, C. W., and B. L. Rogers. 1958. Irrigation studies with certain fruit and vegetable crops in Maryland. Maryland Agric. Exp. Stn. Bull. 463.

Scarsbrook, C. E., O. L. Bennett, L. J. Chapman, R. W. Pearson, and D. G. Sturkie. 1961. Management of irrigated cotton. Alabama Agric. Exp. Stan. Bull. 332.

Sneed, R. E., and C. K. Martin. 1969. Response of corn, cotton and peanuts to irrigation. Soil Sci. Sco. North Carolina Proc. 12:81-93.

Sopher, C. D., R. J. McCracken, and D. D. Mason. 1973. Relationships between drought and corn yields on selected South Atlantic Coastal Plain soils. Agron. J. 65:351-354.

Spooner, A. E. 1961. Effects of irrigation timing and length of flooding periods on soybean yields. Arkansas Agric. Exp. Stn. Bull. 644.

Unrath, C. R. 1972a. The evaporative cooling effects of overtree sprinkler irrigation on 'Red Delicious' apples. J. Amer. Soc. Hort. Sci. 97:55-58.

Unrath, C. R. 1972b. The quality of 'Red Delicious' apples as affected by overtree sprinkler irrigation. J. Amer. Soc. Hort. Sci. 97:58-61.

Unrath, C. R., and R. E. Sneed. 1974. Evaporative cooling of 'Delicious' apples—The economic feasibility of reducing environmental heat stress. J. Amer. Soc. Hort. Sci. 99:372-375.

Van Bavel, C. H. M., and F. J. Verlinden. 1956. Agricultural drought in North Carolina. North Carolina Agric. Exp. Stn. Tech. Bull. 122.

Van Bavel, C. H. M., and J. H. Lillard. 1957. Agricultural drought in Virginia. Virginia Agric. Exp. Stn. Tech. Bull. 128.

Van Bavel, C. H. M., L. A. Forrest, and T. C. Peele. 1957. Agricultural drought in South Carolina. South Carolina Agric. Exp. Stn. Bull. 447.

Van Horn, A. G., W. M. Whitaker, R. H. Lush, and J. R. Carreker. 1956. Irrigation of pastures for dairy cattle. Tennessee Agric. Exp. Stn. Bull. 248.

Vittum, M. T., R. B. Alderfer, B. E. Jones, C. W. Reynolds, and R. A. Struchtemeyer. 1963. Crop response to irrigation in the Northeast. New York Agric. Exp. Stn. Bull. 800.

Wesley, W. K. 1979. Irrigated corn production and moisture management. Georgia Agric. Ext. Ser. Bull. 820.

White, J. H. 1961. Gravity irrigation on cotton and soybeans in central Arkansas. Arkansas Agric. Exp. Stn. Report Series 98.

Whitt, D. M., and C. H. M. Van Bavel. 1955. Irrigation of tobacco, peanuts and soybeans. p. 376-381. In Water, U.S. Dept. of Agric. Yearbook. U.S. Government Printing Office, Washington, D.C.

Wilson, T. V., and A. W. Snell. 1968. Irrigation of corn and cotton in South Carolina. South Carolina Agric. Exp. Stn. Bull. 540.

Woodhouse, W. W., Jr., J. L. Moore, R. K. Waugh, and W. H. Pierce. 1958. Supplemental pasture irrigation for lactating dairy cows. North Carolina Agric. Exp. Stn. Bull. 404.

13

# PHOTOSYNTHETIC ADAPTATION TO WATER STRESS AND IMPLICATIONS FOR DROUGHT RESISTANCE

**P. E. Kriedemann and H. D. Barrs**

Water supply probably limits field crop growth more frequently than any other physical input, and yet irrigation scheduling and general management of soil/water resources is still beset with difficulty. Both dryland and irrigated field crops encounter moisture stress of varying frequency, intensity, and duration. The ability of a plant in either situation to maintain positive carbon balance with minimum disruption, despite drought will be advantageous.

Crop plants may have a multiplicity of structural adaptations or physiological and biochemical adjustments that confer some measure of drought resistance. Inevitably, some loss in photosynthetic capacity, or at least in instantaneous rates, will be associated with these adaptations and adjustments; but, component processes have been identified, and their "physiological cost" forms a basis for our discussion.

## STRUCTURAL ADJUSTMENT FOR DROUGHT RESISTANCE

### Whole Plants

Arid-zone vegetation is generally sparse, and because of micrometeorological factors, individual plants usually have diminutive features. Tall vegetation tends not to occur because potential evapotranspiration, compared with photosynthesis, increases with height. This pattern is accentuated under heavy insolation and strongly advective conditions so that tall plants are at a disadvantage. For example, in a corn crop, the ratio of potential transpiration to potential photosynthesis increased from 10 g·cal/cm$^2$/day at a height of 100 cm to 18 g·cal/cm$^2$/day at 265 cm above the soil (Allen et al., 1964). Tall plants receive heavier radiation as well as extract more momentum from wind; hence, the evapotranspiration can increase relatively more than photosynthesis. This roughness component also finds expression in the transfer coefficient for heat and water vapor (K), which increases almost logarithmically with height. Lemon's (1966) comparisons of red clover and orchard grass with bulrush millet and corn provide K values of 10 to 300 sec/cm$^2$ for the short crop and 300 to 3000 sec/cm$^2$ for the tall crops.

While chlorophyll-containing stems on drought-deciduous plants such as *Cercidium floridum* can complement leaf photosynthesis during periods of moisture stress (Adams et al., 1967), further specialization in assimilatory organs also is evident in mesophytic crop plants. For example, Evans et al. (1972) and Kaul (1974) describe the photosynthetic contribution from awns in wheat ears towards kernel growth under dry conditions. Awns and related organs are far more desiccation tolerant than flag leaves, so that drought increases the proportion of assimilate contributed by ear photosynthesis. Evans et al. (1972) put the contribution at 43% in awned wheat ears under stress conditions and 13% in an awnless variety under irrigation. Similar compensatory responses have been reported by Pasternak and Wilson (1976) for *Sorghum bicolor.* When irrigation was withheld, photosynthesis in leaves virtually ceased. The value dropped from 89.3 $\mu$g $CO_2$/cm$^2$/sec in turgid leaves to 2.0 $\mu$ $CO_2$/cm$^2$/sec at a relative water content of 80%. However, photosynthetic activity of heads remained unchanged during this same period. Consequently the relative importance of head photosynthesis for grain filling increased, and in terms of whole plant carbon fixation, $CO_2$ assimilation by heads increased from 12% of the total in a well-watered plant to about 88% under drought.

#### Heat Budgets for Individual Leaves

Adaptive responses for drought resistance become apparent in single leaves as processes which minimize thermal load but maximize photosynthesis. Analysis of heat budgets for leaves in a simulated arid-zone climate (Gates et al., 1968; Vogel, 1970) have demonstrated 10 to 15 C increases in leaf temperature for broad-leafed plants over ambient air temperature. However, smaller leaves are cooler by virtue of better convective transfer of heat. In addition to moderating influences on transpiration, the temperature of smaller leaves and photosynthetic stems may be closer to the optimum for photosynthesis. These principles are exemplified in *Larrea divaricata* where small, thin leaves (4 to 10 mm long and 2 to 4 mm wide) show little increase over air temperature despite strong insolation and high stomatal resistance (Oechel et al., 1972).

The interplay of leaf size and leaf orientation, with respect to incident sunlight, has a major bearing on gas exchange. Morrow (1971), and Morrow and Mooney (1974) compared the two evergreen plants *Arbutus menziesii* and *Heteromeles arbutifolia* which have almost identical photosynthetic characters but widely differing leaf orientation according to native habitat. *A. menziesii* from the cooler habitat had larger and more or less horizontal leaves which tended to remain above ambient temperature because of strong radiant flux and lower convection transfer. The warmer leaf thereby derived some photosynthetic benefit in the cooler habitat. Conversely, *H. arbutifolia* encounters warmer conditions where air temperature is close to the photosynthetic optimum and occurs in habitats with longer periods of drought. In that case, smaller and vertically oriented leaves permit photosynthetic gain while preserving leaf moisture.

Sun tracking and leaf cupping movements can be seen as further refinements in regulation of thermal load. The desert lupin *Lupinus arizonicus,* (Wainwright, 1977), and *Stylosanthes humilis,* (Begg and Torsell, 1974), exhibit both forms of movement. Control over laminar orientation was seen as an active process, analogous to stomatal guard cell operation, and subject to regulation by water potential ($\psi_w$). Cupping movement predominated in response to moisture stress and constituted an adaptive response to drought conditions. Photosynthetic consequences of such leaf movements can be inferred from measurements of radiation climate and the incident angle of direct radiation ($\theta$). As a leaf is rotated from a plane normal to the beam to one parallel to the beam, incident flux will be proportional to the cosine of $\theta$ and will decrease linearly (Kriedemann et al., 1964). Heat load, and hence potential transpiration, will drop accordingly (Hadfield, 1968), but photosynthesis will not be similarly affected in plants such as *Camellia sinensis* where photosynthesis becomes light saturated at a fraction of full sun. Consequently, leaves held parallel to the beam under full sun can still achieve high rates of photosynthesis, that are up to 50% of the light-saturated rate (Went, 1958), but at considerable saving on potential transpiration. Rawson (1979) makes a similar case for sunflower leaves, and points out the additional benefit to improved water use efficiency from simultaneous reduction in leaf temperature as leaf orientation changes from horizontal to vertical during wilting.

One further avenue to reduce thermal load is by increased reflection combined with faster heat dissipation. Specialized coatings and surface structures facilitate the loss. For example, spines on cacti moderate daily temperature variation and ribs reduce daytime temperatures by providing a greater surface area for conductive heat loss (Lewis and Nobel, 1977). Pubescent leaves utilize a similar principle to regulate gas exchange and heat budgets, and higher plants show a tendency for increased pubescence along gradients of increasing aridity. Studies by Ehleringer (1977) on the effect of leaf pubescence on light adsorptance and photosynthesis in the drought-deciduous desert shrub *Encelia farinosa* provide a nice illustration. The presence of leaf hairs reduced absorptance of photosynthetically active radiation by more than 50% compared with a non-pubescent relative from moist regions. Furthermore, the degree of pubescance increases as the season advances (lower $\psi_w$), and photosynthetic attributes change accordingly. Light-saturated rates decreased from 4 to 2.5 nmol $CO_2/cm^2/sec$, and the initial slope of light-response curves for photosynthesis versus incident radiation was depressed; but quantum yield (based on absorbed quanta) were unchanged at 0.048 and 0.050 for non-pubescent and fully pubescent leaves, respectively. Significantly, the reduction in photosynthetic activity was attributable to decreased light absorption rather than increased diffusive resistance. The strong positive correlation between light absorptance and precipitation at sampling sites noted by Ehleringer (1976) is further evidence that pubescence is an adaptive response that confers a selective advantage under arid conditions.

**Differential Control Over Water and $CO_2$**

Leaf characteristics which limit loss of water to a greater extent than their effect on gain of $CO_2$ have obvious benefit for water use efficiency (hence drought resistance), and their effects can be analyzed in terms of gaseous diffusive resistances.

Bloom on sorghum leaf surfaces, which is evident as a powdery waxy layer, has the reputation of improving productivity, compared with bloomless varieties, under dry-land conditions. Higher water use efficiency seemed a likely explanation, and Chatterton et al. (1975) have provided relevant data. Their ratio of net $CO_2$ fixation to transpiration showed significantly higher values for bloomed than bloomless varieties.

Leaf pubescence and gas exchange characteristics of grasses from North Central USA (Table 1) also demonstrate the beneficial effects of leaf hairs on water use efficiency. These data, taken from Frank and Barker's (1976) study of six $C_3$ grasses, emphasize the importance of epidermal versus internal factors in water use efficiency. Western wheatgrass had the fastest photosynthesis and highest transpiration because of low stomatal diffusion resistance ($r_s$). By contrast, reed canarygrass had the lowest transpiration and highest ($r_s$); however, mesophyll resistance ($r'_m$) was even higher so that photosynthesis was adversely affected. Consequently, the potential water use efficiency, as indicated by the ratio $r_s/r'_m$ in Table 1, was only intermediate. Beneficial effects of the high $r_s$ associated with pubescence and low $r'_m$ (hence, faster photosynthesis) are exemplified in pubescent wheatgrass (Table 1). In this case, gas exchange has been selectively influenced so that potential water use efficiency ($r_s/r'_m$) is maximized and contributes to

**Table 1. Leaf pubescence and gas exchange of grasses at the tillering stage of growth. (Adapted from Rank and Barker, 1976)**

| Species | Photosynthesis | Transpiration | Diffusive Resistance $r_s$ | Diffusive Resistance $r'_m$ | Ratio $r_s/r'_m$ |
|---|---|---|---|---|---|
| | *– mg $CO_2/dm^2/hr$ –* | *– $g/dm^2/hr$ –* | *– sec/cm –* | | |
| Western wheatgrass *(Agropyron smithii)* | 26.2 | 3.6 | 0.6 | 5.4 | 0.11 |
| Pubescent wheatgrass *(A. intermedium* var. *trichlophorum)* | 18.1 | 1.5 | 3.7 | 3.4 | 1.08 |
| Reed canarygrass *(Phalaris arundinacea)* | 10.2 | 1.1 | 4.6 | 10.2 | 0.45 |

drought resistance despite a modest reduction in photosynthesis. Increased water use efficiency as stomates close in response to water stress can only be anticipated as long as $r'_m$ does not increase, or at least does not increase faster that $r_s$. Although Rawson (1979) has shown this to be true for hardened sunflower plants, the general relevance of this mechanism in other crop plants needs to be gauged.

Stomatal modification can be another adaptive feature of drought-resistant plants. Development of epistomatal cavities in *Eucalyptus* (Hallam and Chambers, 1970) and *Acacia* (Hellmuth, 1969), due to specially sculptured cuticles, often are associated with xeromorphic foliage. Cuticular resistances ($r_c$) range from 20 to 400 sec/cm depending upon species. As a general indication, $r_c$ is approximately 50 sec/cm for mesophytes and 200 sec/cm for xerophytes (Burrows and Milthorpe, 1976). Such structures contribute to enormous cuticular resistance and are important for retention of leaf moisture. For example, *Acacia* phyllodes have $r_c$ values well in excess of 100 sec/cm and are beyond the measurement range of standard instruments (Tunstall and Connor, 1975).

Further development of epidermal structures, and an unusual arrangement of photosynthetic tissues, are shown in the Australian aridzone $C_4$ species *Triodia irritans.* All species within this genus are perennial evergreen grasses which form large tussocks, are capable of exploiting a wide range of soils, and can endure high temperatures and moisture stress. Leaves are highly lignified and rolled, with chlorophyllous tissue flanking deep grooves on inner and outer surfaces of the rolled laminae. Stomata are located only in these grooves and provide direct gaseous access to photosynthetic mesophyll, but are protected themselves by interlocking papillae. Consequently, stomatal resistance is high. McWilliam and Mison (1974) cite minimum values of 14.7 sec/cm, so that photosynthetic activity has relatively low rates of

3 to 9 mg $CO_2/dm^2$/hour despite levels of PEP carboxylase comparable to other $C_4$ plants. High stomatal resistance is exacerbated by small, densely packed cells within the photosynthetic mesophyll which result in virtually no intercellular air spaces for $CO_2$ diffusion (Craig and Goodchild, 1977). Consequently, inward diffusion of $CO_2$ along a presumably steep concentration gradient (high PEP carboxylase activity) is impeded still further due to the limited "face" through which adsorption can occur. Nevertheless, the foliar anatomy of *T. irritans,* as characterized by recessed photosynthetic tissue, sunken stomata, and heavily thickened epidermal cell walls, make the grass ideally suited to arid conditions, since the reduced photosynthetic capacity is secondary to the ability to persist under extreme conditions. Mesophyll fine structure is of additional interest in this context, as Craig and Goodchild (1977) have associated the virtual absence of vacuolated mesophyll cells with the xeromorphic nature of the plant. It is suggested that an unusually high ratio of bound to unbound protoplasmic water would reduce cell damage due to tissue shrinkage on desiccation.

Cuticular waxes, resinous coatings, and epistomatal structures clearly amplify gaseous diffusive resistance, but anatomical features in the substomatal chamber can lead to selective reduction in transpiration. Jeffree et al. (1971) describe such a situation in Sitka spruce where wax-filled antechambers beneath the stomatal apparatus act as excellent antitranspirants. Wax tubules reduce both cross-sectional area and void space so that free diffusion no longer applies. Since the sum of resistances to diffusion of $CO_2$ ($\Sigma r'$) of 7 sec/cm is substantially larger than the sum of resistances to diffusion of water vapor ($\Sigma r$) of 2 sec/cm, the additional resistance imposed by a wax-filled antechamber causes a selective reduction in transpiration. Jeffree et al. (1971) calculate that an additional 0.73 sec/cm contributed by wax tubules would reduce $CO_2$ fixation by 32%, but transpiration would fall by 66%.

A selective influence of $r_s$ over transpiration compared with photosynthesis also can be inferred from a series of observations on amphistomatous leaves. In potato, tomato, and snap beans (Ackerson et al., 1977b; Duniway, 1971; Kanemasu and Tanner, 1969) adaxial stomata show a higher $\psi_w$ threshold for closure of approximately -8 bar compared to stomata on the abaxial (shaded) surface where $\psi_w$ had to fall to about -10 to -12 bar before $r_s$ increased to any extent. Earlier restriction of a potentially greater rate of transpiration from the adaxial surface thereby would improve overall water use efficiency by individual leaves as plants encountered drought.

By contrast, container-hardened sunflower appears to lack a critical leaf water potential for stomatal closure, but nevertheless achieves some improvement in water use efficiency during stress via alteration in leaf pores. Rawson (1979) showed that a change in orientation from horizontal to vertical as a result of wilting reduced water consumption from about 77 to 54 g

$H_2O$/g $CO_2$. Photosynthetic rates on wilted leaves were still about 50% of peak values.

## FUNCTIONAL ADJUSTMENT FOR DROUGHT RESISTANCE

Plant responses to stress that provide some measure of drought resistance include both short-term photosynthetic adjustment to brief episodes of water shortage and long-term accommodation of recurrent, and often intensifying periods, of moisture stress. This section will deal with long-term aspects that lead to some displacement in photosynthetic performance, while the more dynamic aspects of photosynthetic response during stress and recovery cycles are covered in the next section.

### Growth

Virtually every facet of the plant's metabolism is dislocated to some degree by moisture stress, but turgor-dependent processes such as root extension, leaf cell enlargement, and stomatal opening are especially vulnerable and contribute directly to the reduced growth during stress. Constraints on assimilation rate contribute further, but this effect becomes compounded by a decrease in assimilatory surface. The interactive effects of leaf area and photosynthetic rate in determining growth during stress can be deduced from the simple relationship RGR = NAR x LAR. Watson (1952) discusses the derivation and application of these indices of growth analysis, but in summary, the relative growth rate (RGR = dry weight increment/unit weight/per unit time) can be resolved into two primary determinants: net assimilation rate (NAR = dry weight increment/unit leaf area or leaf weight/unit time) and leaf area ratio (LAR = total leaf area/whole plant dry weight).

Recurring cycles of short term and mild stress cause reduced RGR in tomato (Gates, 1955a, 1955b) and *Panicum maximum* (Ng et al., 1975). The reduced RGR in tomato was attributed largely to a reduction in NAR from 0.3 to 0.1 g/g leaf/day after a brief stress to permanent wilting point. However, leaf expansion was adversely affected in *P. maximum* once the base level of water potential fell below about –11 bar, so that reduced LAR also had a significant influence on RGR as the experiment progressed.

Restriction of leaf expansion, which lowers LAR, is the more common determinant to reduced RGR, even during mild stress (Hsiao, 1973). Data of Kanemasu and Tanner (1969) on snap beans support this. Field-grown material showed reduced NAR once $\psi_w$ fell towards –8 bar. Stomatal closure was initiated at about the same $\psi_w$, but from parallel experiments in a growth chamber, extension growth would have been affected adversely within days.

Initiation, expansion, and retention of foliage are all influenced by $\psi_w$ and can feature in the response of a plant to water stress and, hence, drought resistance. Comprehensive reviews of Boyer (1976a, 1976b), Hsiao and Acevedo (1974), and Fisher and Hagan (1965) emphasize the greater sensitivity of leaf expansion to reduced turgor than to either stomatal aperture or photosynthesis. Gas exchange generally is affected only at lower $\psi_w$. While brief interruptions to leaf enlargement by crop plants are reversible, expansion is impaired permanently if turgor is not regained for several days and leads to the observed reduction in LAR on droughted plants (Gates, 1955a, 1955b). While leaf area decreases with water stress, specific leaf weight (SLW) can show an associated increase as demonstrated by Rawson et al. (1977b) for cereal leaves. After recurring cycles of moisture stress, SLW increased from 3.51 to 5.13 mg/cm$^2$ in wheat and from 3.24 to 4.87 mg/cm$^2$ in barley. Smaller, thicker leaves maintain photosynthesis at lower $\psi_w$ and, thereby, contribute to drought resistance.

In photosynthetic terms, droughted plants are especially disadvantaged because the effect of lower NAR is compounded by reduced LAR; thus, some ability to sustain leaf growth despite lower $\psi_w$ confers the potential for renewed RGR once stress is relieved. This principle would at least apply under recurring episodes of mild stress. Since leaf growth is turgor dependent, this form of drought resistance will depend on the ability of the plant to generate turgor in foliar zones of cell enlargement. Data in the review by Begg and Turner (1976) bear on this point. One particular case from Watts (1974) shows the relationship between leaf extension and $\psi_w$ for corn growth in the field and controlled environments. Expansion in light at 30 C for controlled-environment corn started to decline once $\psi_w$ dropped below -2 bar, and fell to zero at $\psi_w$ of -7 bar. By contrast, field-grown corn showed little impairment down to -9 bar. Leaf elongation continued at about 3.5 mm/hour at $\psi_w$ of -8 bar in field-grown corn compared to 4.5 mm/hour at $\psi_w$ of -2 bar for corn grown in controlled environments. Osmotic adjustment to environmental stress offsets the effects of reduced $\psi_w$ on leaf expansion as well as fostering a resurgence in extension growth once drought is relieved. This is a characteristic of drought-resistant species of cotton and sorghum (Ackerson et al., 1977a) that contrasts with the less resistant potato which appears to have very limited capacity for osmotic adjustment. Consequently, growth and yield of potato are adversely affected by even mild water stress (Ackerson et al., 1977b). Similarly, young unemerged wheat leaves are capable of large osmotic adjustment of -12 to -40 bar during water stress and can resume active growth promptly when stress is relieved, but exposed leaves show comparactively small osmotic adjustment and fail to recover on rewatering (Munns et al., 1979).

The ability of a plant to maintain positive turgor via osmotic adjustment is an important adaptation to stress. As demonstrated by Johnson and Brown (1977), response of turgor pressure over the wide range in $\psi_w$ of

−2 to −28 bars is positively correlated with field assessment of drought resistance. Osmotic adjustment by roots constitutes a useful form of that adaptation which was shown in *Capsicum frutescens* and *Gossypium hirsutum* by Bernstein (1961) and more recently in small grain crop species by Bower and Tamimi (1979) in response to root zone solutes. Reduction in soil matric potential (Hsiao, 1973) or increased mechanical resistance (Greacen and Oh, 1972) can elicit a similar response, although Russell and Goss (1974) have queried the latter effect.

Nevertheless, as Hsiao and Acevedo (1974) point out, root growth is generally favored over shoot growth under water stress, even leading in some cases to an absolute increase in root growth (see Figure 8, Hsiao and Acevedo, 1974). They speculate that mild water stress leads to reduced shoot growth, permitting roots to receive an increased share of assimilates, undergo osmotic adjustment, and thereby achieve extra growth which leads to more extensive exploration of soil. Their hypothesis deserves wider evaluation over a range of crop species and environments.

Despite virtues attributable to osmotic adjustment, an alternative and distinctive process occurs in the broad-leaved tropical legume *Macroptilium atropurpureum.* Rather than accumulate osmotic solutes, this plant uses its limited energy resources to sustain production of new leaves which can adapt morphologically to withstand stress (J. R. Wilson et al., 1980). This species derives no advantage from large osmotic adjustment to keep stomates open because water potentials below −20 bar are lethal.

**Net Photosynthesis**

Drought effects on NAR are primarily due to reduced photosynthesis rather than increased respiration by the whole plant. Moreover, conversion efficiency by new assimilates into growth in grain crops was not adversely affected, despite photosynthetic reduction, under stress (D. R. Wilson et al., 1980). Drought resistance in the present context therefore will relate to the capability of a plant in maintaining a positive carbon balance under stress.

Reduction in $\psi_w$ leads to a massive displacement in processes underlying gas exchange. Invariably, $CO_2$ fixation is impaired, whereas relative rate of $CO_2$ production can be stimulated. Notwithstanding short-term osmotic adjustments which may have buffered turgor-dependent processes against adverse changes in $\psi_{leaf}$, increased stomatal resistance ultimately will occur and can be accompanied by higher $r'_m$ because of the combined influences of greater vapor and liquid phase resistances (Gale et al., 1967; Gaastra, 1959) as well as reduced photochemical (Boyer, 1971b; Keck and Boyer, 1974) and carboxylating (O'Toole et al., 1977) efficiencies. Regardless of the relative magnitudes in the displacement of carbon assimilation versus dissimilation under stress, increased $CO_2$ compensation point at low $\psi_w$ is reported widely and carries serious implications for net photosynthesis.

For example, the $CO_2$ compensation concentration of wheat leaves changes from 60 ppm $CO_2$ at $\psi_w$ of -5 bar to about 80 ppm at -16 bar and undergoes a sharp increase at lower potentials reaching 320 ppm at $\psi_w$ of - 22 bar (Lawlor, 1976). Leaves in that condition can show $CO_2$ evolution even under illumination. Shearman et al. (1972) have reported a similar $CO_2$ efflux for water-stressed sorghum leaves which extends Meidner's (1967) earlier observations on corn leaves showing an increase in $CO_2$ compensation concentration from 0 to about 26 ppm $CO_2$ under severe water stress. Clearly, $C_4$ plants would not necessarily maintain the same advantage over $C_3$ plants under drought as under well-watered conditions, if the balance between assimilation and dissimilation was solely responsible for drought resistance. Boyer (1970) provides some support for this view from his comparative studies where soybean photosynthesis was unaffected by desiccation until $\psi_w$ was below -11 bar, whereas corn photosynthesis was inhibited once $\psi_w$ dropped below -3.5 bar. Although corn was capable of faster photosynthesis during initial desiccation down to $\psi_w$ of about -8 bar, stomatal factors supervened by dictating early reponse patterns of $\psi_w$. In Boyer's comparison, tissue tolerance to desiccation was of more significance for drought resistance than the timecourse of photosynthetic response. Furthermore, $C_3$ plants are not disadvantaged universally by lower rates of photosynthesis compared with $C_4$ plants. Mooney et al. (1976) encountered particularly high rates in *Camissonia claviformis,* a winter annual of Death Valley. The desert ecosystem appears to have selected, in this case, for a $C_3$ species with high photosynthetic rate.

In summary, the $C_4$ syndrome cannot be regarded as intrinsically responsible for drought resistance by altering the form of photosynthetic response to stress or by altering the threshold $\psi_w$ at which stomatal closure is initiated. Variation among plants is more commonly attributable to previous environmental conditions than to their photosynthetic mode (Ludlow, 1976).

If the $CO_2$ concentration gradient from air to chloroplast remains unchanged despite some reduction in $\psi_w$, then reduced flux becomes attributable to increased $\Sigma r'$, and depending upon the extent of stomatal control, $r'_m$ can become a key factor. Although derivations of $r'_m$ from gas exchange measurements in normal air carry the disadvantage that photorespiratory factors are not excluded, such data do provide some insight into differences between mesic and xeric plants in their photosynthetic response to moisture stress. Regardless of stomatal factors, maintenance of relatively low $r'_m$, despite reduced $\psi_w$, would be advantageous for carbon balance. Data from Bunce (1977) shown in Table 2 support that inference. From a comparison of gaseous diffusive resistance in mesophytic trees versus arid-zone shrubs,

**Table 2.** **Mesophyll resistance to $CO_2$ update ($r'_m$) in relation to leaf water potential ($\psi_w$). (Adapted from Bunce. 1977.)**

| Species | Leaf Water Potential, Bars | | | | | | | |
|---|---|---|---|---|---|---|---|---|
| | -5 | -10 | -15 | -20 | -25 | -30 | -35 | -40 |
| | | | *— mesophyll resistance, sec/cm —* | | | | | |
| **Meosphytes** | | | | | | | | |
| *Fraxinus pennsylvanica* spp. *velutina* | 5 | 10 | 18 | 25 | 33 | 40 | 50 | |
| *Acer saccharum* | - | - | 60 | 90 | 120 | | | |
| *Alnus oblongifolia* | 4 | 14 | 24 | | | | | |
| **Xerophytes** | | | | | | | | |
| *Simmondsia chinensis* | - | 12 | 14 | 16 | 18 | 20 | 22 | 24 |
| *Acacia greggii* | - | - | - | - | 14 | 19 | 23 | - |
| *Larrea divaricata* | - | - | 2 | 6 | 10 | 14 | 18 | 22 |
| *Vauquelinia californica* | - | 11 | 12 | 14 | 16 | 18 | 20 | 22 |

it is obvious that xerophytic shrubs have a photosynthetic advantage at low water potential. The progressive increase in apparent $r'_m$ for the meophytic group as water potential diminishes might stem from the combined effects of a shallower $CO_2$ concentration gradient, as well as higher diffusive resistance per se. Unfortunately, since Bunce (1977) provides no additional information on the magnitude of low $O_2$ enhancement or on values of $CO_2$ compensation concentration at different water potenials, the issue remains unresolved. Nevertheless, Slatyer (1973) also refers to species differences in $r'_m$ sensitivity to decreased $\psi_w$ that relate to provenance. Drought-adapted wheat, millet, and salt bush showed a more limited $r'_m$ response to low $\psi_w$ than did corn and cotton which were considered to be less drought resistant.

The relevance of $r'_m$ sensitivity to $\psi_w$ for drought resistance is also apparent in the comparative studies of Dubé et al. (1974). They compared two lines of corn which differed in phenotypic response to water stress. One line 'Q-188' was a wilting inbred and the other 'DR-1' was heat and drought resistant. In both lines, stomatal closure was triggered between -8.5 and -9.5 bar, and $r'_m$ started to increase within this same narrow range. However, the absolute magnitude and degree of $r'_m$ increase differed substantially. Under adequate moisture ($\psi_w$ of -2 to -8 bar), $r'_m$ for 'Q-188' ranged between 1.0 and 1.4 sec/cm, but increased to a maximum value of 124 sec/cm at $\psi_w$ of -11 bar. By contrast, 'DR-1' showed $r'_m$ between 0.8 and 1.1 sec/cm at high $\psi_w$ with a maximum value of 32 sec/cm when $\psi_w$ fell to -12 bar. Then lower $r'_m$, especially under stress, was a likely contributor to the greater drought resistance of 'DR-1.' Early wilting in 'Q-188' under field conditions was attributed to a higher xylem resistance which no doubt amplified the photosynthetic disadvantage of high $r'_m$ in this line.

Drought resistant plants vary as to whether their assimilatory organs endure or avoid low $\psi_w$. Avoidance necessitates either close control over $r_s$ to preserve leaf moisture or a highly efficient system for acquisition and distribution of water within the plant. These two forms are discussed below.

*Photosynthetic Avoidance of Low* $\psi_w$. Early stomatal closure during the onset of stress can contribute towards drought resistance (Grace et al. 1975; Kozlowski, 1964), especially when combined with high cuticular resistance. Superior drought resistance in ponderosa pine compared to grand fir has been ascribed to such an effect by Lopushinsky (1969) who followed the time course of transpiration decline in 3-year old, potted seedlings. Pine needles showed an abrupt stomatal closure within 6 to 7 days, whereas stomata on fir showed a more gradual decline and took twice as long to close. Short-term comparisons of stomatal closure on excised twigs confirmed this distinction. Nighttime $r_s$ was also lower in fir so that transpiration was proportionately higher, and hence, drought resistance was lower. Stomatal sensitivity to increased leaf temperature and associated increase in transpiration represents a further refinement in regulation of gas exchange for drought resistance. As demonstrated by Wuenscher and Kozlowksi (1971a, 1971b), *Quercus* species normally occupying a xeric habitat showed greater increase in $r_s$ from 20 to 40 C leaf temperature than the mesophytic group. Ironically, for some arid-zone plants, a lower resistance with increased leaf temperature also can constitute an adaptive mechanism which is partly related to transpirational cooling, but such prodigal water use is confined both geographically and temporally (Schulze et al., 1973; Bjorkman et al., 1972; Pearcy et al., 1972).

Vapor pressure gradient from leaf to air is obviously the primary determinant of transpiration so that stomatal sensitivity to this gradient and associated flux will figure prominently in preservation of $\psi_w$. The dry land plant *Sesamum indicum* has this characteristic (Fig. 1) and shows increased $r_s$ with steepening water vapor pressure gradient from leaf to air. Significantly, $r'_m$ was not affected similarly, so that the stomatal response of *Sesame* leaves to potential transpiration offers the dual advantage of preserving both leaf moisture and carbon balance. Drought-susceptible perennial plants such as citrus lack this characteristically flat response in $r'_m$ to vapor pressure gradient; instead, $r'_m$ increases in step with $r_s$ (Kriedemann, 1971), although some adaptation can occur whereby $r'_m$ increments diminish after repeated exposure to desiccating conditions (Khairi and Hall, 1976).

Stomatal response to leaf/air vapor pressure gradient in some drought-resistant plants is further accentuated by a slight reduction in $\psi_w$ (Schulze et al., 1972, 1974) so that $r_s$ sensitivity to transpirational flux is heightened and a greater measure of protection for the photosynthetic apparatus is ensured. Such plants tend to be drought-escaping or opportunistic individuals.

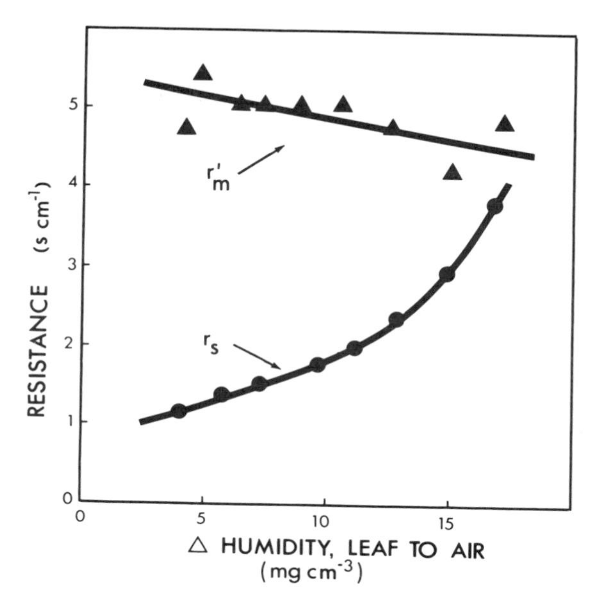


**Figure 1. Differential effects of leaf to air vapor pressure gradient on stomatal ($r_s$) and mesophyll ($r'_m$) resistances to gas exchange in *Sesamum indicum* L. (Adopted from Hall and Kaufmann, 1975).**

Sheriff and Kaye (1977) illustrate such an association in their stomatal comparisons of two leguminous plants from a semi-arid tropical region of Australia. *Macroptilium atropurpureum* avoids low $\psi_w$ by early closure and is highly drought resistant compared to *Desmodium uncinatum,* which inhabits high rainfall areas, is virtually insensitive to vapor pressure gradient, and has low drought resistance.

Another form of physiological adaptation that confers drought resistance by obviating low $\psi_w$ occurs in the $C_4$ amaranth *Tidestromia oblongifolia.* As a low growing perennial herb, it combines $C_4$ photosynthesis with thermal stability and low resistance to water flow (Troughton et al., 1974). In sharp contrast to most arid-zone inhibitants, *Tidestromia* grows during the hot summer. Maximum photosynthesis occurs at leaf temperatures of around 47 C and does not reach light saturation even under full sun. Gas exchange is prodigious. Fixation rates of $CO_2$ up to 58 mg/dm$^2$/hour and transpiration rates of 12.6 g water/dm$^2$/hour have been measured (Björkman et al. 1972). Despite such flux, $\psi_w$ ranges from −10 bar at dawn and −25 bar at midday. The water requirements of the plant are serviced by a highly efficient vascular network which draws on water supplies at soil depths of 25-40 cm and below.

### Photosynthetic Endurance of Low $\psi_w$

Maintenance of photosynthesis despite reduction in $\psi_w$ constitutes an alternative adaptation for drought resistance which is highly developed in sclerophyllous xerophytes and also is present to varying degrees in mesophytes. For photosynthetic endurance, osmotic adjustment can act as a counter measure to reductions in $\psi_w$. Begg and Turner (1976) cite many instances where photosynthetic and stomatal conductance decline at lower $\psi_w$ in plants growing in the field than in controlled environments. In cotton, for example, gas exchange declines rapidly at $\psi_w$ below -16 bar in plants grown in controlled environments, whereas field-grown specimens were influenced only marginally at -27 bar. Similarly, Ludlow (1976) found no reduction in photosynthesis of field-grown *Panicum maximum* down to $\psi_w$ of -18 bar, but in plants grown in controlled environments found restricting gas exchange starting at -3 bar and zero net photosynthesis at -12 bar. One facet of the widely reported yield difference between corn and sorghum during drought also has been attributed to osmotic adjustment. As outlined by Hsiao et al. (1976), sorghum either is more capable of osmotic adjustment or else contains more osmotically active solutes than corn, so that turgor pressure reaches zero at a lower $\psi_w$ threshold. The net result is that during the onset of moisture stress sorghum maintains photosynthesis for a longer time than corn. Beadle et al. (1973) report on growth-chamber studies where sorghum photosynthesis was still substantial at $\psi_w$ -12 bar, whereas corn leaf assimilation was virtually zero at this same $\psi_w$.

Specific loci for osmotic adjustment need further resolution. Jones and Rawson (1979) suggest that bulk leaf changes do not explain lower stomatal sensitivity to $\psi_{leaf}$ in hardened sorghum. By imposing stress at different rates, plants were obtained with differing $\psi_{leaf}$ to stomatal conductivity relationships despite similar bulk leaf osmotic adjustment. Plants stressed more slowly were thought to have greater opportunity for localized solute accumulation in guard cells; hence, turgor generation potential of the stomatal apparatus was shifted via-a-vis adjacent tissue.

## DYNAMICS OF GAS EXCHANGE DURING DROUGHT AND RECOVERY

Long-term photosynthetic adjustment for drought resistance is one consequence of moisture stress that occurs either during the life cycle of a particular plant or during the evolution of the species or cultivar. Recurring experiences of stress and recovery cycles therefore seem to be a prerequisite for such adaptation. Our concerns in this section are an analysis of such short-term photosynthetic adjustments to stress which ultimately lead to adaptation and an assessment of how early stomatal response to $\psi_w$, rapid recovery of gas exchange after watering, or even a major photosynthetic modification such as $C_3$ to CAM contribute towards drought resistance.

**Stomatal-Leaf Water Potential Interrelations**

Stomatal operation is closely coupled "hydraulically" to $\psi_w$ (Raschke, 1970) but hormonal systems can exert an overriding influence. Whenever transpirational demand outstrips water supply and $\psi_w$ falls substantially, stomatal resistance is subject to the combined influences of reduced guard cell turgor, increased epidermal abscisic acid (ABA) and elevated $CO_2$ within the stomatal cavity. Since $CO_2$ potentiates stomatal response to ABA (Raschke et al., 1976) and drought sensitizes stomata to $CO_2$ (Meidner and Mansfield, 1968), interaction of these three factors will ensure the close coupling among $r'_s$, environmental conditions, and $\psi_w$ that is associated with drought resistance. For example, early stomatal closure during the onset of drought can afford some protection to the leaf's photosynthetic apparatus against debilitating effects of low $\psi_w$.

A higher threshold for stomatal closure in drought-resistant Mexican corn containing the Latente germplasm might well derive from higher endogenous ABA (Larque-Saavedra and Wain, 1974). There was an increase in ABA from 13 to 1149 μg/kg fresh weight in the drought-resistant line '$M_{35\text{-}1}$' held 24 hours under moderate stress, but in the drought-sensitive cultivar 'Shallu' ABA increased from 13 to 441 μg/kg fresh weight. The more massive increase in ABA was thought to make an additional contribution to drought resistance in '$M_{35\text{-}1}$' by arresting synthetic events and by counter-acting the effect of growth stimulating hormones during dry periods.

Studies on tomato provide further evidence of ABA involvement in plant water relations. The mutant *flacca,* derived by irradiation from the normal cultivar 'Rheinlands Ruhm', is characterized by very low levels of ABA-like substances and a propensity to wilt because stomata do not close despite severe reduction in $\psi_w$ (Tal, 1966). Levels of ABA-like substances are an order of magnitude lower in *flacca* than in 'Rheinlands Ruhm' (Tal and Imber, 1970), but in response to external application of ABA to *flacca* parental-type stomatal activity was restored and the wilting syndrome disappeared. Reversion to the mutant phenotype occured as soon as ABA treatment was terminated.

The $\psi_w$ threshold which instigates stomatal closure varies enormously among species, and this plasticity in stomatal $\psi_w$ relations appears to derive from different drought-resistance mechanisms, as well as from recent environmental history. Osmotic adjustment, for example, lowers the $\psi_w$ threshold for stomatal closure, and it might be significant, in this context, that previously stressed plants also show diminished sensitivity to exogenous ABA (Kriedemann et al., 1972). Moreover, the correlations between $\psi_w$, $r_s$ and endogenous ABA also become more tenuous in hardened material (Loveys and Kriedemann, unpublished).

ABA synthesis is of additional relevance to drought resistance in situations where hormonal modulation of stomatal aperture offers greatest advantage, viz. unhardened mesophytes whose assimilatory organs are exposed to occasional droughts of varying frequency and intensity. In that case, a high $\psi_w$ threshold for increased $r_s$ would protect the photosynthetic apparatus from periods of low $\psi_w$ and, in effect, would amount to a trade-off between early cessation of carbon gain with the onset of stress and early restoration of photosynthetic activity following relief from stress. As discussed previously, irrigated crop plants experiencing strong isolation exert such control.

ABA metabolism holds less relevance for plants in the arid zone where drought resistance requires judicious use of available water and an early resumption of photosynthesis following rain. In that case, structural and physiological adaptations for drought resistance that are far less transitory than ABA generation appear to prevail. Indeed, a massive accumulation of ABA could be a serious disadvantage if dissimilation of the inhibitor was a prerequisite for resumption of gas exchange.

Rates of gas exchange do not reach prestress levels despite restoration of $\psi_w$ (Fischer et al., 1970; Hsiao, 1973). These after-effects of drought on stomatal physiology have been ascribed in specific instances to stress related accumulation of ABA (Allaway and Mansfield, 1970; Richmond, 1976). Nevertheless, two sets of observations are difficult to reconcile with this supposed role for ABA. First, as discussed previously, foliar concentrations of ABA can reach prestress levels within hours of irrigation while recovery of gas exchange is more protracted (Beardsell and Cohen, 1975). Thus, hormonal change and $r_s$ might show close correlation as plants enter stress, but they are not coincident during recovery. The inhibition of leaf photosynthesis during stress often embodies a nonstomatal component, especially under severe conditions, which becomes even more apparent during recovery. Since ABA effects on gas exchange are mediated via stomatal changes, rather than directly on photosynthesis (Kriedemann et al., 1972), additional factors must be involved. A discussion of these follows.

### Leaf Moisture and Diffusive Resistance

Unhardened mesophytes encountering stress for the first time (e.g. Brix, 1962), or leaves undergoing cyclic oscillations in gas exchange (e.g. Barrs, 1968), can show parallel changes in transpiration and photosynthesis to convey the impression that $r_s$ is solely responsible for regulation of photosynthesis. However, a correlation does not necessarily imply causation, especially one with such "bothersome exceptions" (Boyer, 1976a). For example, both spruce and birch trees entering a period of stress show greater reductions in photosynthesis than can be accounted for by stomatal factors alone (Beadle and Jarvis, 1977; Hari et al., 1975) because of additional limitations imposed by some internal factor. Similarly, Regehr et al. (1975) showed a continuation of photosynthesis despite a reduction in transpiration during early stages of stress induced by stomatal closure. These authors simultaneously measured fluxes of $CO_2$ and water vapor from four tree species with varying degrees of drought resistance. Red cedar *(Juniper virgaurea)* was especially noteworthy because photosynthesis continued at 10% of the maximal rate despite $\psi_w$ reduction to -35 bar and minimal transpiration. The winged elm *(Ulmus alata)* showed peak photosynthesis at around -10 to -12 bar even though transpiration had started tc decline once $\psi_w$ fell below -6 bar.

Such differential effects on $CO_2$ and water vapor exchange carry obvious implications for drought resistance, which have been emphasized in the preceding section. The extent and dynamics of such nonstomatal control are our present concern. The importance of such factors was foreshadowed prior to the advent of diffusive resistance analysis in the suble approach of Pisek and Winkler (1956). Working with leaves on *Asarum europaeum,* they measured photosynthesis and stomatal opening on leaves of different moisture deficit. For a given stomatal aperture, e.g. 0.5 $\mu$ wide, photosynthetic rate at low water deficit was 9.5 mg $CO_2/dm^2$/hour compared to 3.1 mg $CO_2/dm^2$/hour for the same stomatal aperture but higher water deficit. Presumably, some form of mesophyll resistance to $CO_2$ assimilation had increased with loss of leaf moisture, but no clear resolution between physical, biochemical, or photochemical processes was presented.

With the advent of Gaastra's (1959) methodology for derivation of diffusive resistances, and the subsequent use of low-$O_2$ gas streams to eliminate photorespiratory complications, further differentiation between stomatal and nonstomatal components became feasible. Numerous laboratories reported that under severe stress, increased $r_s$ was accompanied by a substantial increase in $r'_m$ (Boyer, 1976a, 1976b). During initial phases of stress/recovery cycles in short-term experiments with cotton where leaf stress was achieved by chilling the roots, increased $r_s$ is primarily responsible for the fall in photosynthesis because $r'_m$ stays the same (Troughton, 1969). Stomatal closure was initiated at relative leaf water contents (RLWC) below 85 to 80%, but $r'_m$ in air of normal $O_2$ content did not increase until the RLWC was below 70% at $\psi_w$ of -15 bar.

In subsequent work on cotton (Troughton and Slatyer, 1969), $r'_m$ was derived from measurements of photosynthetic response to $CO_2$ concentration in a gas stream forced through the amphistomatous leaf. Mesophyll resistance in $O_2$-"free" air was unaffected by RLWC between 56 and 92% or by variation in temperature between 22.5 and 38 C. Short-term water stress was therefore without effect on liquid phase diffusion of $CO_2$ implying that photorespiration increased in normal air (Hsiao, 1973).

This difference between $r_s$ and $r'_m$ in sensitivity to changes in leaf moisture was examined further by Pasternak and Wilson (1974) in attached and excised sorghum leaves. Both threshold $\psi_w$ for stomatal closure and the form of photosynthesis-stomatal relationships were altered. Leaves held in an assimilation chamber and attached to a turgid, but recently decapitated plant, showed an increase in $r_s$ once RLWC fell below 95% at $\psi_w$ of −1 bar. Stomatal closure was complete after 2 hours, and over this period the decline in photosynthesis was largely accounted for by gradual closure of stomata. In leaves excised adjacent to the chamber, an initial hydropassive opening for 7 minutes was followed by rapid closure which was complete within another 10 minutes. In that case, RLWC had to fall to 70% at $\psi_w$ of −24 bar before closure was instigated while photosynthetic rate for a given stomatal resistance was substantially lower than on slowly wilting leaves, e.g. 15 versus 20 mg $CO_2/dm^2$/hour at $r_s$ of 3.2 s/cm. Clearly, nonstomatal inhibition became manifest only during the accelerated wilting.

If stress develops suddenly, and $\psi_w$ of an unhardened plant falls in the space of minutes, then $r'_m$ might show an upward inflexion due to structural collapse of the leaf's mesophyll (Slatyer, 1973). Withdrawal of moisture from walls lining intercellular spaces ultimately would lead to increased liquid phase resistance to $CO_2$ (Nobel, 1977; Redshaw and Meidner, 1972) while the greater tortuosity of pathways for gaseous diffusion within a flaccid leaf would contribute further to a higher $r_m$ and $r'_m$ (Levitt and Ben Zaken, 1975). Biochemical and photochemical factors of impeded $CO_2$ fixation by chloroplasts would complement physical factors by having a specific effect on $r'_m$. Correlations between transpiration and photosynthesis as plants approach the wilting point, taken earlier as a priori evidence for exclusively stomatal control over photosynthesis, now warrant reassessment (Boyer, 1976a).

**Stomatal and Photosynthetic Adjustments**

During drought, stomatal physiology becomes subject to stomatal and $\psi_w$ interactions as well as to hormonal systems. Both avenues of control relate to drought resistance, either through osmotic adjustment to offset low $\psi_w$, or via ABA sensitization which lifts $\psi_w$ threshold for stomal closure. Drought-resistant sorghums that generate massive quantities of ABA when under stress were cited previously to illustrate drought resistance via early stomatal closure. Since the locus for ABA effects is stomatal closure rather than the photosynthetic apparatus per se, hormonal regulation implies additional benefit for drought resistance because $r_s$ will increase without commensurate increase in $r'_m$.

Despite these protective measures, a leaf's photosynthetic apparatus will ultimately become debilitated if drought intensifies. As a result, photosynthetic recovery will be impaired. Photochemical components seem especially labile, or at least more prone to inhibition, at low $\psi_w$ than many enzymatic processes.

The extent of stomatal and photosynthetic disruption following drought is a function of both the duration and intensity of the previous stress. For example, sunflower leaves subjected to only partial desiccation at $\psi_w$ not below -12 bar show complete restitution in 3 to 5 hours with $r_s$ controlling gas exchange during both stress and recovery. If $\psi_w$ falls below -12 bar, then photosynthetic recovery is incomplete due to increases in both $r_s$ and $r'_m$, while more intense drought stress at $\psi_w$ of -20 bar is lethal (Boyer, 1971a). Spring wheat provides an additional illustration where the $\psi_w$ threshold for stomatal closure decreases with plant development; viz. -12 bar to -13 bar at tillering, -16 to -18 bar at heading, and -31 bar during grain filling. The corollary of lower $\psi_w$ for $r_s$ increase was a more protracted photosynthetic recovery. In wheat plants subjected to drought during grain filling, $CO_2$ assimilation showed no sign of recovery even 48 hours after rewatering, despite full, and virtually immediate restoration of $\psi_w$ and $r_s$. Photosynthetic disruption occurred well in advance of general physiological deterioration, because senescence was not accelerated until $\psi_w$ was reduced to -44 bar (Frank et al., 1973).

Although the components of photosynthetic inhibition during drought have been resolved in some cases into diffusive resistances, photochemical activity, and biochemical processes (Boyer, 1976b), the most unequivocal demonstrations of nonstomatal inhibition of photosynthesis occur during recovery. The time course shown in Figure 2 for $\psi_w$, stomatal conductance, and net photosynthesis in normal air in the Concord grapevine *(Vitis labruscana)* provides such an example. This mesophytic vine had grown 2 years with regular irrigation in an 85 liter container under field conditions as part of a larger experiment. Stress developed gradually and 14 days elapsed after irrigation ceased before $\psi_w$ fell from -6 to -16 bar. Water was withheld another 3 days and minimum $\psi_w$ on the vine was -18 bar. Restoration of $\psi_w$ was complete within 2 hours of irrigation (Figure 2), while stomatal recovery was virtually immediate. The time course was even more compressed when droughted shoots were excised and recut under water. Nevertheless, photosynthesis in either attached (Figure 2) or excised shoots showed only slight recovery even after 4 hours. Additional photosynthetic recovery occurred over the next 24 hours, and at the end of 3 days, light saturated rates of 22 mg $CO_2/dm^2$/hour exceeded prestress levels of 26 mg $CO_2/dm^2$/hour.

Since photosynthesis and stomatal conductance had shown a parallel decline as vines encountered drought, the outstanding question is what led to the specific inhibition of photosynthesis during recovery? Accumulation

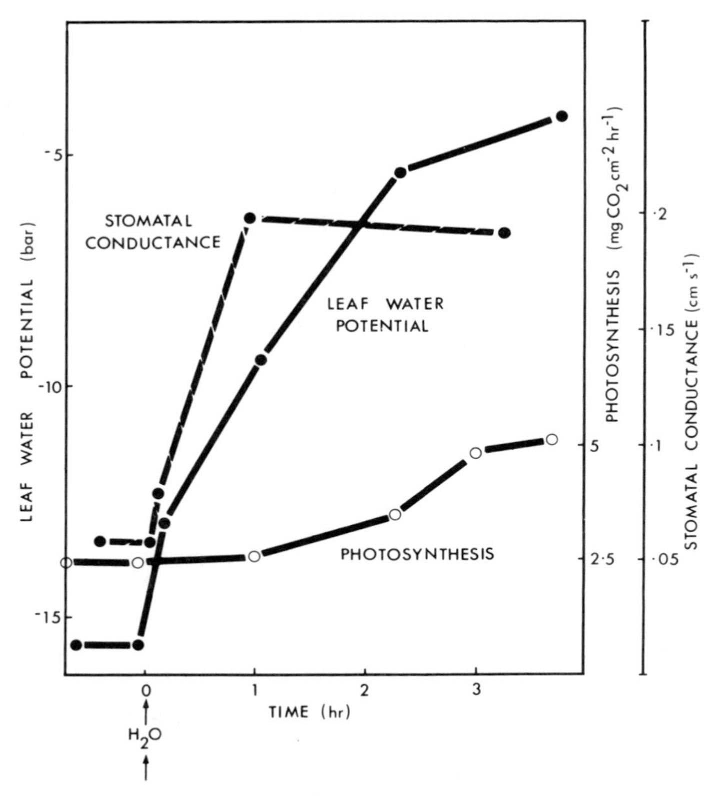


**Figure 2. Time course of recovery in $\psi_w$, stomatal conductance, and photosynthesis in *Vitis labruscana* following drought. (From Liu, Wenkert, and Kriedemann, unpublished data, 1973).**

of ABA must be discounted because that inhibitor is $r_s$ specific (Kriedemann et al., 1972) and stomatal recovery was virtually immediate and independent of photosynthesis. Light response curves imply some loss of photochemical efficiency, and inhibition by some other form of bioregulation seems more likely. For example, one metabolic derivation of ABA, viz. phaseic acid (PA), shows an upsurge in concentration as ABA level falls with rewatering (Kriedemann et al., 1975). Furthermore, plant extracts containing PA have been shown to cause a specific inhibition of photosynthesis independent of stomatal closure, with photochemical processes as the most likely site of action (Kriedemann et al., 1976; Loveys, unpublished data). However, additional compounds will almost certainly be involved because PA purified by crystallization proved ineffectual as a sole regulatory compound when applied to detached leaves, epidermal strips, or isolated mesophyll cells (Sharkey and Raschke, 1980). Existence of stress-related photosynthetic inhibitors within leaves nevertheless has been confirmed (Kriedemann et al., 1980), and one component in the complex tentative identified as malonic acid (Loveys, unpublished). Decreased photosynthetic efficiency in other droughted plants (Boyer, 1976b) may well depend in part on such naturally-occurring inhibitors.

Drought resistance, in the present context, therefore relates to the speed and extent of this post-drought photosynthetic recovery and especially to the resumption of transpiration relative to $CO_2$ assimilation. Allaway and Mansfield (1970) view the aftereffect of stress on $r_s$ in *Rumex sanguineus* as a fail-safe mechanism to prevent excessive transpiration when soil moisture could still be limiting, but really judicious water usage necessitates minimal $r'_m$ compared with $r_s$, especially after stress. Drought-resistant Mitchell grass *(Astrebla lappacea,* $C_4$) embodies this characteristic, and as described by Doley and Trivett (1974), its leaves combine photosynthetic endurance of low $\psi_w$ with rapid recovery. Only 2 hours were required for 70% restitution of photosynthesis after intense drought-stress to minimum $\psi_w$ of −48 bar, and total recovery was complete in a week. Significantly, transpiration at a given $\psi_w$ was 15% lower after drought than during the drying phase. Such resilience with respect to $\psi_w$, combined with a photosynthetic optimum in excess of 40 C, makes Mitchell grass ideally suited to summer growth under the sporadic rainfall that typifies its native habitat.

Nonstomatal inhibition of photosynthetic activity, subsequent to drought, is frequently observed as a "hysteresis loop" when net photosynthesis is plotted as a function of $r_s$ over the full course of a stress-recovery cycle. Drought-resistant plants, however, are less prone to such secondary inhibition. This contrast is self-evident in Hinkley's (1973) comparison between tomato and black locust (Figure 3). Black locust *(Robinia pseudoacacia)* is a shade intolerant leguminous perennial which is frost and drought resistant, whereas tomato is susceptible to either form of stress. Experimental plants growing in vermiculite were desiccated to the permanent wilting point then rehydrated. The threshold $\psi_w$ for stomatal closure was −7 bar in tomato compared to −11 bar in black locust. Photosynthetic differences were even more definitive. While dehydration/rehydration curves for black locust were identical, for tomato leaves they differed significantly between the two phases. The discontinuity in photosynthetic rate of tomato leaves in the closing stages of dehydration coincided with appearance of necrotic patches. Although tomato leaves had a greater capacity for gas exchange at high $\psi_w$, their photosynthetic apparatus was less resilient. Black locust leaves had withstood $\psi_w$ down to −24 bars without aftereffect, whereas tomato was adversely affected to −16 bars. Photosynthetic endurance of low $\psi_w$ by

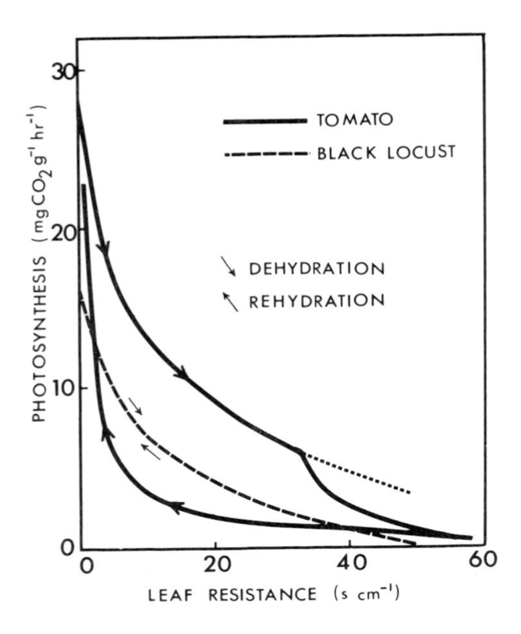


**Figure 3. Photosynthetic response to dehydration and rehydration in drought-sensitive tomato and drought-resistant black locust. (Adapted from Hinkley, 1973).**

black locust, combined with immediate recovery upon rehydration, was a feature of this plant's drought resistance.

## SUMMARY AND SUGGESTIONS FOR FUTURE RESEARCH

In photosynthetic terms, drought resistance has been equated with a crop plant's ability to maintain positive carbon balance under desiccating conditions. Synthesis, under stress, was achieved via a number of devices that ranged from the stabilizing influences of osmotic adjustment on stomatal physiology to structural and physiological modifications which decoupled the leaf's photosynthetic apparatus from adverse microclimate variations.

Given the multifaceted nature of drought resistance and the physiological plasticity of drought resistant plants (J. R. Wilson et al., 1980), future research needs to be extended well beyond soft laboratory-grown material. Hardened field-plants now are known to show drought adaptation at all levels of organization from photochemical events in relation to $\psi_w$ (Mooney et al., 1978) up to stomatal and growth response as a function of plant moisture status (Begg and Turner, 1976). Specific areas that might require further attention include:

(a) Naturally occuring substances which may be associated with early stomatal closure during drought, and subsequent nonstomatal inhibition of photosynthesis when moisture stress is relieved should be identified. Sesquiterpenoids such as abscisic acid, or other stress-related inhibitors, could hold additional relevance for drought resistance.

(b) The effects of $\psi_w$ on $r'_m$ in drought-resistant material exposed to a progressive, rather than sudden, reduction in $\psi_w$ should be resolved. The consequences of osmotic adjustment, or other forms of adaptation to moisture stress, then might become apparent. Rawson et al. (1978) have already made a significant contribution to this context.

(c) The interactions of intercellular $CO_2$ concentrations and endogenous regulators in modulating the $\psi_w$ threshold for stomatal response should be investigated. Fine control over gas exchange at low $r_s$ appears crucial in maximizing water use efficiency during early drought (Cowan and Farquhar, 1977).

(d) The "productivity-cost" in undertaking osmotic adjustment and other forms of photosynthetic adaptation to moisture stress should be determined in closely planted summer row crops. The possibility that such a cost might exist, at least for salinized plants, was recognized almost 20 years ago by Bernstein (1961). However, progress in quantifying such effects has been slow, especially where osmotic adaptation occurs in response to lowered soil matric potential. There is also a need to evaluate the value for productivity or survival of the extra carbon assimilation due to osmotic adjustment (J. R. Wilson et al., 1980). A beneficial trade-off may exist if frugal watering, and hence hardening in the vegetative phase, confers a subsequent advantage during later developmental and grainfilling stages when conditions of strong insolation and evaporative demand commonly prevail. Such a regime would be consistent with water use patterns in crops such as sorghum where water conservation through stomatal control is manifest to some extent prior to reproductive development. Plants appeared more prodigal during grain filling as photosynthetic activity took precedence over water conservation (Ackerson et al., 1980).

(e) Other aspects of osmotic adjustment require further study in a wide range of crop species and cultivars. These include evaluation of its role: in countering water stress induced reduction in fertility by minimizing associated increases in ABA (Morgan, 1980); in displacing relationships between stomatal conductance and leaf water potential (Jones and Rawson, 1979); and in facilitating greater root growth and exploration of the soil in adverse physical environments (Hsiao and Acevedo, 1974).

Given the multifacted nature of drought resistance in crop plants, and the diversity of environmental situations already utilized, plant breeders have an onerous task in adapting crops to either arid or humid environments.

Recognition by physiologists and biochemists of the various mechansims which confer some measure of drought resistance have already proved useful by at least reducing the plant breeders' earlier dependence on pure empiricism.

## NOTES

P. E. Kriedemann and H. D. Barrs, CSIRO Division of Irrigation Research, Griffith, New South Wales, 2680, Australia.

## LITERATURE CITED

Ackerson, R. C., D. R. Krieg, T. D. Miller, and R. E. Zartman. 1977a. Water relations of field grown cotton and sorghum: Temporal and diurnal changes in leaf water, osmotic and turgor potentials. Crop Sci. 17:76-80.

Ackerson, R. C., D. R. Krieg, T. D. Miller, and R. G. Stevens. 1977b. Water relations and physiological activity of potatoes. J. Amer. Soc. Hort. Sci. 102:572-575.

Ackerson, R. C., D. R. Krieg, and F. J. M. Sung, 1980. Leaf conductance and osmoregulation of field-grown sorghum canopies. Crop Sci. 20:10-14.

Adams, M. S., B. R. Strain, and I. P. Ting. 1967. Photosynthesis in chlorophyllous stem tissue and leaves of *Cercidium floridum:* Accumulation and distribution of $^{14}C$ from $^{14}CO_2$. Plant Physiol. 42:1797-1799.

Allaway, W. G., and T. A. Mansfield. 1970. Experiments and observations on the after affect of wilting on stomata of *Rumex sanguineus.* Can. J. Bot. 48:513-521.

Allen, L. H., C. S. Yocum, and E. R. Lemon. 1964. Photosynthesis under field conditions. VII. Radiant energy exchanges within a corn crop canopy and implications in water use efficiency. Agron. J. 56:253-259.

Barrs, H. D. 1968. Effect of cyclic variations in gas exchange under constant environmental conditions on the ratio of transpiration to net photosynthesis. Physiol. Plant. 21:918-929.

Beadle, C. L. K. R. Stevenson, H. H. Neumann, G. W. Thurtell, and K. M. King. 1973. Diffusive resistance, transpiration, and photosynthesis in single leaves of corn and sorghum in relation to leaf water potential. Can. J. Plant Sci. 53:537-544.

Beadle, C. L., and P. G. Jarvis. 1977. The effect of short water status on some photosynthetic partial processes in Sitka spruce. Physiol. Plant. 41:7-13.

Beardsell, M. F., and D. Cohen. 1975. Relationships between leaf water status, abscisic acid levels, and stomatal resistance in maize and sorghum. Plant Physiol. 56:207-212.

Begg, J. E., and B. W. R. Torssell. 1974. Diaphotonastic and parahelionastic leaf movements in *Stylosanthes humilis* H. B. K. (Townsville Stylo), p. 277-283. In R. L. Bieleski, A. R. Ferguson and M. M. Cresswell (eds.) Mechanisms of regulation of plant growth. Bulletin 12, Royal Soc. N. Z., Wellington.

Begg, J. E., and N. C. Turner. 1976. Crop water deficits. Adv. Agron. 28:161-217.

Bernstein, L. 1961. Osmotic adjustments of plants to saline media. I. Steady state. Amer. J. Bot. 48:909-918.

Bjorkman, O., R. W. Pearcy, A. T. Harrison, and H. Mooney. 1972. Photosynthetic adaptation to high temperatures: A field study in Death Valley, California. Science 175:785-789.

Boussiba, S., and A. E. Richmond. 1976. Abscisic acid and the aftereffect of stress in tobacco plants. Planta 129:217-219.

Bower, C. A., and Y. N. Tamimi. 1979. Root adjustments associated with salt tolerance in small grains. Agron. J. 71:690-693.

Boyer, J. S. 1970. Differing sensitivity of photosynthesis to low leaf water potentials in corn and soybeans. Plant Physiol. 46:236-239.

Boyer, J. S. 1971a. Recovery of photosynthesis in sunflower after a period of low leaf water potential. Plant Physiol. 47:816-820.

Boyer, J. S. 1971b. Nonstomatal inhibition of photosynthesis in sunflower at low leaf water potentials and high light intensities. Plant Physiol. 48:532-536.

Boyer, J. S. 1976a. Photosynthesis at low water potentials. Phil. Trans. Royal Soc. Lond. Bull. 273:501-512.

Boyer, J. S. 1976b. Water deficits and photosynthesis. p. 153-190. In T. T. Kozlowski (ed.) Water deficits and plant growth. Vol. IV. Soil water measurement, plant responses, and breeding for drought resistance. Academic Press, New York.

Brix, H. 1962. The effect of water stress on the rates of photosynthesis and respiration in tomato plants and loblolly pine seedlings. Physiol. Plant. 15:10-20.

Bunce, J. A. 1977. Nonstomatal inhibition of photosynthesis at low water potentials in intact leaves of species from a variety of habitats. Plant Physiol. 59:348-350.

Burrows, F. J., and F. L. Milthorpe. 1976. Stomatal conductance in the control of gas exchange. p. 103-152. In T. T. Kozlowski (ed.) Water deficits and plant growth. Vol. IV. Soil water measurement, plant responses, and breeding for drought resistance. Academic Press, New York.

Chatterton, N. J., W. W. Hanna, J. B. Powell, and D. R. Lee. 1975. Photosynthesis and transpiration of bloom and bloomless sorghum. Can. J. Plant Sci. 55:641-643.

Cowan, I. R., and G. D. Farquhar. 1977. Stomatal functions in relation to leaf metabolism and environment. p. 471-505. In D. H. Jennings (ed.) Integration of activity in the higher plant. Vol. 31. Symposia of the Society of Experimental Biology, Camb. Univ. Press, Cambridge.

Craig, S., and D. J. Goodchild. 1977. Leaf ultrastructure in *Triodia irritans:* a $C_4$ grass possessing an unusual arrangement of photosynthetic tissues. Aust. J. Bot. 25: 277-290.

Doley, D., and N. B. A. Trivett. 1974. Effects of low water potentials on transpiration and photosynthesis in Mitchell grass *(Astrebla lappacea).* Aust. J. Plant Physiol. 1:539-550.

Dube, P. A., K. R. Stevenson, and G. W. Thurtell. 1974. Comparison between two inbred corn lines for diffusive resistance, photosynthesis and transpiration as a function of leaf water potential. Can. J. Plant Sci. 54:765-770.

Duniway, J. M. 1971. Water relations of *Fusarium* wilt in tomato. Physiol. Plant Pathol. 1:537-546.

Ehleringer, J. 1976. Leaf absorptance and photosynthesis as affected by pubescence in the genus *Encelia.* Carnegie Inst. Wash. Year Book 75:413-418.

Ehleringer, J. 1977. Adaptive value of leaf hairs in *Encelia farinosa.* Carnegie Inst. Wash. Year Book 76:367-369.

Evans, L. T., J. Bingham, P. Jackson, and J. Sutherland. 1972. Effect of awns and drought on the supply of photosynthate and its distribution within wheat ears. Ann. Appl. Biol. 70:67-76.

Fischer, R. A., and R. M. Hagan. 1965. Plant water relations, irrigation management and crop yield. Exp. Agric. 1:161-177.

Fischer, R. A., T. C. Hsiao, and R. M. Hagan. 1970. After-effect of water stress on stomatal opening potential. J. Exp. Bot. 21:371-385.

Frank, A. B., and R. E. Barker. 1976. Rates of photosynthesis and transpiration and diffusive resistance of six grasses grown under controlled conditions. Agron. J. 68: 487-490.

Frank, A. B., J. F. Power, and W. O. Willis. 1973. Effect of temperature and plant water stress on photosynthesis, diffusion resistance, and leaf water potential in spring wheat. Agron. J. 65:777-780.

Gaastra, P. 1959. Photosynthesis of crop plants as influenced by light, $CO_2$, temperature and stomatal diffusion resistances. Meded. Landbhoogesch. Wageningen 13:1-68.

Gale, J., A. Poljakoff-Mayber, and I. Kahane. 1967. The gas diffusion porometer technique and its application to the measurement of leaf mesophyll resistance. Israel J. Bot. 16:187-204.

Gates, C. T. 1955a. The response of the young tomato plant to a brief period of water shortage. I. The whole plant and its principal parts. Aust. J. Biol. Sci. 8:196-214.

Gates, C. T. 1955b. The response of the young tomato plant to a brief period of water shortage. II. The individual leaves. Aust. J. Biol. Sci. 8:215-230.

Gates, D. M., R. Alderfer, and E. Taylor. 1968. Leaf temperatures of desert plants. Science 159:994-995.

Grace, J., D. C. Malcolm, and I. K. Bradbury. 1975. The effect of wind and humidity on leaf diffusive resistance in Sitka spruce seedlings. J. Appl. Ecol. 12:931-940.

Greacen, E. L., and J. S. Oh. 1972. Physics of root growth. Nature. 235:24-25.

Hadfield, W. 1968. Leaf temperature, leaf pose and productivity of the tea bush. Nature 219:282-284.

Hall, A. E., and M. R. Kaufmann. 1975. Stomatal response to environment with *Sesamum indicum* L. Plant Physiol. 55:455-459.

Hallam, N. D., and T. C. Chambers. 1970. The leaf waxes of the genus *Eucalyptus* L' Heritier. Aust. J. Bot. 18:335-386.

Hari, P., O. Luukkanen, P. Pelkonen, and H. Smolander. 1975. Comparisons between photosynthesis and transpiration in birch. Physiol. Plant. 33:13-17.

Hellmuth, E. O. 1969. Eco-physiological studies on plants in arid and semi-arid regions of Western Australia. II. Field physiology of *Acacia craspedocarpa* F. Muell. J. Ecol. 57:613-634.

Hinckley, T. M. 1973. Responses of black locust and tomato plants after water stress. Hort. Sci. 8:405-407.

Hsiao, T. C. 1973. Plant responses to water stress. Ann. Rev. Plant Physiol. 24:519-570.

Hsiao, T. C., and E. Acevedo. 1974. Plant responses to water deficits, water-use efficiency, and drought resistance. Agric. Meterol. 14:59-84.

Hsiao, T. C., E. Acevedo, E. Fereres, and D. W. Henderson. 1976. Stress metabolism: Water stress, growth, and osmotic adjustment. Phil. Trans. Royal Soc. Lond. Bull. 273:479-500.

Jeffree, C. E., R. P. C. Johnson, and P. G. Jarvis. 1971. Epicuticular wax in the stomatal antechamber of Sitka spruce and its effects on the diffusion of water vapour and carbon dioxide. Planta. 98:1-10.

Johnson, D. A., and R. W. Brown. 1977. Psychrometric analysis of turgor pressure response: A possible technique for evaluating plant water stress resistance. Crop Sci. 17:507-510.

Jones, M. M., and H. M. Rawson. 1979. Influence of rate of development of leaf water deficits upon photosynthesis, leaf conductance, water use efficiency and osmotic potential in grain sorghum. Physiol. Plant. 45:103-111.

Kanemasu, E. T., and C. B. Tanner. 1969. Stomatal diffusion resistance of snap beans. I. Influence of leaf-water potential. Plant Physiol. 44:1547-1552.

Kaul, R. 1974. Potential net photosynthesis in flag leaves of severely drought-stressed wheat cultivars and its relationship to grain yield. Can. J. Plant Sci. 54:811-815.

Keck, R. W., and J. S. Boyer. 1974. Chloroplast response to low leaf water potentials. III. Differing inhibition of electron transport and photophosphorylation. Plant Physiol. 53:474-479.

Khairi, M. M. A., and A. E. Hall. 1976. Temperature and humidity effects on net photosynthesis and transpiration of citrus. Physiol. Plant. 36:29-34.

Kozlowski, T. T. 1964. Water metabolism in plants. Harper Row Co., New York.

Kriedemann, P. E. 1971. Photosynthesis and transpiration as a function of gaseous diffusive resistances in orange leaves. Physiol. Plant. 24:218-225.

Kriedemann, P. E., B. R. Loveys, and H. M. van Dijk. 1980. Photosynthetic inhibitors from *Capsicum annum* (L.)—Extraction and bioassay. Aust. J. Plant Physiol. 7: 629-633.

Kriedemann, P. E., T. F. Neales, and D. H. Ashton. 1964. Photosynthesis in relation to leaf orientation and light interception. Aust. J. Biol. Sci. 17:591-600.

Kriedemann, P. E., B. R. Loveys, G. L. Fuller, and A. C. Leopold. 1972. Abscisic acid and stomatal regulation. Plant Physiol. 49:842-847.

Kriedemann, P. E., B. R. Loveys, and W. J. S. Downton. 1975. Internal control of stomatal physiology and photosynthesis. II. Photosynthetic responses to phaseic acid. Aust. J. Plant Physiol. 2:553-567.

Kriedemann, P. E., B. R. Loveys, J. V. Possingham, and M. Satoh. 1976. Sink effects on stomatal physiology and photosynthesis. p. 401-414. In I. F. Wardlaw and J. B. Passioura (eds.) Transport and transfer processes in plants. Academic Press, New York.

Larque-Saavedra, A., and R. L. Wain. 1974. Abscisic acid levels in relation to drought tolerance in varieties of *Zea mays* l. Nature 251:716-717.

Lawlor, D. W. 1976. Water stress induced changes in photosynthesis, photorespiration, respiration and $CO_2$ compensation concentration of wheat. Photosynthetica 10: 378-387.

Lawlor, D. W. 1979. Effects of water and heat stress on carbon metabolism of plants with $C_3$ and $C_4$ photosynthesis. p. 303-326. In H. Mussell and R. C. Staples (eds.) Stress physiology in crop plants. John Wiley and Sons, New York.

Lemon, E. R. 1966. Impact of the atmospheric environment on the integument of the organism. p. 57-69. In Proc. Fourth Int. Biometeorology Congress, Rutgers University.

Levitt, J., and R. BenZaken. 1975. Effects of small water stresses on cell turgor and intercellular space. Physiol. Plant. 34:273-279.

Lewis, D. A., and P. S. Nobel. 1977. Thermal energy exchange model and water loss of a barrel cactus, *Ferocactus acanthodes.* Plant Physiol. 60:609-616.

Lopushinsky, W. 1969. Stomatal closure in conifer seedlings in response to leaf moisture stress. Bot. Gaz. 130:258-263.

Ludlow, M. M. 1976. Ecophysiology of $C_4$ grasses. p. 364-386. In O. L. Lange, L. Kappen, and E. D. Schulze. Ecological studies. Vol. 18. Water and plant life. Springer-Verlag, Berlin.

McWilliam, J. R., and K. Mison. 1974. Significance of the $C_4$ pathway in *Triodia irritans* (Spinifex), a grass adapted to arid environments. Aust. J. Plant Physiol. 1:171-175.

Meidner, H. 1967. Further observations on the minimum intercellular space carbon-dioxide concentration ($\Gamma$) of maize leaves and the postulated roles of "photorespiration" and glycolate metabolism. J. Exp. Bot. 18:177-185.

Meidner, H., and T. A. Mansfield. 1968. Physiology of stomata, McGraw-Hill, New York.

Mooney, H. A., J. Ehleringer, and J. A. Berry. 1976. High photosynthetic capacity of a winter annual in Death Valley. Science 194:322-324.

Mooney, H. A., O. Bjorkman, and G. J. Collatz. 1978. Photosynthetic acclimation to temperature in the desert shrub, *Larrea divaricata.* Plant Physiol. 61:406-410.

Morrow, P. A. 1971. The ecophysiology of drought adaptation of two mediterranean climate evergreens. Ph.D. Thesis. Stanford University. Univ. Microfilms, Ann Arbor, Mich.

Morrow, P. A., and H. A. Mooney. 1974. Drought adaptation in two Californian evergreen sclerophylls. Oecologia 15:205-222.

Munns, R., C. J. Brady, and E. W. R. Barlow. 1979. Solute accumulation in the apex and leaves of wheat during water stress. Aust. J. Plant Physiol. 6:379-389.

Ng, T. T., J. R. Wilson, and M. M. Ludlow. 1975. Influence of water stress on water relations and growth of a tropical ($C_4$) grass, *Panicum maximum* var. trichoglume. Aust. J. Plant Physiol. 2:581-595.

Nobel, P. S. 1977. Internal leaf area and cellular $CO_2$ resistance: Photosynthetic implications of variations with growth conditions and plant species. Physiol. Plant. 40:137-144.

Oechel, W. C., B. R. Strain, and W. R. Odening. 1972. Photosynthetic rates of a desert shrub, *Larrea divaricata* Cav., under field conditions. Photosynthetica 6:183-188.

O'Toole, J. C., J. L. Ozbun, and D. H. Wallace. 1977. Photosynthetic response to water stress in *Phaseolus vulgaris.* Physiol. Plant. 40:111-114.

Pasternak, D., and G. L. Wilson. 1974. Differing effects of water deficit on net photosynthesis of intact and excised sorghum leaves. New Phytol. 73:847-850.

Pasternak, D., and G. L. Wilson. 1976. Photosynthesis and transpiration in the heads of droughted grain sorghum. Aust. J. Exp. Agric. Anim. Husbandry 16:272-275.

Pearcy, R. W., J. A. Berry, and B. Bartholomew. 1972. Field measurements of the gas exchange capacities of *Phragmites communis* under summer conditions in Death Valley. Carnegie Inst. Wash. Year Book 71: 161-164.

Pisek, A., and E. Winkler. 1956. Wassersattigungsdefizit, spaltenbewegung und photosynthese. Protoplasma 46:597-611.

Raschke, K. 1970. Stomatal response to pressure changes and interruptions in the water supply of detached leaves of *Zea mays* L. Plant Physiol. 45:415-423.

Raschke, K., M. Pierce, and C. C. Popiela. 1976. Abscisic acid content and stomatal sensitivity to $CO_2$ in leaves of *Xanthium strumarium* L. after pretreatments in warm and cold growth chambers. Plant Physiol. 57:115-121.

Rawson, H. M. 1979. Vertical wilting and photosynthesis, transpiration, and water use efficiency of sunflower leaves. Aust. J. Plant Physiol. 6:109-120.

Rawson, H. M., A. K. Bagga, and P. M. Bremner. 1977. Aspects of adaptation by wheat and barley to soil moisture deficits. Aust. J. Plant Physiol. 4:389-401.

Rawson, H. M., N. C. Turner, and J. E. Begg. 1978. Agronomic and physiological response of soybean and sorghum crops to water deficits. IV. Photosynthesis, transpiration and water use efficiency of leaves. Aust. J. Plant Physiol. 5:195-209.

Redshaw, A. J., and H. Meidner. 1972. Effects of water stress on the resistance to uptake of carbon dioxide in tobacco. J. Exp. Bot. 23:229-240.

Regehr, D. L. F. A. Bazzaz, and W. R. Boggess. 1975. Photosynthesis, transpiration and leaf conductance of *Populus deltoides* in relation to flooding and droughts. Photosynthetica 9:52-61.

Russell, R. S., and M. J. Goss. 1974. Physical aspects of soil fertility—The response of roots to mechanical impedance. Netherlands J. Agric. Sci. 22:305-318.

Schultze, E.-D., O. L. Lange, U. Buschbom, L. Kappen, and M. Evenari. 1972. Stomatal responses to changes in humidity in plants growing in the desert. Planta 108: 259-270.

Schulze, E.-D., O. L. Lange, L. Kappen, U. Buschbom, and M. Evenari. 1973. Stomatal responses to changes in temperature at increasing water stress. Planta. 110:29-42.

Schultze, E.-D., O. L. Lange, M. Evenari, L. Kappen, and U. Buschbom. 1974. The role of air humidity and leaf temperature in controlling stomatal resistance of *Prunus armeniaca* L. under desert conditions. I. A simulation of the daily course of stomatal resistance. Oecologia 17:159-170.

Sharkey, T. D., and K. Raschke. 1980. Effects of phaseic acid and dihydrophaseic acid on stomata and the photosynthetic apparatus. Plant Physiol. 65:291-297.

Shearman, L. L., J. D. Eastin, C. Y. Sullivan, and E. J. Kinbacher. 1972. Carbon dioxide exchange in water stressed sorghum. Crop Sci. 12:406-409.

Sheriff, D. W., and P. E. Kaye. 1977. Responses of diffusive conductance to humidity in a drought avoiding and a drought resistant (in terms of stomatal response) legume. Ann. Bot. 41:653-655.

Slatyer, R. O. 1973. The effect of internal water status on plant growth, development and yield. p. 171-191. In R. O. Slatyer (ed.) Plant response to climatic factors. Proceedings of the Uppsala Symposium, UNESCO, Paris.

Tal, M. 1966. Abnormal stomatal behavior in wilty mutants of tomato. Plant Physiol. 41:1387-1391.

Tal, M., and D. Imber. 1970. Abnormal stomatal behavior and hormonal imbalance in *flacca,* a wilty mutant of tomato. II. Auxin and abscisic acid-like activity. Plant Physiol. 46:373-376.

Troughton, J. H. 1969. Plant water status and carbon dioxide exchange of cotton leaves. Aust. J. Biol. Sci. 22:289-302.

Troughton, J. H., and R. O. Slatyer. 1969. Plant water status, leaf temperature, and the calculated mesophyll resistance to carbon dioxide of cotton leaves. Aust. J. Biol. Sci. 22:815-827.

Troughton, J. H., S. E. Comacho-B., and A. E. Hall. 1974. Transpiration rate, plant water status and resistance to water flow in *Tidestromia oblongifolia.* Carnegie Inst. Wash. Year Book 73:830-835.

Tunstall, B. R., and D. J. Connor. 1975. Internal water balance of brigalow *(Acacia harpophylla* F. Muell.) under natural conditions. Aust. J. Plant Physiol. 2:489-499.

Vogel, S. 1970. Convective cooling at low air speeds and the shapes of broad leaves. J. Exp. Bot. 21:91-101.

Wainwright, C. M. 1977. Sun-tracking and related leaf movements in a desert lupine *(Lupinus arizonicus).* Amer. J. Bot. 64:1032-1041.

Watson, D. J. 1952. Physiological basis for variation in yield. Adv. Agron. 4:101-145.

Watts, W. R. 1974. Leaf extension in *Zea mays.* III. Field measurements of leaf extension in response to temperature and leaf water potential. J. Exp. Bot. 25:1085-1096.

Went, F. W. 1959. The physiology of photosynthesis in higher plants. Preslia 30:225-249.

Wilson, J. R., M. M. Ludlow, M. J. Fisher, and E.-D. Schulze. 1980. Adaptation to water stress of the leaf water relations of four tropical forage species. Aust. J. Plant Physiol. 7:207-220.

Wilson, D. F., C. H. M. van Bavel, and K. J. McCree. 1980. Carbon balance of water-deficient grain sorghum plants. Crop Sci. 20:153-159.

Wuenscher, J. E., and T. T. Kozlowski. 1971a. The response of transpiration resistance to leaf temperature as a desiccation resistance mechanism in tree seedlings. Physiol. Plant. 24:254-259.

Wuenscher, J. E., and T. T. Kozlowski. 1971b. Relationship of gas exchange resistance to tree-seedling ecology. Ecology 52:1016-1023.

14

# USE OF SIMULATION AS A TOOL IN CROP MANAGEMENT STRATEGIES FOR STRESS AVOIDANCE

R. B. Curry and A. Eshel

Simulation of plant growth is one strategy available to crop scientists in the pursuit of increased productivity through improved stress avoidance and adaptation of the plants. In order to describe how this relatively new tool can be used, it is necessary to discuss the nature of the crop simulation and explain how it was used in application to a particular crop. Some suggestions for application of crop simulation to stress studies will be presented.

## DEFINITION OF SIMULATION IN THIS CONTEXT

Simulation of a system is a representation of its behavior by a computer program, and may be defined as the act of minicry. Using dynamic modeling as a technique of solving problems by following changes over time is another definition of system simulation.

In simulation, a system is described by a set of mathematical equations. The status of the system is defined at every instant by a complement of state variables and its environment by a set of driving variables. Simultaneous solution of this set of equations is carried out by the computer at every time step of the simulation process.

In recent years, new attempts have been made to extend the scope of simulation to areas which previously were considered too complex and difficult to be simulated. Modelers involved in agricultural research long have known that they deal in their everyday work with these complex systems which exhibit varied behavior governed by external stimuli of the natural environment and the internal controls of plant physiology. Therefore, agricultural researchers recently have become interested in systems science, a discipline devoted to the study of such complex systems (Witz, 1973). They found out rather early that our understanding of the plant system leaves a lot to be desired, and it is impossible to fully describe the plant and its interaction with the environment in exact mathematical terms.

It is not surprising, under these circumstances, that development of crop simulation has been slow. In addition to the complex natural systems, varying environments make management schemes difficult, and lack of information concerning interactions of different parts of the system create problems in structuring the model. Scientists are faced with the reality that agricultural systems, especially plant systems, are inherently more complicated than standard statistical methods would lead people to believe and that they therefore present new problems. These systems involve highly interactive subsystems, which require description by non-linear functions with changing thresholds and other complex response functions.

Another difficulty arises from the complexity of the biosystem which includes a large number of subsystems. Accurate modeling of these numerous subsystems is impractical.

Developing a simulator is considered to be an iterative process. Several stages are involved in the modeling sequence. These include: (1) problem definition, (2) statement of initial assumptions and model structure, (3) construction of preliminary models and assessment of assumptions, (4) determination and availability of input data requirements, (5) determination of output needs, (6) precise specifications of final model, including programming language, and model plans, and (7) re-examination of previous steps.

Perhaps, researchers have been over-optimistic about the future of agricultural system models. Passioura (1973) points out that simulation of complex physical systems, such as rocket technology, has been highly successful. Even though it is tempting to extrapolate this success to agricultural systems, Passioura cautions that while all the underlying physics on which the design of a rocket is based has been thoroughly worked out, the same cannot be said for crop systems. Even when the individual processes are known, difficulties arise because of lack of understanding of interactions among them.

## VERIFICATION AND VALIDATION

The terms "verification" (calibration) and "validation" of a model often are used synonymously in the literature, but a useful distinction can be drawn between the two. In a literal context "to verify" means "to test the truth or correctness of." Verification is then a test of the logic of the model and the reasonableness of parameter values.

In contrast to verification, validation is concerned with the usefulness in comparison with real world situations, rather than the truthfulness, of a model. In practice, validation is likely to be based upon statistical tests where possible, although problems of the appropriate test and the interpretation of results will be even more complex than for verification.

## REVIEW OF THE APPLICATION OF SIMULATION TO PLANT SYSTEMS

Since the time of the earliest application of simulation to crop growth systems, the correction approach to the problem and the selection of language and simulation technique have been topics of discussion. Two examples of the philosophical approach to the problem can be pointed out from the literature. DeWit (1970) discussed the application of dynamic concepts in the study of biology, and Baker and Curry (1976) presented a philosophical view of the structure of agricultural simulators. DeWit suggested that the level of complexity must provide an explanation of physiological processes and the code written so that it is easily understood and transferable. Baker and Curry stressed the principle of a systematic analyses of the crop system and the fact that simulation is a research and management tool.

The early simulation studies were done using FORTRAN and CSMP as languages. Use of CSMP was encouraged by deWit and Goudriaan (1974). Some simulation studies have been done using GASP IV (Miles et al., 1979). Most of the current work in crop simulation is being done using FORTRAN IV. The one reason for this return to FORTRAN has been transferability.

Simulation has been used to study individual plant processes, such as photosynthesis, stomatal action, and water uptake by roots. Simulation of photosynthesis involves simulation of the energy received, absorbed and reflected, the temperature of system based on an energy balance, the storage and transport of the photosynthetic products, as well as the biochemical process itself. This submodel may be used as the driving force in an overall plant simulator. Stomatal action also can be simulated as a series of the interrelated processes of water transport to and from the epidermis, water transport to and from the guard cells, ion transport to and from the guard cells, and biochemical control processes occurring inside the guard cells. This simulator may serve as a controlling submodel for exchange of gases and water between plant leaves and the atmosphere. Simulation of the water uptake by roots involves the simulation of movement of water in the soil, the uptake by the root, and the transport through the plant system to the leaves. Since the supply of water to the leaf has an effect on stomatal action, one can see that all three of these examples exhibit interactions and feedback. These form part of the control mechanism for behavior of the total plant system.

Simulation also has been used to mimic the growth and development of the whole plant. Plant models of varying complexities have been developed. Early efforts were those of deWit (1970) on general crop modeling, Duncan (1971) on cotton and corn, and Stapleton (1973) on cotton. Later, other simulators have been developed for cotton, corn alfalfa, soybeans, peanuts, sugarbeets, sorghum and wheat, and other crops. Baker (1980) listed a total of 23 different models as examples of the current status of crop growth simulation.

A few examples of current simulation research in crop systems have been summarized from published research. They are described below.

RHIZOS was developed by J. R. Lambert at Clemson University in co-operation with several other people (Lambert et al., 1976). This simulator of the dynamic processes in the Rhizosphere includes physical, biological, and microbiological processes which occur in the root zone of the crop. This simulator has been designed to form a module in a row crop simulator. It has been implemented in a whole plant simulator of cotton and is now being modified for use in a soybean model.

GOSSYM, developed by D. N. Baker and others in Mississippi (Baker, et al., 1976), is a dynamic simulator of materials balance for cotton growth and development. In the plant model, nitrogen, which is taken up from the soil via the transpiration stream, and labile carbohydrates, which are produced by the photosynthesis process, are represented by a system of pools. These materials are distributed to the leaves, stems, fruit and roots. Losses as results of insect damage and the natural plant processes of senescence and abscission in response to physiological stress are taken into account. The model also depicts the redistribution of nitrogen within the plant. The initiation of organs on the plant occurs as a series of discrete events. Rates of development of these organs depend on temperature and the physiological status of the plant. GOSSYM has been validated and is being used to study how possible changes in the physiology and structure of the cotton plant affect growth and yield.

SIMED was developed at Purdue University by Holt et al. (1975) to simulate the growth of alfalfa as a crop model in an overall simulator of the alfalfa pest management system. SIMED has been developed around the concept of a balance between carbon production and consumption. Whenever the amounts of carbon entering the system as carbon dioxide and retained in the system are known, the total dry matter accumulated is estimated by assuming that the proportions of the various elements in carbohydrates and proteins tend to be constant. SIMED considers the alfalfa field as a whole and not as a group of individual plants; therefore, all calculations are made on the basis of land area. Rates of material flows into and from plant compartments are depicted by the physical environment and the physiological state of the crop. Verification and validation of SIMED were done with data sets from a large pool of research data on alfalfa at Purdue. A major use of SIMED outside the research area has been as part of a pest management model. SIMED has been combined with on-line weather data to provide real-time guidance of scheduling control pest management practices in Indiana.

SOYMOD/OARDC was developed at the Ohio Agricultural Research and Development Center to simulate the growth and development of the soybean plant (Curry et al., 1975; Meyer et al., 1979). Since the principle experience of the authors with simulation has been with SOYMOD, it will be described in greater detail and used as a vehicle to suggest how simulation can be used as a tool in analysis of various stress avoidance strategies.

## DESCRIPTION OF SOYMOD

SOYMOD/OARDC was developed to serve as a tool to further the understanding of the growth and development of the soybean plant and as an aid to research on improving the production and management of the soybean crop. The objectives of this project were to include as much of the known physiological processes as necessary to produce a simulator which responds to a simulated environment in the same way as field-grown soybeans respond to a real environment. This simulator then would be useful in answering questions regarding response of plants to particular modifications in their structure, physiology, or environment. Once the accuracy of the predicted responses has been checked in field experiments, the simulator would reduce greatly the amount of field work by enabling a rapid screening of various combinations of these modifications and pointing out the most promising ones.

SOYMOD/OARDC is composed of a system of dynamic partial-differential equations describing the mass and energy balance within the soybean plant. The equations describe the soybean as an open system with import, export, and internal control processes. The simulator employs dynamic simulation techniques at a detailed physiological level. One of the underlying hypotheses used in building the simulator was that the processes described are based mathematically on the currently most acceptable theory and the most logical hypotheses derived from available data. The equations which describe various physiological processes were organized into a computer program in such a way that improvements in the simulator could be made easily when new data or new theories become available. The processes simulated include photosynthesis, respiration, carbohydrate translocation and partitioning, nitrogen assimilation and partitioning, leaf senescence, and transpiration. Nodal formation, flowering, initiation of pod-fill, and leaf and petiole abscission are considered as discrete events.

The input data required for SOYMOD can be summarized into four categories: planting parameters, soil-water parameters, operational parameters, and climatic parameters. *Planting parameters* required as input are variety, emergence date, row spacing, and plant spacing in the row; *soil-water parameters* are soil water retention curve, soil bulk density, hydraulic conductivity, initial soil water content, and irrigation schedule; *operational parameters* are choice of output forms, interval between time of printout, special tests such as defoliation and depodding, and maximum number of days simulated; *climatic parameters* are daylength, daily maximum air temperature, daily minimum air temperature, daily dew point temperature, total daily solar radiation, daily rainfall, and total daily wind-run. All of these parameters can be specified by the user. The climatic parameters are supplied automatically from disc storage beginning with date of emergence. By changing any of these input parameters, its effect on the simulator can be tested. Also, the effects on the soybean crop of agronomical practices, changes in weather, and soil type can be studied.

Three categories of output data are generated by SOYMOD. *Log,* which is a chronology of dry matter accumulation and physiological events on a whole plant basis, includes dry matter for each plant part (leaf blades, stems plus petioles, roots, and fruits), plant height, maximum stem diameter, number of fruits, flowering and podfill events, leaf area, irrigation events, leaf abscission events, and seed yield. *Partitioning summary* provides once per day, for the specific days requested and on the basis of each node, leaf area; number of fruits; total dry matter and the nitrogen, available carbohydrate, storage starch, and structural carbohydrate components of dry matter for each plant part; rates of net photosynthesis, respiration, phloem loading, translocation, and carbohydrate storage; and growth rate for each plant part. *Short summary,* at the end of the season, gives location, date, variety, soil type, and planting configuration; total rainfall, total irrigated water, and total solar radiation and heat units from emergence to maturity; and a crop summary of number of nodes attained, maximum plant dry matter, total evapotranspiration from emergence to maturity, weight per 100 seeds, seed yield, total fruit dry matter per plant.

The water update system used in the present SOYMOD/OARDC version is rather elementary. It requires modification to include a root generating system which would provide a water uptake function based on demand, availability, and location of water in the soil profile. In order for these systems to work satisfactorily for the row crop, the soil profile in the model must be layered and grided to show root development and where water is added and removed. Work has continued in Ohio towards development and validation of a soil-root-submodel for SOYMOD/OARDC. This soil-root submodel is based on RHIZOS (Lambert et al., 1976), but has an hourly time step which is compatible with the canopy part of SOYMOD. Simulator sections for soil-temperature and soil-moisture regime have been developed to provide routines which depend on soil characteristics rather than on empirical functions that were derived from historic location-dependent data. In order to reduce computation time, a finite-difference integration scheme with a variable time-step has been used for the temperature simulator. Details of this submodel are given by Eshel and Curry (1980). The soil-root-submodel will be interconnected with the main simulator SOYMOD and validated in the near future.

The root growth function in the submodel, as well as water uptake activities by various parts of the root system, depend upon soil moisture status and soil temperature in every location within the root zone. The simulator thus will include several interconnected feed-back loops. One such feed-back loop is the effect of soil moisture on photosynthesis. Through its effect on photosynthesis, soil moisture affects plant growth, which in turn provides carbohydrates for future growth of roots. The growth of roots is necessary to supply the canopy with required water. Another example of a control function which is exerted upon the canopy by the root water supply is its effect on leaf growth. These leaves absorb radiation and thus reduce heating and evaporation from the soil surface. This, in turn, has an effect on root growth and water uptake.

## APPLICATIONS OF SOYMOD

Uses of SOYMOD/OARDC include both studies to further the understanding of the soybean plant and tests of the effect of change in plant structure or management. For example, one can study the effect of planting density. Table 1 shows the variation in yield with planting density for four years. Also shown is the effect of changing moisture regime from year-to-year. These results, although not totally validated, confirm the importance of high canopy density for high reproductive yields generally accepted by agronomists. However, the effect of soybean branching and stand compensation is not solved completely in SOYMOD.

In addition to or in combination with plant spacing tests, the simulator could be used to study the effect of plant structural changes such as leaf arrangement, leaf size, and light distribution. By testing different forms of canopy architecture and its effect on light absorption, photosynthesis, transpiration, and yields, the simulator would reduce the research time and cost necessary to select the best attributes for the field. This example indicates how engineering changes in plants and their effects on stress avoidance could be studied by simulation.

Similarly, other hypothetical modifications of plant morphology and physiology can be studied. An example is the root-growth-photosynthesis

**Table 1.** Simulated performance of 'Beeson' soybeans under various planting configurations and moisture regimes at Wooster, Ohio. (Adapted from Curry et al., 1980.)

| Year | Spacing Row by Plant | Density | Max. Leaf. Area | | Fruits Weight | | Fruit Number | | Seed Yield | |
|---|---|---|---|---|---|---|---|---|---|---|
| | | | Irrigated | Not Irrigated | Irrigated | Not Irrigated | Irrigated | Not Irrigated | Irrigated | Not Irrigated |
| | –cm– | –plant/m– | –cm²/plant– | | –g/plant– | | –no./plant– | | –kg/ha– | |
| 1975 | 13x13 | 62 | 1154 | 688 | 13.8 | 8.9 | 22 | 20 | 4422 | 2849 |
| | 76x 5 | 26 | 1730 | 1316 | 15.0 | 10.4 | 35 | 30 | 2016 | 1391 |
| | 91x15 | 7 | 2867 | 1941 | 45.4 | 33.0 | 71 | 61 | 1687 | 1230 |
| 1976 | 13x13 | 62 | 1135 | 1085 | 12.2 | 12.8 | 20 | 22 | 3938 | 4132 |
| | 76x 5 | 26 | 1699 | 1591 | 19.3 | 21.1 | 34 | 33 | 2587 | 2822 |
| | 91x15 | 7 | 2846 | 2405 | 47.2 | 32.5 | 67 | 57 | 1754 | 1210 |
| 1977 | 13x13 | 62 | 1127 | 1053 | 7.5 | 17.1 | 13 | 14 | 2399 | 5484 |
| | 76x 5 | 26 | 1702 | 1574 | 13.0 | 22.5 | 29 | 34 | 1734 | 3010 |
| | 91x15 | 7 | 2652 | 2224 | 40.6 | 28.1 | 70 | 58 | 1512 | 1048 |
| 1978 | 13x13 | 62 | 1162 | 902 | 15.2 | 11.7 | 12 | 20 | 4939 | 3750 |
| | 76x 5 | 26 | 1721 | 1338 | 16.0 | 12.1 | 28 | 34 | 2144 | 1626 |
| | 91x15 | 7 | 2552 | 1919 | 43.3 | 21.6 | 69 | 47 | 1613 | 806 |

interaction mentioned previously. Thus, various possible strategies of stress avoidance can be evaluated before any of them actually have to be tested in the field.

Finally, work can now be started on coupling SOYMOD/OARDC to an overall crop-management model, which would include such management practices as planting time, spraying, harvesting, etc. Such a model would be used by a farmer on almost any interactive basis to guide day-to-day or week-to-week decisions. In order for this to happen, real-time weather data would have to be available. This also will require modification of the simulator and possible simplification in order to keep the overall model in a practical size. These modifications and adaptations must be developed and validated before a simulator such as SOYMOD is used in crop-management strategy studies.

We have confined our example to simulation related to soybeans. The same approach could be used on other crops provided the simulator utilized is based on a set of sound physiological principles and developed to respond to dynamic environmental changes.

Irrigation is a technique of stress avoidance. Both Baker (1980) and Jones and Smajstria (1980) cite examples of how simulation can be used to guide irrigation management. Baker discusses an example for cotton in which a farmer would have to make a decision relative to balancing irrigation and increased pest control cost versus return from increased yield. The example by Jones and Smajstria involves the use of models to study the potential for increasing the yield of soybeans in the humid southeastern U.S. by irrigation. Both authors stress the need for a plant simulator which includes a dynamically simulated soil-water balance and a soil-root-plant system that is sensitive to internal plant water deficits.

## RESEARCH NEEDS TO MAKE SIMULATION A MORE USEFUL TOOL

In order that the simulation may be a useful tool in stress studies, further research is needed in several areas. There are four areas that come to immediate attention as priority research.

Improved understanding of the physiological response of plants to stress conditions is needed to improve the ability of the simulator for correctly mimicking the response of the plant to stress. How does stress effect the physiological process and how do these processes compensate for various levels of stress? In order for this information to be used in a dynamic simulator, these questions must be answered quantitatively whereby process response is defined by differential equations.

Another area requiring increased knowledge is a better understanding of the timing of phenological events in order to simulate the response of such events to stress with the correct phase relationship. For example, in most crop simulators the transfer from vegetative to reproductive state is timed by empirical relationships. In order to adequately study time-varying stress impacts on this event, the description of timing must be represented by a set of differential equations internally controlled by the interaction of other physiological processes in the simulator.

As mentioned earlier, an improved soil-root-environment subsystem is another necessity for stress studies. It is the soil-root system that initiates the changes in moisture stress levels; therefore, shoot-root relationships may determine the plant response to the aerial environment. A step in this direction has been made with soybeans (Eshel and Curry, 1980), but further improvements will be needed as new knowledge is developed.

In order to incorporate a plant model into an overall management model, further work should be done on companion models of tillage, management, and environment. These are under development and refinement. Extensions will be needed to make sure the overall simulator can mimic the tillage, management, and environmental interactions with the crop.

## CONCLUSIONS

The authors have tried to describe simulation as a tool for identifying strategies with the grest potential for manipulating crop management systems to provide stress avoidance.

As one looks over the work on simulators of crop growth, several points stand out. There is considerable similarity between the various crop simulators discussed. Probably this is because there are a lot of common physiological principles involved in each of the crops discussed, and many of the developers have been in communication with each other over the years. Also, there are many differences, types of approaches, languages used, depth of detail included, and amount of validation. The use of the model should guide the development.

An important point, which has been emphasized many times in discussion of simulation, is that simulation is just a tool to further knowledge of a system and is not an end in itself. The SOYMOD system is very close to being ready to be used in stress avoidance studies. Other physiologically based crop models also are adaptable to stress studies. Several research imperatives to make simulation a more useful tool in studies of stress avoidance management are an improved understanding of the physiological responses of plants to stress, improved knowledge of phenological timing, an improved soil-root-subsystem simulator, and further development of tillage management environmental models. As these areas are developed, simulation will be a useful tool in future research in this important field of crop production.

## NOTES

R. B. Curry, Department of Agricultural Engineering, Ohio Agricultural Research and Development Center, Wooster, Ohio 44691; Amram Eshel, Department of Botany, Tel Aviv University, Tel Aviv, Israel.

This paper was approved for publication as journal article number 192-80 of the Ohio Agricultural Research and Development Center, Wooster, Ohio 44691. The research was supported in part by PL89-106 grant number 801-15-51.

## LITERATURE CITED

Baker, D. N. 1980. Simulation for research and crop management p. 533-546. In F. T. Corbin (ed.) World soybean research conference II: Proceedings. Westview Press, Boulder, Colorado.

Baker, C. H., and R. B. Curry. 1976. Structure of agricultural simulators: A philosophical view. Agric. Systems 1:201-218.

Baker, D. N., J. R. Lambert, C. J. Phene, and J. M. McKinion. 1976. GOSSYM: A simulator of cotton crop dynamics. p. 100-133. In Computers applied to the management of large scale agricultural enterprises. Proceedings of the US/USSR Seminar, Moscow, Riga, Kishinev. National Science Foundation, Washington, D.C.

Curry, R. B., C. H. Baker, and J. G. Streeter. 1975. SOYMOD I: A dynamic simulator of soybean growth and development. Trans. Amer. Soc. Agric. Eng. 18:964-968.

Curry, R. B., G. E. Meyer, J. G. Streeter, and H. J. Mederski. 1980. Simulation of the vegetative and reproductive growth of soybeans. p. 557-569. in F. T. Corbin (ed.) World soybean research conference II: Proceedings. Westview Press, Boulder, Colorado.

Duncan, W. G. 1971. SIMBOT: A simulator of cotton growth and yield. p. 115-118. In C. Murphy (ed.) Proceedings of the workshop on tree growth dynamics and modeling. Duke University, Durham, North Carolina.

Eshel, A., and R. B. Curry. 1980. SMASH—soil moisture and soil heat simulator. Users' information guide. Ohio Agric. Res. Dev. Center, Agric. Eng. Series 103.

Holt, D. A., R. J. Bula, G. E. Miles, M. M. Schreiber, and R. M. Peart. 1975. Environmental physiology, modeling and simulation of alfalfa growth I. Conceptual development of SIMED. Indiana Agric. Exp. Stn. Bull. 907.

Jones, J. M., and A. G. Smajstria. 1980. Applications of modeling to irrigation management of soybeans. p. 571-599. In F. T. Corbin (ed.) World soybean research conference II: Proceedings. Westview Press, Boulder, Colorado.

Lambert, J. R., D. N. Baker, and C. J. Phene. 1976. Dynamic simulation of processes in the soil under growing row crops: RHIZOS. In Computers applied to the management of large scale agricultural enterprises. Proceedings of the US/USSR Seminar, Moscow, Riga, Kishinev. National Science Foundation, Washington, D.C.

Meyer, G. E., R. B. Curry, J. G. Streeter, and H. J. Mederski. 1979. SOYMOD/OARDC: A dynamic simulator of soybean growth, development and seed yield. Ohio Agric. Res. Dev. Center Res. Bull. 1113.

Miles, G. E., R. M. Peart, D. A. Holt, A. A. B. Pritsker, and R. J. Bula. 1979. CROPS: A gasp IV based crop simulation language. p. 177-185. In EPPO/IOBC conf. on system modeling in modern crop protection. EPPO Bull. 9(3). Paris.

Passioura, J. B. 1973. Sense and nonsense in simulation. J. Aust. Inst. Agric. Sci. 39: 181-183.

Stapleton, H. N., D. R. Buxton, F. L. Watson, D. J. Nolting, and D. N. Baker. 1973. Cotton: A computer simulation of cotton growth. Arizona Agric. Exp. Sta. Tech. Bull.

Wit, C. E. de. 1970. Dynamic concepts in biology. p. 17-23. In Prediction and measurement of photosynthetic productivity. Proceedings of IBP/PP Technical Meeting, Trebon, Czechoslovakia. Pudoc, Wageningen, The Netherlands.

Wit, C. T. de, and J. Goudriaan. 1974. Simulation of ecological processes. Centre for Agricultural Publishing and Documentation, Wageningen,The Netherlands.

Witz, J. A. 1973. Integration of systems science methodology and scientific research. Agric. Sci. Rev. 11(2):37-48.

## SECTION V

# REDUCTION OF INJURY BY PLANT BREEDING

*This section opens with a discussion by Csonka et al. on genetic engineering in relation to osmoregulation. They first discuss the mechanisms of osmoregulation in bacteria, caused chiefly by increase in proline. They propose that introduction of the plasmid for proline production into Rhizobium, which has weak osmotic regulation, might make nitrogen fixation in legumes more tolerant of water stress. However, other research indicates that although proline accumulation is common in stressed plants it is doubtful if it increases tolerance of water stress in seed plants.*

*Blum reminds us that nature selects for survival, man selects for productivity, and that drought tolerance is not a specific factor. He still does not know the best methods of testing for tolerance but uses greenhouses and growth chambers in combination with field tests to determine yield and stress tolerance. He doubts if yield is negatively correlated with drought tolerance. It was pointed out during the discussion that dwarf forms of grain have less drought tolerance than tall forms because they have less reserve food stores in their stems. Blum thinks that considerable drought tolerance exists among cultivated populations of wheat and sorghum and it is not necessary to search for it in wild forms.*

*Brim and others agree that there also is considerable stress tolerance among soybean varieties and some do much better than others under stress. In a presentation not reproduced in this volume, Brim indicated that he regards differences in time from emergence to flowering, fruiting, and seed filling as important characteristics. Brim suggests planting a mixture of two or three varieties with the hope that not all will be stressed at a critical state of development. He, like Blum, believes that varieties which yield well in the absence of stress will yield well under stress.*

*Castleberry emphasizes the need for good criteria for selection of temperature and drought tolerance in corn. He has used "firing" of leaves, synchrony of silk and tassel development, growth, and ear development, and finds leaf water potential and leaf firing closely related. A better definition of the physiological characters associated with improved stress tolerance is needed, also a better understanding of environments. For example, it may be unwise to use the low humidity of Colorado to test for drought tolerance in the Southeast where the humidity usually is high. It appears that the approach used to achieve stability of yield differs among the three plant breeders. Blum selects for cultivars with relatively high yields under stress. Castleberry selects for cultivars which demonstrate tolerance of dehydration and ability to resume growth after stress is removed. Brim proposes exploitation of varietal diversity in timing of reproductive growth stages to develop planting mixtures in which not all components would be at critical growth stages during a stress episode.*

*Burton's most drought-tolerant varieties of grasses owe their success to deep root systems, but he finds tolerance in pearl millet associated with leaf characters which reduce transpiration. He suggests the use of deep sands in desert areas, accompanied by irrigation, to screen for drought tolerance, but the low humidity and high irradiance of deserts are regarded by some as objectionable for testing cultivars intended for humid climates. During discussion, Ritchie proposed the use of large scale rain shelters for producing the water stress needed for screening plants in humid climates.*

*There was some discussion of the energy cost to plants of adaptations that increase stress tolerance. Some regarded it as important and others as relatively unimportant. Perhaps the cost in terms of yield depends on the type of adaptation and age of plants. While an increase in root:shoot ratio during seedling growth might enhance later stress tolerance and yield potential, an increase during reproductive growth might divert metabolites from seed or other economic products. Osmotic adjustment would be less expensive than increased root:shoot ratio regardless of plant age, and more responsive stomata would cost little or nothing.*

15

# GENETIC ENGINEERING FOR OSMOTICALLY TOLERANT MICROORGANISMS AND PLANTS

L. Csonka, D. Le Rudulier, S. S. Yang, A. Valentine, T. Croughan, S. J. Stavarek, D. W. Rains, and R. C. Valentine

Genetic engineering of osmotic tolerance (osmoregulation) is the application of genetic techniques, both conventional and recombinant DNA, for enhancing osmotic tolerance in microorganisms and plants. Osmoregulation is the mechanism(s) utilized by cells to balance their internal osmotic strength with that of their surroundings. The genes governing this process(es), referred to as the *(osm)* osmoregulation genes in microorganisms (Andersen et al., 1980) play an important role in tolerance to water stress brought about by salinization or drought.

For a summary of the state of knowledge of the mechanism of osmoregulation see Reference 1, a recent symposium volume dealing with this subject (D. W. Rains et al., eds., 1980).

The first section of this article draws heavily on recent work on the *osm* genes from our laboratory (R.C. Valentine and coworkers) since there is very little literature on this subject (see Epstein and Schultz, 1965 and Rhoads et al., 1976, for role of $K^+$ transport genes in osmoregulation). In addition the little information currently available on the molecular genetics of *osm* genes is found primarily in the microbial literature. Nevertheless, we feel that the picture of the nature of the *osm* genes now emerging from microbial studies is very relevant to the subject of this symposium and provides information for discussions of osmotic stress genes and their manipulation in higher plants.

The second portion (D. W. Rains and coworkers) deals with the use of pioneering techniques of plant cell tissue culture for selection of saline resistant cell lines, studies which may provide badly needed basic information on the mechanism of osmoregulation in higher plants and with the potential of providing "stress tolerant mutants" of major crop plants for use in programs of plant breeding (see several articles in this volume on current strategies for plant breeding of stress tolerance).

In examining the potential application of cell culture to plant breeding for stress tolerance, an area showing considerable potential for contributing to the genetic variability available to plant breeders is the selection of salt tolerant mutant individuals from cultures of plant cells. This use of cell-culture is based upon the application to cultured higher plant cells of techniques of mutant isolation developed in microbial and fungal systems (Bottino, 1975 and see selections employed below). Since the totipotency of individual cells has been firmly established (Nickell, 1977), each plant cell within a culture can be considered an individual organism. It is possible then, with appropriate techniques, to recover a mutation in any one of these large number of cultured plant cells. The general selective strategy used is to grow cells that have or have not been treated with a mutagen on a medium containing toxic levels of salt (NaCl) at a concentration that normally inhibits the growth of all cells. Any cell that can proliferate under these conditions is considered a presumptive mutant. The proliferating cells are continually challenged on the media, and survivors are eventually regenerated into plants where subsequent genetic analysis can be done. The important point here is that mutations are selected at the cellular level (Bottino, 1975). The development of selective procedures for salt tolerant mutants of microorganisms recently reported and as described in the next section provides additional incentive for work based on plant cell cultures.

## MICROORGANISMS: TOWARD GENETIC ENGINEERING OF ORGANISMS WITH ENHANCED OSMOTOLERANCE

As outlined in Figure 1, molecular cloning of the *osm* genes is proceeding in several steps, with recent progress being summarized in this section (see Mielenz et al., 1979, for discussions of some strategies using recombinant techniques for cloning *osm* genes in microbes).

### Mechanism of Osmoregulation

The first step, shown as A in Figure 1, summarizes current information on the mechanism of adaptation of bacteria such as *E. coli, Salmonella* spp. and *Klebsiella* spp. to osmotic stress. Osmotically active molecules surrounding the cell are shown as $X^-$ and $Y^+$ for charged molecules and Z for uncharged species. For simplicity the composite osmotic strength is considered to be equivalent to about 0.5 M NaCl. As illustrated in Figure 1A, cells adapt by using some remarkable biochemical machinery including a potassium ($K^+$) accumulation system (indicated as *i* in Figure 1A) which is somehow triggered

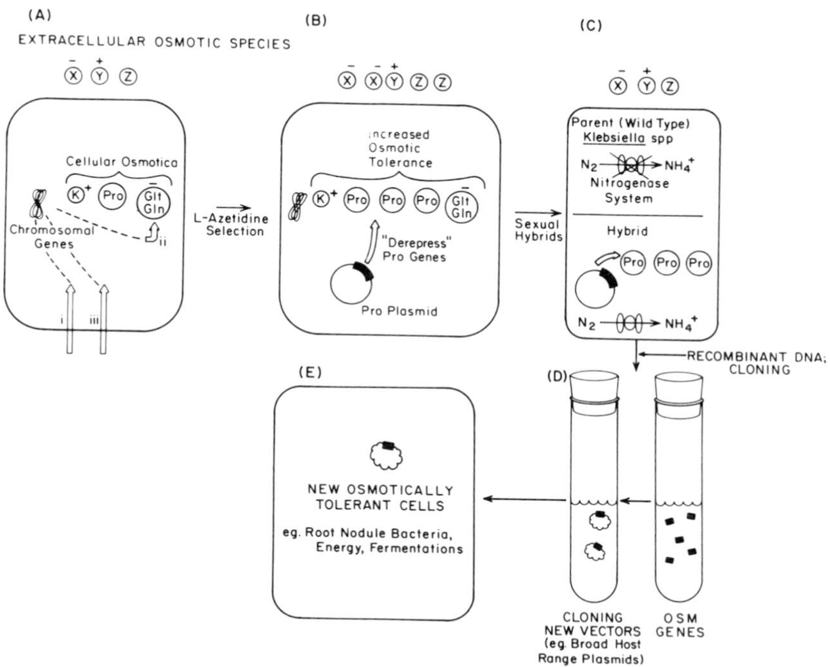


**Figure 1. Genetic engineering for proline over-producing plasmids for osmotic tolerance. See text for details.**

in response to increasing osmotic strength in the environment. This osmotically stimulated system appears already to be present in the cell since only a few seconds are sufficient to permit active accumulation of $K^+$ (probably too short a time scale to allow new rounds of protein synthesis to occur). Accumulation of $K^+$ appears to be a general response to increased osmotic stress caused by both neutral organic compounds (e.g., sucrose), as well as inorganic (charged) molecules (e.g., NaCl).

As pointed out by Epstein and Schultz (1965) and Christian and Waltho (1966), inorganic ions such as $K^+$, commonly available in most environments, may be energetically the cheapest form of osmoregulators. In plant systems there has been considerable emphasis on the role of ions whereas there is a much smaller literature on bacteria. Epstein and coworkers have carried out a comprehensive study of the role of potassium ($K^+$) in osmoregulation in Enteric bacteria, and their conclusions are of considerable interest here. In their 1965 paper Epstein and Schultz say, "The ability of *E. coli* to maintain its internal osmotic activity equal to, or somewhat greater than that of the surrounding environment, through regulation of its cell $K^+$ content, is of profound functional significance for an organism which may be subjected to a wide range of growth conditions." They further conclude that:

"1. Under a variety of conditions including those suitable for optimal growth, the osmolality of the growth medium is a major determinant of the cell K content."

"2. The growing cell responds to abrupt changes in the surrounding osmolality with rapid changes in cell K content in the direction necessary to minimize the osmotic difference."

"3. The conclusion that the bulk of the intracellular K in *E. coli* exists in an unbound, osmotically active form, though not directly established, is strongly suggested."

In studies reported in 1976 Epstein and coworkers (Rhoads et al., 1976) described the interesting finding that certain classes of $K^+$ uptake mutants, which they have analyzed extensively, are diminished in their capacity to grow in medium of high osmotic strength. In other words, such mutants behave as osmotically sensitive strains, which is interesting indirect evidence for a key role of potassium accumulation in osmotic adaptation. This important area deserves further work.

The finding by Epstein and Schultz (1965) that the increase in cellular $K^+$ raises the cellular osmolarity by about half as much as the increase in medium osmolality is of particular interest since each mole of $K^+$ accumulated must be associated with an equivalent of intracellular anion. The increase in cellular osmolality thus would be equal to the increase in medium osmolality if the anions accumulated with K were univalent and osmotically active.

With this as background, it is interesting to speculate that the elusive anion mentioned by these workers is in fact glutamic acid harboring a negative charge at neutral pH. Experiments to determine whether the kinetics of accumulation of $K^+$ and glutamate coincide are discussed next.

In continuing the discussion of some of the key features of the mechanisms of osmoregulation as outlined in Figure 1A, Measures (1975) found that many varieties of bacteria respond to osmotic stress by accumulating high internal pools of glutamate which he proposes as playing an important role as a counterion for $K^+$. However, since these observations were obtained using a rich broth medium containing large amounts of preformed glutamate, it is not possible from these data to distinguish between accumulation of exogenous glutamate and *de novo* biosynthesis of glutamate. We have answered this question by growing cells in defined medium containing no exogenous glutamate and following glutamate pools in response to osmotic stress. The major point here is that glutamate levels increase markedly during osmotic stress to levels approximately equivalent to that of $K^+$ and on a time scale essentially synchronized with $K^+$ accumulation. These findings, to be presented in detail elsewhere, are summarized in Figure 2.

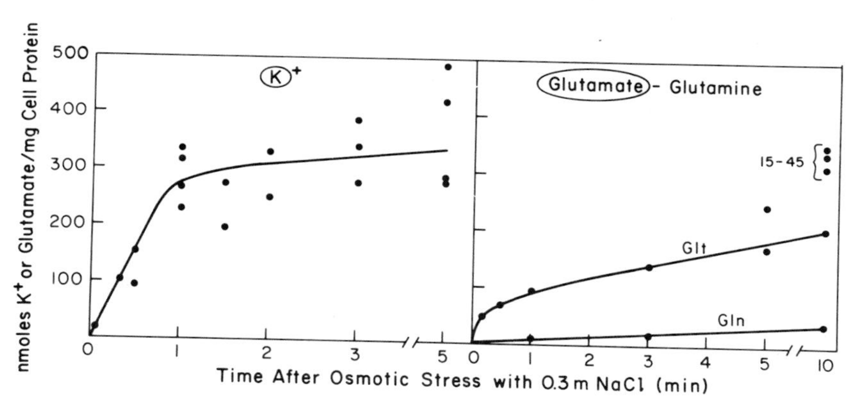


**Figure 2.** Cellular accumulation of potassium and glutamate-glutamine following osmotic stress with 0.3 M NaCl. For experimental details of measurement of cellular $K^+$ see Epstein and Schultz (1965). For details regarding intracellular pools of amino acids see Tempest, Meers, and Brown (1970).

It is interesting to speculate that the sophisticated control system(s) which have been evolved for modulating glutamate and glutamine biosynthesis in bacteria may play a vital role in osmoregulation. Indeed, the control system for the glutamine-dependent route of glutamate biosynthesis (glutamate synthase-glutamine synthetase) is among the most elaborate yet discovered in bacteria, and the glutamate dehydrogenase route of glutamate biosynthesis may respond to osmotic conditions (see Measures, 1975). The network of control of the glutamine-glutamate pathway includes not only the now classic examples of induction and repression at the genetic level and feedback or end product inhibition, but also covalent modification (adenylylation) occurring at specific sites of glutamine synthetase (see Chock et al., 1980). Covalent modification of glutamine synthetase resulting in raising or lowering in catalytic activity for glutamine biosynthesis and consequently glutamate production is mediated in turn by an elaborate enzyme cascade that is triggered by signals from the environment. In this manner the demand of the cell for glutamate is intimately linked to changes in the environment. It seems clear that the elegant glutamine synthetase cascade is suited for, and is capable of, rapid amplification of environmental signals into biochemical language. Experiments are in progress to determine whether modulation of glutamate and glutamine biosynthesis represent a part of the trigger or sensing mechanism of osmoregulation. The "mutant approach" is being utilized to determine which step(s) is essential (Dendinger et al., 1980).

The role of proline (pro) as an organic osmoticum (referring still to Figure 1A) is discussed in detail in an earlier paper (Csonka, 1980) and will be summarized here only briefly. Christian (1955a,b) discovered more than

25 years ago that exogenously added proline could alleviate osmotic inhibition in *Salmonella orianenburg.* We have confirmed and further characterized this phenomenon with *Salmonella typhimurium* (Csonka, 1980). The basic observation is that the addition of exogenous proline at concentrations as low as 0.5 mM markedly stimulates NaCl-inhibited growth. A similar result is seen where growth is inhibited by sucrose. Proline also has stimulatory effects in the presence of inhibitory concentrations of a number of other solutes, including $(NH_4)_2SO_4$, $K_2SO_4$, and $KH_2PO_4$. Proline is unique in this regard since, in agreement with the observations of Christian (1955 a), we found that none of the other 19 common amino acids caused a comparable stimulation. The concentrations of proline required for the stimulation are slight in comparison to those of the inhibitory solutes. Maximal stimulation is produced at 1 mM proline, and approximately 0.1 mM is sufficient for half maximal effect. Catabolism of proline is not necessary because a putA+ mutation, which blocks the only known proline catabolic pathway of the organism (Ratzkin et al., 1978), does not diminish the stimulatory effect.

It is important to clarify another point regarding the choice of proline as probable osmoregulator in bacteria. Measures (1975), in his comprehensive survey of the amino pools found in bacteria during osmotic stress, provided exogenous proline in the rich medium that he used to grow all his cultures. Under these circumstances, proline was found specifically to accumulate to high levels; however, we have found that, when several of the same species used by Measures are grown in chemically defined medium (salts and a single carbon source), the picture is dramatically changed with proline levels being very low and glutamine and glutamate elevated (see Table 1). This finding has influenced our interpretation of events occurring during osmotic adaptation. For example, note in Figure 1A that the dark area representing glutamine and glutamate synthesis is depicted as being the major source of gluta-

**Table 1. Amino acid levels in *S. typhimurium* LT2.**

| Amino Acid | Minimal | Minimal + 0.65 M NaCl |
|---|---|---|
| | *— nmoles/mg protein —* | |
| Glutamate | 253 | 762 |
| Glutamine | 31 | 139 |
| Alanine | 15 | 13 |
| Glycine | 10 | 15 |
| Leucine | 5 | 5 |
| Lysine | 1 | 3 |
| Proline & others | $< 1.4$ | $< 1.3$ |

mine and glutamate (some exogenous glutamate also may be available). In contrast, the exogenously supplied proline appears to be the major source of proline during osmotic stress. Undoubtedly this scheme is oversimplified and will require further changes as more knowledge of the mechanism is elucidated.

All of the processes mentioned above appear to be governed by chromosomal genes in contrast to, say, naturally occurring plasmids. This is indicated by the dotted lines connecting the chromosome to the respective process.

The studies up to this point have provided a picture of how cells adapt to osmotic stress and lead naturally to the next step illustrated as Figure 1B.

### The Isolation of Proline Over-Producing Mutants with Increased Osmotolerance

Proline over-producing mutants were selected as strains resistant to a toxic proline analogue, L-azetidine-2-carboxylate. The rationale for the methodology was that proline, produced at high levels, could antagonize the analogue. However, only a minority (1%) of the L-azetidine-2-carboxylate resistant mutants proved to be proline overproducers. The same phenotype also could be conferred by a much more frequent type of a mutation, *putP*$^-$, inactivating the major proline permease that functions in the uptake of the analogue (Ratzkin et al., 1978). Previously, Condamine (1971) isolated mutations resulting in proline over-production, some of which proved to be closely linked to the *proBA* genes. This fact provided us with a short cut in the isolation of a large number of additional proline over-producing strains.

We have used a strain carrying the *proB*$^+$*A*$^+$ genes on a self-transmissible plasmid (shown in Figure 1B as a circular DNA molecular with *pro* gene as darker area), because derivatives that became resistant to L-azetidine-2-carboxylate due to a mutation in the *proBA* region could be readily identified on the basis of their ability to transfer the mutation to other *Salmonella typhimurium* strains (and Enteric bacteria) that carried a mutation of the *proBA* genes. Since this work will be described in detail elsewhere, only a brief summary is given here.

The major finding is that it is possible to generate a class of simple mutations which enhance the osmotolerance of *Salmonella typhimurium.* These mutations were obtained as alterations resulting in proline over-production, suggesting that the enhanced osmotolerance might be a consequence of the high intracellular levels of proline. The mutation is mapped in the *proB*$^+$*A*$^+$ genes which, upon transfer to other *Salmonella* strains, confer the phenotype found with the original hosts.

Solid media containing glucose, leucine, and 0.65 M NaCl was used to test 107 proline over-producing derivatives for enhanced osmotolerance. After five days of incubation, six of these were judged to give rise to larger colonies than their parental strains.

In order to verify whether enhanced osmotolerance is correlated with the intracellular proline levels, we have determined the free amino acid content of the mutants and their parental strains grown in minimal medium and also under conditions of osmotic stress. The results are in Table 2. There are three major points to be made about these data. First, in TL128 ("wild-type" with respect to proline production and osmotolerance) the levels of glutamine and glutamic acid increase 14-fold and 2-fold, respectively, over the control values as a consequence of growth in the presence of 0.65 M NaCl. Similar results, which may be part of the osmoregulatory response of the organisms, were observed with other *Salmonella typhimurium* LT-2 strains (L. Csonka, unpublished data) and with *E. coli* (Munroe et al., 1972). Second, the mutations in TL124 and TL126, originally selected to confer L-azetidine-2-carboxylate resistance, result in proline over-production, for the proline levels in these two strains are at least 12 times greater than in TL128. Moreover, the proline levels in the markedly osmotolerant strain TL126 are higher than the corresponding levels in the slightly osmotolerant TL124. Third, both in TL124 and in TL126, the intracellular proline levels are higher in the presence of NaCl than in its absence. This is especially true of TL126, in which under conditions of osmotic stress proline constitutes 61% of the ninhydrin positive small molecules. In Figure 1B the term "derepressed Pro genes" is used to illustrate this point although the nature of this mutation remains unknown.

One of our goals is to develop a procedure to obtain osmotolerant mutants of other bacteria, including *Rhizobia.* As a first step in testing whether

**Table 2. Doubling times and proline levels for strains of *Salmonella typhimurium* grown in minimal medium and exposed to osmotic stress.**

| Strain | Doubling Time | | Proline Level | |
|---|---|---|---|---|
| | Minimal | Minimal +0.65 M NaCl | Minimal | Minimal +0.65 M NaCl |
| | *– hours –* | | *– nmoles/mg protein –* | |
| **Wild Type:** | | | | |
| TL128 | 1.0 | 6.6 | 1.2 | 1.9 |
| **Osmotolerant Mutants:** | | | | |
| TL126 | 1.2 | 2.4 | 36 | 787 |
| TL123 | 1.3 | 5.1 | 19 | 98 |
| TL124 | 1.3 | 5.6 | 24 | 75 |
| TL125 | 1.2 | 6.0 | 19 | 22 |
| TL120 | 1.2 | 6.3 | 4 | 39 |
| TL119 | 1.2 | 6.5 | 19 | 55 |

the selection of proline over-producing mutants might be a practical approach, we transferred into a nitrogen fixing strain of *Klebsiella pneumoniae* one of the F's that conferred enhanced osmotolerance on *S. typhimurium,* and determined the effect proline over-production had on nitrogenase activity under osmotic stress (Le Rudulier et al., 1981, see Figure 1C).

First, we examined the effect of osmotic inhibition, achieved by the addition of NaCl, on the growth rate (Table 3) and nitrogenase activity (Table 4) of *Klebsiella pneumoniae.* Nitrogenase activity was much more sensitive to osmotic inhibition than the overall growth rate of the cells. For instance, in the absence of proline, 0.4 M NaCl caused a ten-fold decrease in the nitrogenase activity of strain M5A1 (Table 4, column 1), whereas it caused only a

**Table 3. The effect of exogenously added proline and of proline over-production on the growth rate of *K. pneumoniae* under conditions of osmotic inhibition. Growth was under anaerobic conditions at room temperature.**

| | Growth Rate | | | | |
|---|---|---|---|---|---|
| | M5AI (Wild Type) | | XY1 (F'*pro-74)* | XY2 (F'$proB^+A^+$) | |
| NaCl | – Proline | +0.5 mM proline | –Proline | –Proline | +0.5 mM proline |
| *M* | | | *– generation/hr –* | | |
| 0.0 | 0.33 | 0.33 | 0.31 | 0.33 | 0.33 |
| 0.3 | 0.251 | 0.26 | 0.24 | 0.25 | 0.26 |
| 0.4 | 0.17 | 0.24 | 0.17 | 0.17 | 0.24 |
| 0.5 | 0.15 | 0.16 | 0.14 | 0.14 | 0.17 |
| 0.6 | 0.054 | 0.096 | 0.079 | 0.033 | 0.15 |

**Table 4. The effect of exogenously added proline and of proline over-production on nitrogenase activity under conditions of osmotic inhibition.**

| | Nitrogenase Activity | | | | |
|---|---|---|---|---|---|
| | M5AI (Wild Type) | | XY1 (F'*pro-74)* | XY2 (F'*proB+A+)* | |
| NaCl | – Proline | +0.5 mM proline | –Proline | –Proline | +0.5 mM proline |
| *M* | | *– µmoles ethylene produced/hour/mg protein –* | | | |
| 0.0 | 2.55 | 2.66 | 2.64 | 2.71 | 2.59 |
| 0.3 | 0.42 | 1.26 | 1.55 | 0.60 | 1.31 |
| 0.4 | 0.25 | 0.96 | 1.53 | 0.14 | 0.74 |
| 0.5 | 0.02 | 0.94 | 0.42 | 0.04 | 0.73 |
| 0.6 | 0.02 | 0.12 | 0.26 | 0.01 | 0.19 |

two-fold decrease in the growth rate (Table 2, column 1). Similarly, in the absence of proline, 0.6 M NaCl decreased the nitrogenase activity of strain M5A1 over a hundred-fold, while it caused only about a six-fold reduction in growth rate. Analogous results were obtained with strain KY2 *(pro-3/*$F'_{128}$ *proB*$^+$*A*$^+$*)*.

The stimulatory effect of proline on the growth rate is manifested only under conditions of extreme osmotic inhibition ($\leq$ 0.6 M NaCl with *K. penumoniae,* unpublished results). However, proline exerted a much greater stimulatory effect on nitrogenase at lower osmolarities. Thus, in the presence of 0.4 M NaCl, 0.5 mM proline caused approximately a four-fold enhancement of the nitrogenase activity of strain M5A1 and approximately fifty-fold enhancement in the presence of 0.5 M NaCl (Table 4, columns 1 and 2).

The effect of the mutation resulting in proline over-production in strain KY1 ($F'_{128}$ *pro-74)* was similar to that seen when proline was supplied exogenously, in that the growth rate was stimulated only under extreme osmotic inhibition (Table 4; compare strain KY1, column 2, with KY2 without proline, column 4, at 0.6 M NaCl). The mutation, however, had a much more pronounced stimulatory effect on nitrogenase activity. In the absence of proline, the nitrogenase activity of strain KY1 was at least ten times greater at 0.4 and 0.5 M NaCl than that of strain KY2. At 0.6 M NaCl, the stimulatory effect of the mutation was greater than 25-fold.

The final steps of Figure 1 (D and E) represent future strategies for genetic engineering of the *osm* genes using recombinant DNA technology. The availability of *osm* genes on a plasmid vector should hasten these experiments.

It should also be pointed out that these microbial systems should not be regarded as merely models for higher plants. Indeed, osmotic tolerance is an essential trait in the utilization (fermentation) of biomass yielding ethanol and numerous other essential fuels and chemicals. In short, the cardinal importance of osmoregulation as an essential cellular process provides added incentive for research in this area, knowledge which may have applications for selection and breeding of stress tolerant plants.

## APPLICATIONS OF PLANT TISSUE CULTURE

In this section we are concerned with the potential for selection of osmotic tolerant cell lines and ultimately whole plants. Cell-culture techniques have proved useful in a broad spectrum of plant research areas. These applications range from basic physiological studies on enzymes and biochemical pathways at the cellular level to large-scale vegetative multiplication of desirable genotypes at the commercial level (Scowcroft, 1977). The major areas in which the cell-culture approach has been used effectively are plant biochemistry, breeding, plant development, disease control, and plant propagation.

Although several experiments illustrate the tremendous potential of cell-culture as a plant breeding technique, this approach has certain limitations and problems. For example, only certain types of characters can be selected, and at present there appears no way to directly select for increased yields with cell-culture methods. Nonetheless, if it is established that the alteration or increase in quantity of a particular enzyme (a biochemical marker) will boost yields, cell culture could be used to select for mutants at this enzymatic level. Other problems include a difficulty in distinguishing between truly mutant cells and cells that have not changed genetically but have epigenetically adapted to the selection pressure applied. The final proof of a genetic basis for a particular cellular trait requires regeneration of whole plants from the selected cell lines and subsequent physiological and genetic analysis at the whole-plant level.

The development of salt tolerant crop plants may ameliorate the growing problem of salt in the agricultural environment, and the techniques of cell culture appear applicable to developing these salt tolerant crops. Cell culture has several advantages as a technique for studying salt tolerance. It allows the processes or markers involved in salt tolerance to be characterized at the cellular level. The relative lack of differentiation in the cultured cells eliminates complications arising from the morphological variability and highly differentiated state of the various tissues of whole plants. The culturing of plant cells on rigidly defined culture media also permits a relatively uniform and precise treatment of these tissues with salt.

It is now well established that cell culture techniques can be used to select salt tolerant lines from nontolerant agricultural plants. Nabors et al. (1975) obtained tobacco cells that showed superior tolerance to 0.16% NaCl (w/v) and, subsequently, to 0.52% NaCl. Dix and Street (1975) selected lines of tobacco and pepper cells that grew in liquid medium containing up to 2% NaCl. Both studies measured growth as an increase in cell number or change in packed cell volume. Dix and Street observed that the packed cell volume tended to decrease after exposure to salt despite an increase in cell number. This was attributed to a decrease in cell size. The results of these studies demonstrate that selection for salt tolerance at the cellular level is feasible.

A number of plant species have been selected for salt tolerance through procedures of cell selection. Some examples are presented in Table 5.

In our laboratory we have obtained data from a comparative study in which a salt-selected line of alfalfa callus cells grew better in the presence of salt than did the unselected cell line from which the variants had been selected (Croughan et al., 1978). Exposing cultured alfalfa cells to an agar-solidified nutrient medium containing a typically lethal concentration of NaCl permitted identification of a cell line with an elevated resistance to salt toxicity. The tissue to be screened was increased in quantity through growth

**Table 5. Examples of salt-tolerant cell lines selected through tissue culture.**

| Species | Callus or Suspension | NaCl Level Tolerated | Reference |
|---|---|---|---|
| | | *% (w/v)* | |
| *Nicotiana tabacum* | Suspension | 0.52 | Nabors et al. (1975) |
| *Nicotiana sylvestris* | Callus | 1.0 | Dix and Street (1975) |
| | Suspension | 2.0 | |
| *Capsicum annum* | Callus | 1.0 | Dix and Street (1975) |
| | Suspension | | |
| *Medicago sativa* | Callus | 1.0 | Croughan et al. (1978) |
| *Oryza sativa* | Callus | 1.5 | Rains et al. (1980) |

as a suspension culture. Inoculating salt-containing media in petri dishes with a slurry of these suspension-grown cells allowed this extremely large number of individuals to be screened for salt tolerance. The screening consisted of visual identification of cells that maintained both growth and a healthy appearance despite exposure to high salinity. Consistent subculture to saline nutrient medium permitted identification of the most tolerant cell lines, ultimately resulting in isolation of the few salt-tolerant types present in the large original population.

The salt-selected line of alfalfa cells grew better than unselected cells at a high level of salt (1.0% NaCl), indicating that the selection isolated variant cells with increased capacity for growth in the presence of high levels of NaCl. This selected line additionally displayed other characteristics suggesting that the tolerance was the consequence of a shift toward a true halophytic nature. Besides tolerating high levels of salinity, halophytes actually require some salt, as evidenced by poor growth in its absence and growth stimulation when it is added (Black, 1960; Brownell, 1965). This same pattern is displayed by the salt-selected line of alfalfa, which grows optimally at 0.5% (85 mM) NaCl and poorly in the absence of salt.

A similar pattern was observed with rice cells selected for tolerance to salt. Two populations of rice cells, one selected in the presence of 1.5% salt and the other unselected, were grown for 42 days on media containing various levels of NaCl. The salt-selected line required the presence of 0.5% NaCl for optimal growth and successfully grew at 1.5% NaCl, a concentration that was lethal to the unselected line (Figure 3).

Another characteristic of halophytes is their capacity to accumulate certain ions to high internal concentrations (Flowers et al., 1977). Comparing the ionic content to the two alfalfa cell lines indicates that in the presence of salt the salt-selected line accumulated more $Cl^-$ and $K^+$ than the unselected line (Figures 4 and 5). Particularly noteworthy is this capacity to maintain

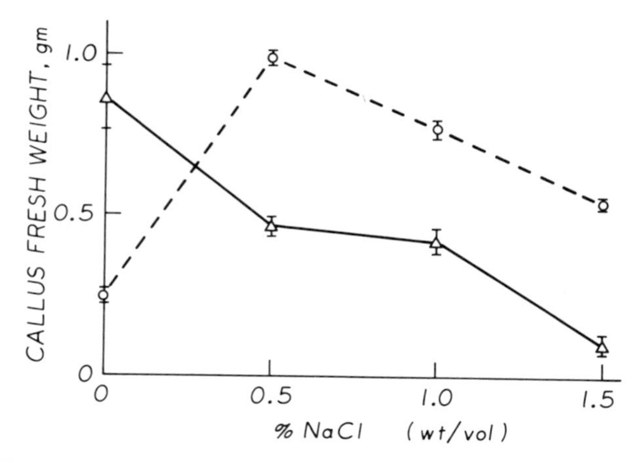


**Figure 3. Rice callus fresh weight as a function of salt concentration in the medium, comparing the nonselected cell line (△—△) to the salt-selected cell line (○—○).**

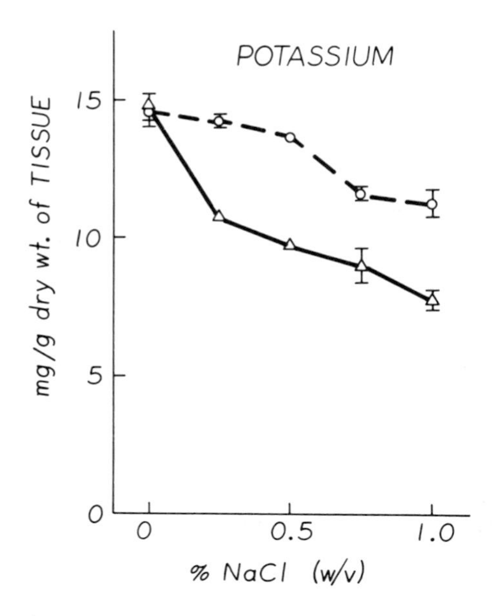


**Figure 4. Potassium concentration in alfalfa callus as a function of salt concentration, comparing the nonselected cell line (△—△) to the salt-selected cell line (○—○).**

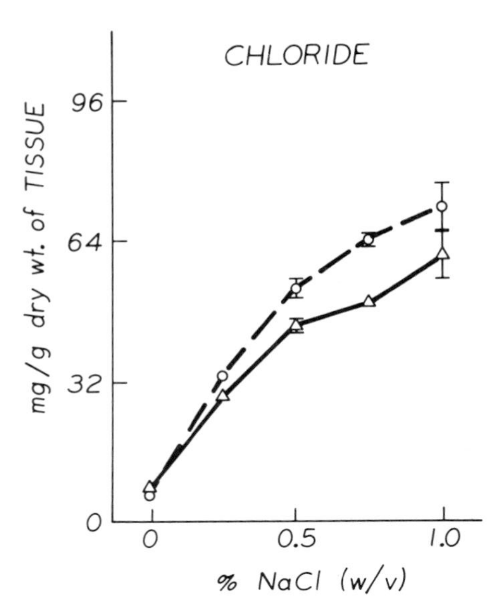


**Figure 5. Chloride concentration in alfalfa callus as a function of salt concentration, comparing the nonselected cell line (△—△) to the salt-selected cell line (○—○).**

higher levels of $K^+$ in the presence of high levels of $Na^+$, since this trait is strongly correlated with salt tolerance in both halophilic bacteria and halophytic plants (Rains, 1972, 1979; Rains and Epstein, 1967; Brown, 1964). An elevated level of $NO_3^-$ within the salt-selected line at the low salt levels is also consistent with observations that halophytic plants tend to accumulate $NO_3^-$ when grown in the absence of $Cl^-$ (Flowers et al., 1977).

Data on the ionic contents of the two cell lines suggest that the salt-selected cell line may differ from the unselected line in ionic transport systems. The salt-selected line generally accumulated more $NO_3^-$ and $Cl^-$ from the medium. The affinity for the substrate or specificity of the transport mechanisms for these anions may differ, or perhaps the quantity of transport mechanisms is greater in the salt-selected line (Rains, 1972). As to the Na and K ions, the transport mechanisms in the salt-selected line may show a greater specificity for $K^+$, permitting higher internal levels of this ion despite high external (and internal) levels of $Na^+$. This resulted in a higher ratio of $K^+/Na^+$ within the salt-selected cell line at all levels of additional NaCl.

Figure 6 is a flow sheet which provides the basic steps in selecting salt tolerant plants using cell selection procedures. This approach could be applied to a number of stress factors.

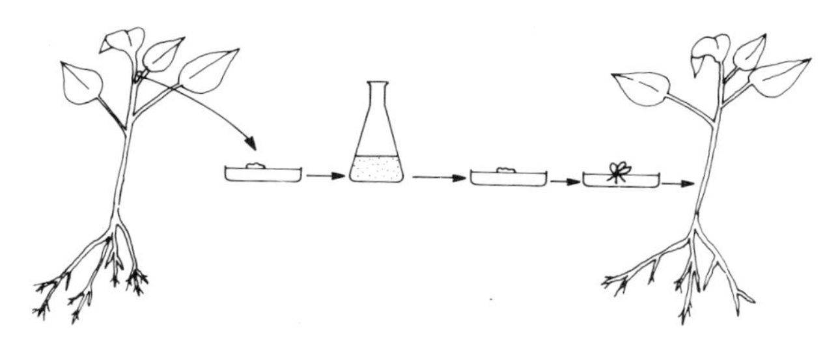

**Figure 6. Tissue culture selection for salinity tolerant plants. The flow diagram illustrates the procedures for selection. An explant is cultured as callus (a); callus placed in liquid suspension to raise millions of cells (b); suspensions plated on salt media and tolerant cells selected (c); selected cell lines regenerated to plants (d).**

## NOTES

L. Csonka, D. LeRudulier, S. S. Yang, A. Valentine, T. Croughan, S. J. Stavarek, D. W. Rains, and R. C. Valentine, Plant Growth Laboratory and Department of Agronomy and Range Science, University of Salifornia, Davis, California 95616.

This work was supported by the National Science Foundation under Grant No. PFR 77-07301. Any opinions, findings, and conclusions or recommendations expressed in this publication are those of the authors and do not necessarily reflect the views of the National Science Foundation. We also thank the Kearney Research Foundation for support of work on stress tolerant soil microorganisms. Support for work on selection of salt tolerant plant cells was funded by the United States Department of Agriculture (Cooperative Agreement) and by the California Crop Improvement Association. D. Le Rudulier was a recipient of a fellowship from the North Atlantic Treaty Organization (NATO). S. S. Yang was supported by the Ministry of Education of the People's Republic of China.

## LITERATURE CITED

Andersen, K. T., K. T. Shanmugam, S. T. Lim, L. N. Csonka, R. Tait, H. Hennecke, D. B. Scott, S. M. Hom, J. F. Haury, A. Valentine, and R. C. Valentine. 1980. Genetic engineering in agriculture with emphasis on nitrogen fixation. TIBS 5: 35-39.

Black, R. F. 1960. Effects of NaCl on the ion uptake and growth of *Atriplex vesicaria* Howard. Aust. J. Biol. Sci. 13: 249-266.

Bottino, P. J. 1975. Potential of genetic manipulation of plant-cell culture for plant breeding. Radiat. Bot. 15: 1-16.

Brown, A. D. 1965. Aspects of bacterial response to the ionic environment. Bacteriol. Rev. 28: 296-329.

Brownell, P. F. 1965. Sodium as an essential micronutrient element for a higher plant *(Atriplex vesicaria)*. Plant Physiol. 40: 460-468.

Chock, P. B., S. G. Rhee, and E. R. Stadtman. 1980. Interconvertible enzyme cascades in cellular regulation. Ann. Rev. Biochem. 49: 813-843.

Christian, J. H. B. 1955a. The influence of nutrition on the water relations of *Salmonella orianenburg.* Aust. J. Biol. Sci. 8: 75-82.

Christian, J. H. B. 1955b. The water relations of growth and respiration of *Salmonella orianenburg* at 30°. Aust. J. Biol. Sci. 8: 490-497.

Christian, J. H. B., and J. A. Waltho. 1966. Water relations of *Salmonella orianenburg:* stimulation by amino acids. J. Gen. Microbiol. 43: 345-355.

Condamine, H. 1971. Sur la régulation de la production de proline chez *E. coli* K12. Ann. Inst. Pasteur, Paris 120: 126-143.

Croughan, T. P., S. J. Stavarek, and D. W. Rains. 1978. Selection of a NaCl tolerant line of cultured alfalfa cells. Crop Sci. 18: 959-963.

Csonka, L. N. 1980. The role of L-proline in response to osmotic stress in *Salmonella typhimurium:* selection of mutants with increased osmotolerance as strains which over-produce L-proline. p. 35-52. In D. W. Rains, R. C. Valentine, and A. Hollaender (eds.) Genetic engineering of osmoregulation. Plenum Press, New York.

Dendinger, S., L. G. Patil, and J. E. Brenchley. 1980. *Salmonella typhimurium* mutants with altered glutamate dehydrogenase and glutamate synthase activities. J. Bacteriol. 141: 190-198.

Dix, P. J., and H. E. Street. 1975. Sodium chloride-resistant cultured cell lines from *Nicotiana sylvestris* and *Capsicum annuum.* Plant Sci. Lett. 5: 231-237.

Epstein, W., and S. G. Schultz. 1965. Cation transport in *Escherichia coli.* V. Regulation of cation content. J. Gen. Physiol. 49: 221-234.

Flowers, T. J., P. F. Troke, and A. F. Yeo. 1977. The mechanism of salt tolerance in halophytes. Ann. Rev. Plant Physiol. 28: 89-121.

Le Rudulier, D., S. S. Yang, and L. N. Csonka. 1981. Proline over-production enhances nitrogenase activity under osmotic stress in *Klebsiella pneumoniae.* p. 173-179. In J. M. Lyons, R. C. Valentine, D. A. Phillips, D. W. Rains, and R. C. Huffaker (eds.) Genetic engineering of symbiotic nitrogen fixation and conservation of fixed nitrogen. Plenum Press, New York.

Measures, J. D. 1975. Role of amino acids in osmoregulation in non-halophilic bacteria. Nature 256: 398-400.

Mielenz, J., K. Andersen, R. Tait, and R. C. Valentine. 1979. Protential for genetic engineering of salt tolerance. p. 361-371. In A. Hollaender (ed.) The biosaline concept: an approach to the utilization of underexploited resources. Plenum Press, New York.

Munns, G. F., K. Hercules, J. Morgan, and W. Sauerbier. 1972. Dependence of the putrescine content of *Escherichia coli* on the osmotic strength of the medium. J. Biol. Chem. 247: 1272-1280.

Nabors, M. W., A. Daniels, L. Nadolny, and C. Brown. 1975. Sodium chloride tolerant lines of tobacco cells. Plant Sci. Lett. 4: 155-159;

Nickell, L. G. 1977. Crop improvement in sugarcane: studies using *in vitro* methods. Crop Sci. 17:717-719.

Rains, D. W. 1972. Salt transport by plants in relation to salinity. Ann. Rev. Plant Physiol. 23: 367-388.

Rains, D. W. 1979. Salt tolerance of plants: strategies of biological systems. p. 47-67. In A. Hollaender (ed.) The biosaline concept: an approach to the utilization of underexploited resources. Plenum Press, New York.

Rains, D. W., and E. Epstein. 1967. Preferential absorption of potassium by leaf tissue of the mangrove, *Avicennia marina:* an aspect of halophytic competence in coping with salt. Aust. J. Biol. Sci. 20: 847-857.

Rains, D. W., T. P. Croughan, and S. J. Stavarek. 1980. Selection of salt-tolerant plants using tissue culture. p. 279-292. In D. W. Rains, R. C. Valentine, and A. Hollaender (eds.) Genetic engineering of osmoregulation. Plenum Press, New York.

Ratzkin, B., M. Grabnar, and J. Roth. 1978. Regulation of the major proline permease gene of *Salmonella typhimurium.* J. Bacteriol. 133: 737-743.

Rhoads, D. B., F. B. Waters, and W. Epstein. 1976. Cation transport in *Escherichia coli.* VIII. Potassium transport mutants. J. Gen. Physiol. 67: 325-341.

Scowcroft, W. R. 1977. Somatic cell genetics and plant improvement. p. 39-81. In N. C. Brady (ed.) Advances in agronomy. Vol. 29. Academic Press, New York.

Tempest, D. W., J. L. Meers, and C. M. Brown. 1970. Influence of environment on the content and composition of microbial free amino acid pools. J. Gen. Microbiol. 64: 171-185.

16

# BREEDING PROGRAMS FOR IMPROVING CROP RESISTANCE TO WATER STRESS

A. Blum

Ample evidence has been accumulated in recent years indicating that genetic variation exists within crop plant species in various components of drought resistance. Existence of such variation and the high heritability of given components of resistance (e.g. Roark and Quisenberry, 1977; Williams et al., 1969) are primary requirements for breeding.

When this volume of information is surveyed, a major conclusion may be drawn that drought resistance attributes operate at all plant organization levels. An example with sorghum (Blum, 1979; Sullivan and Eastin, 1974; Sullivan and Ross, 1979) demonstrates that resistance reaction may occur at the cellular level in some genotypes and at the plant community level, involving parameters of energy exchange with the environment, in others. Thus, no singular sorghum genotype was revealed (Blum, 1979) which could be classified as totally drought resistant. This would be expected in cultivated plant cultivars as opposed to natural ecotypes, considering the different modes of selection exerted by man and by nature.

Yield and plant productivity are major selection criteria in most crop plants, while survival is a major criteria in natural vegetation. Natural selection at a given arid site brings about an accumulation of adaptive allels for plant survival. Crop selection processes are designed to accumulate allels that promote the economic productivity of the population in an agricultural ecosystem. Depending on parental materials used in crosses and the selection scheme, drought adaptive allels may occur at random in the breeding population. If selection for yield is exerted in conjunction with selection for stability in yield performance over an array of environments, some drought adaptive genes could be fixed in resulting cultivars. The exact phenotypic-physiological expression of such genes is, however, obscure.

The conclusion that unidentified drought adaptive allels exist at relatively high frequencies in common breeding populations is supported by the fact that empirical breeding programs in many crops continue to produce plant varieties with improved productivity under conditions of drought stress. Indeed, this was repeatedly found in our work with wheat. Consistent large variations in drought avoidance were found among $F_6$ to $F_7$ selections out of populations that were never exposed to stress. Similarly, a range of 10 to 80% injury by desiccation to cell membranes, as a measure of drought tolerance, was revealed among such selections (Blum and Ebercon, 1980). It follows that utilization of exotic genetic resources as a source for drought resistance (Atsmon, 1979) is not a prerequisite at this time for generating the required genetic variation.

The frequency of drought adaptive allels in breeding populations is critical (Daday et al., 1973). Frequency can be increased by identifying and using suitable parental materials in generating the population. We have revealed in wheat, for example, that all crosses having strain 'H574-1-2-6' in their parentage produced a high frequency of dehydration-avoidant progeny. Similarly, a sorghum (Blum, 1974; Blum, 1979; Sullivan and Eastin, 1974) CK60 hybrid parental line performed very well in terms of several drought adaptation responses. It could not be a coincidence that commercial sorghum hybrids having CK60 as a female parent perform relatively well under dryland conditions.

Although numerous efficient selection programs were devised for the manipulation of yield in breeding populations, the genetic control of yield and its components is not fully understood. Yield per unit land area is the target, but selection is often exerted upon yield components in single plants. Yield components determine the yield per unit area of land by way of processes that involve componets' plasticity and interactions (Bradshaw, 1965). The physiological processes that control development of, and interaction among, yield components in time and space are basically not understood. Thus, current selection for yield and yield components even in nonstress environments is basically empirical and relies heavily on the breeder's experience and intuition. At this point, one reaches the thin line drawn between science and art.

It therefore becomes apparent that any discussion of current common breeding programs designed for yield improvement under drought stress may be treated only as a best available approximation. However, judicious incorporation of some physiological work within the framework of a breeding program should result in appreciable feedback of information useful for future improvement of the program's design.

## CURRENT APPROACHES TO BREEDING FOR IMPROVED DROUGHT RESISTANCE

### Selection for Yield Under a Nonstress Environment

Integrated drought resistance in terms of yield can be estimated through tests designed to measure the reduction in yield under conditions of stress, as compared with nonstress conditions. Thus, Mederski and Jeffers (1973) found that old, low potential soybean varieties were the most tolerant to drought, because their yield reduction from nonstress to stress conditions was minimal. Modern, high potential cultivars were most susceptible by this criterion, but their absolute yields under conditions of stress were still the highest.

The relationships between yield potential and its effect on performance under stress has been further explored through the use of stability models (Eberhart and Russell, 1966; Finlay and Wilkinson, 1963; Johnson, 1977). With this analysis, yield of a cultivar tested in an array of environments is regressed against an environmental index (Figure 1). The environmental index consists of the mean yield over all cultivars in each site. Thus, a stable variety

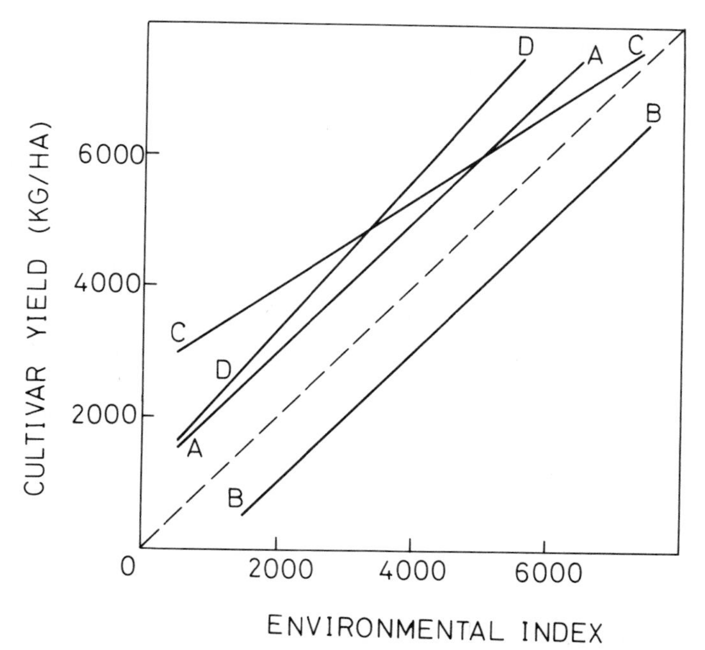


**Figure 1. A schematic example of stability analysis for four typical cultivars. For this example, the major variable in environmental index is water stress.**

is characterized by an equal response in yield over all environments, i.e., regression coefficient 'b' approaches 1.

Stability is independent of yield potential (Finlay and Wilkinson, 1963). Yield potential may affect performance under conditions of stress in stable varieties (e.g., Fischer and Maurer, 1978). Thus, moving from variety B to variety A will increase grain yield under stress conditions, although both varieties possess the same level of drought resistance as indicated by equal rates of reduction in yield from stress to nonstress environments. Even more susceptible varieties may still produce reasonable yields under stress, provided their yield potential is very high (variety D).

Drought susceptibility can be equated with nonstability only in cases where the regression coefficient is larger than 1. Alternatively, nonstability associated with coefficients less than 1 is indicative of drought resistance, provided the intercept 'a' of the linear regression is appreciably greater than 0 (variety C). Thus, while stability analysis commonly involves only the relative values of 'b', the analysis of relative drought resistance in terms of productivity at low environmental indices also requires an evaluation of 'a'.

Thus, we have an analytical tool for classifying varieties according to Reitz's (1974) statement as based on his rich experience: "varieties fall into three categories: (a) those with uniform superiority over all environments, (b) those relatively better in poor environments, and (c) those relatively better in favored environments." It follows that the genetic improvement of the potential yield, as carried out under nonstress conditions, also will bring about yield increases under conditions of stress. Commercial $F_1$ hybrids provide a private case to the rule (Blum, 1979). Hybrid vigor, responsible for the higher yield potential of hybrids, was found to be associated with greater stability over environments (Knight, 1973).

Selection for higher potential yields, carried out under nonstress environments, is an established breeding routine. Under such conditions, genetic variation in and heritability of yield and yield components is large and selection is efficient (Daday et al., 1973; Frey, 1964; Johnson and Frey, 1967; Roy and Murty, 1970). However, as pointed out previously (Blum, 1979), this infers that crop improvement for stress environments will depend on constant efforts to raise the yield plateau.

### Selection for Yield Under Conditions of Stress

Several reports, mainly by Canadian wheat breeders (Hurd, 1969; Townley-Smith and Hurd, 1979), lay out and rationalize a design for yield improvements under the Canadian dryland conditions which uses yield under stress as a major selection criterion. An important problem in using yield and yield components as selection criteria in stressed populations appears to be the low genetic variation and the resultant reduced expected response to selections for these attributes under stress (Daday et al., 1973; Frey, 1964; Johnson and Frey, 1967; Roy and Murty, 1970). One exception, though, has been recently noted in *Brassica* by Richards (1978).

Daday et al. (1973) propose that the depletion of the genetic variation in a population submitted to stress may very well indicate the low frequency of stress-adaptive allels in the given population. This could be taken to indicate that selection for yield under stress can be executed in certain populations, depending on frequencies of adaptive allels. Indeed, Richards (1978) demonstrated that progress in selection under stress was population-specific. Thus, when selection for yield and yield components under conditions of stress is attempted, a major consideration should be given to the genetic makeup of the population. This is indirectly supported by Hurd (1969), who recommends that parental lines used in crosses should originate from varied genetic sources. However, selection of parental lines does not necessarily guarantee the adaptive value of the resulting population. An initial estimate of the genetic variance for yield under stress should probably be obtained for all populations produced by the breeder at an initial phase, such as $F_2$ to $F_3$. Such derived estimates will allow, prior to any serious investments in selection and testing, reduction in the number of populations to only those that have adaptive value. Current biometrical designs for estimating genetic variance in populations are too elaborate for this purpose. Simplified, cheap designs are most warranted, even at some expense of estimate accuracy.

### Selection for Yield and Additional Traits Under Stress

Since the genetic variation in yield under conditions of stress is small, additional selection criteria may improve efficiency of the program. Additional simple criteria, which rely heavily on personal experience and the little scientific proof available at the time, have been developed by breeders. For small grains, additional selection criteria in use are: earliness (Reitz, 1974), limited synchronous tillering (Hurd, 1969), seedling vigor (Roy and Murty, 1970), bearded spikes (Reitz, 1974), narrow leaves (E. I. Smith, personal communication), and leaf rolling or wilting (O'Toole and Chang, 1979). The usefulness of all of these indices as adaptive traits under conditions of moisture deficit were subsequently proven by physiological studies, usually more than a decade after their initial use by breeders.

### Selection for Yield Under Nonstress and for Drought Resistance Under Stress

Selection for yield under nonstress conditions is fairly efficient. Improvement of yield potential also bears some significance toward subpotential environments. Incorporation of appropriate direct selection criteria for drought resistance into a routine program will possibly reveal progeny adapted to stress environments, such as C in Figure 1. Such a method follows the one proposed by Roy and Murty (1970). They selected for yield under nonstress conditions in the early generations. Progeny were tested for yield under both nonstress and stress environments. Selection proceeded in the nonstress environment only, as based on yield performance data derived from both tests.

We have modified the method for wheat. Instead of yield testing under stress, we apply several tests for drought adaptation response carried out under a stress environment. A stress environment in this context is designed in the field, greenhouse, growth chamber or test-tube. Data derived from such tests are used in decision making upon further selection among or within progeny grown and tested for yield in a nonstress environment (Figure 2). This procedure may be repeated twice as the program moves from one generation to another by using an off-season facility to increase seed of desirable progeny.

An appropriate selection strategy, within an agronomically suitable and high yielding genetic pool, has to employ multiple physiological selection criteria within the proposed scheme. The reasons for using multiple physiological selection criteria are beyond the limits of this chapter. The first selection stage (up to $F_3$) will reduce the population into agronomically adapted, reasonably high yielding materials. The second selection stage ($F_4$), which also is carried out under a stress environment, is exerted by applying one or two physiological adaptation tests. Progeny with susceptible reactions are screened out only if their yield potential is anything but very high. The third stage ($F_6$) involves a reduced population tested for both additional drought adaptation criteria and yield potential. At the $F_7$, the population has been screened and sorted out into three groups of lines essentially conforming with Reitz's (1974) classification of: (a) those that have both high yield potential and drought resistance; (b) those that have medium or low potential but are drought resistant; and (c) those that are drought susceptible but have high yield potential.

Although this design has not been in use long enough to present actual results, it can be demonstrated that the variety classes expected to be derived from such a program are existant in the reality of farming. An example with a group of wheat cultivars will conclude this chapter.

## DROUGHT ADAPTATION AND PRODUCTIVITY IN SOME COMMON WHEATS

Seven modern wheat cultivars were compared in terms of yield performance and stability analysis (Table 1). Yield data were derived from 31 locations x year tests. The major variable responsible for site-to-site variation (environmental index variation) was the amount of precipitation as commonly observed for wheat grown in Israel (Lomas and Shashoua,

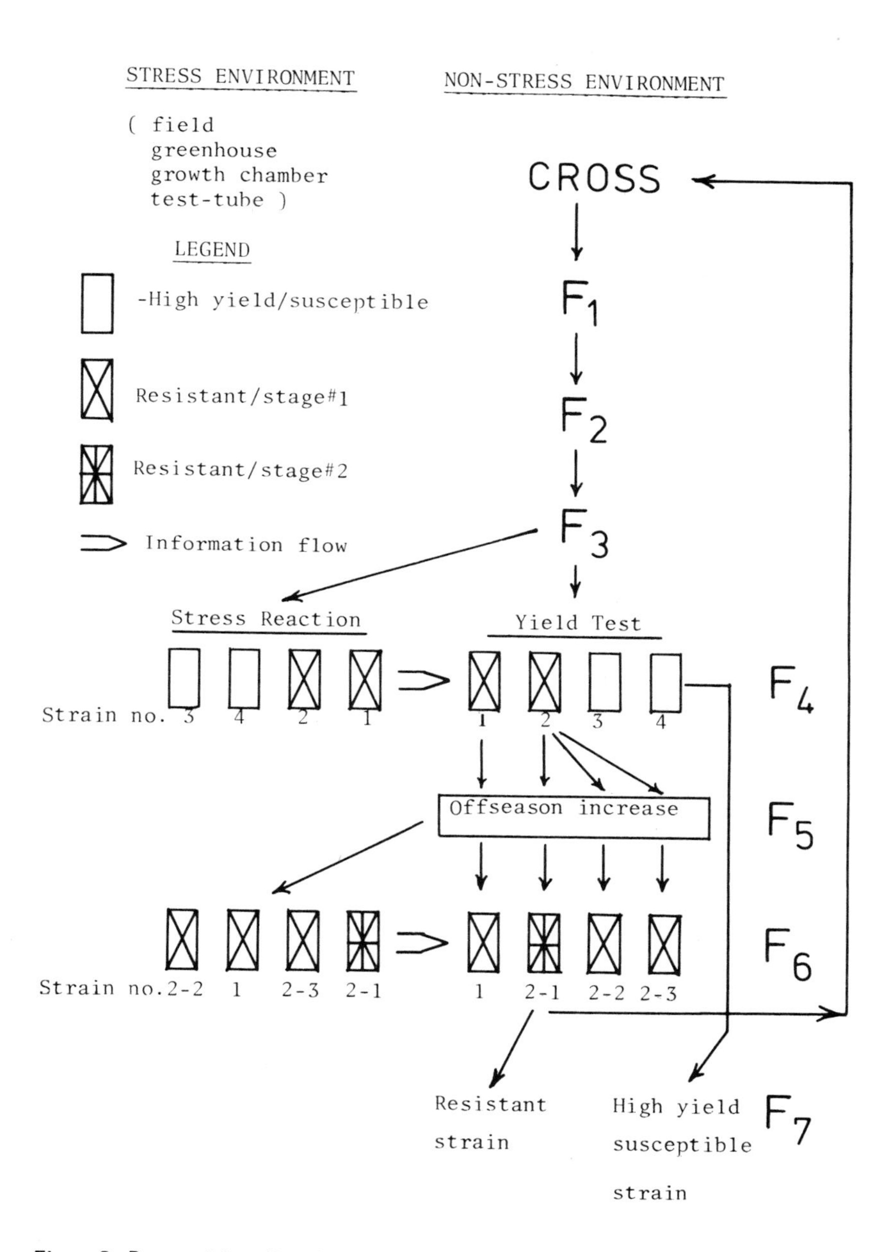


**Figure 2. Proposed breeding design for simultaneous selection for yield under nonstress environments and for drought resistance under stress environments.**

**Table 1. Stability analysis, predicted yield levels, and the resultant classification of seven common wheat cultivars.**

| Cultivar | Regression between Location Mean Yields and Cultivar Yield | Coefficient of Determination | Grain Yield (kg/ha) | | Classification | |
|---|---|---|---|---|---|---|
| | | | Potential[a] | Stress[b] | Yield Potential | Drought Resistance |
| Cajeme-71 | $y = -702 + 1.207x$ | 0.95 | 8950 | 1710 | High | Susceptible |
| BTL | $y = 76 + 0.894x$ | 0.97 | 7230 | 1860 | Low | Susceptible |
| Lakhish | $y = -86 + 1.077x$ | 0.94 | 8540 | 2070 | High | Medium |
| H-895 | $y = 80 + 1,048x$ | 0.95 | 8460 | 2180 | High | Medium |
| Barkaee | $y = 423 + 1.009x$ | 0.90 | 8500 | 2440 | High | Resistant |
| 676 | $y = 405 + 0.958x$ | 0.94 | 8070 | 2320 | Medium | Resistant |
| Miriam | $y = 576 + 0.856x$ | 0.95 | 7420 | 2290 | Low | Resistant |

[a]Cultivar yield at the potential level ($x = 8000$).
[b]Cultivar yield at a stress level ($x = 2000$).

1973). 'Cajeme-71' was revealed as a high potential nonstable cultivar with poor performance under stress. 'BTL' is a low potential nonstable cultivar with poor performance under stress and nonstress environments. 'Barkaee,' which has a high potential yield, also performs reasonably well under stress. 'Miriam' is a "nonstable" cultivar due to relatively superior performance under stress. In this respect, '676' is similar to 'Miriam' but has a slightly better stability. Based on this analysis, and predicted grain yields under stress (Table 1), the cultivars were classified relative to their drought resistance.

Additional cultivar classification for drought response was derived from the relationship between water-use efficiency (WUE) and total annual precipitation for the same 31 location x year tests (Table 2). Typical examples are displayed in Figure 3. 'Miriam' and '676' are typically drought resistant cultivars in terms of production at low moisture levels. Similarly, 'Cajeme-71' and 'Lakhish' are relatively drought susceptible cultivars.

Cultivar classification by the two methods (Table 1 and 2) was fairly consistent. A final classification is presented in Table 3. The relative drought susceptibility of 'Cajeme-71' and 'Lakhish,' in terms of productivity, may be

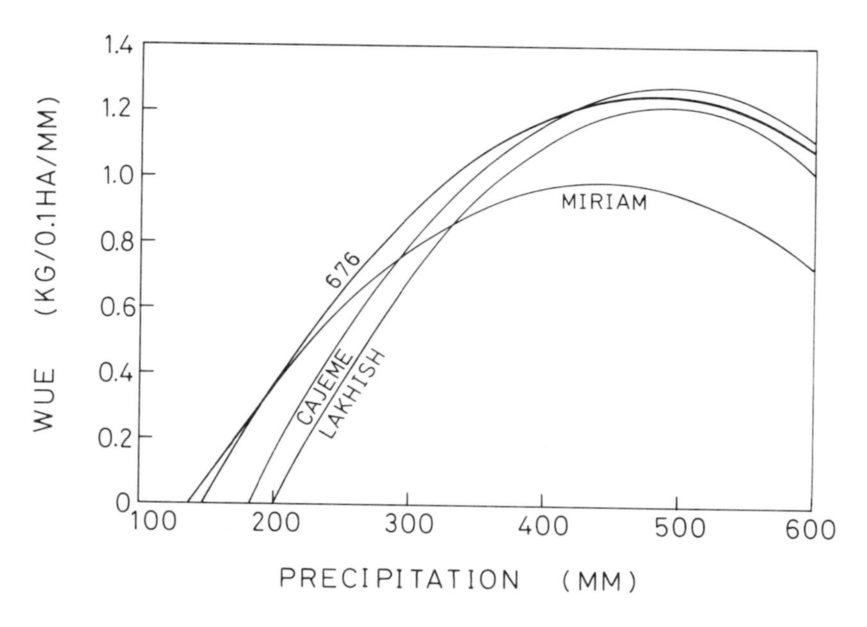


**Figure 3. Second-degree regressions between total annual precipitation (rainfall + irrigation) over 31 location x year tests and water use efficiency (kg/0.1 ha/mm) in several wheat cultivars (functions are presented in Table 2).**

**Table 2.** The regression between annual (seasonal) precipitation (rainfall + irrigation in mm) and water use efficiency (kg/0.1 ha/mm), integrated water use efficiency of given precipitation ranges, and the resultant classification of seven common wheat cultivars.

| Cultivar | Regression | Coefficient of Determination | Simpson's[a] Integral Over | | Classification | |
|---|---|---|---|---|---|---|
| | | | 350 to 550 mm | 100 to 350 mm | Yield Potential | Drought Resistance |
| Cajeme-71 | $y = -1.89 + 0.0129x - 0.0000131x^2$ | 0.86 | 224 | 60 | High | Susceptible |
| BTL | $y = -1.92 + 0.0126x - 0.0000131x^2$ | 0.81 | 186 | 48 | Low | Susceptible |
| Lakhish | $y = -2.35 + 0.0147x - 0.0000151x^2$ | 0.82 | 210 | 39 | Medium | Susceptible |
| H-895 | $y = -1.81 + 0.0130x - 0.0000134x^2$ | 0.86 | 234 | 77 | High | Medium |
| Barkaee | $y = -1.95 + 0.0140x - 0.0000148x^2$ | 0.79 | 233 | 87 | High | Medium |
| 676 | $y = -1.37 + 0.0110x - 0.0000115x^2$ | 0.80 | 217 | 101 | Medium | Resistant |
| Miriam | $y = -1.10 + 0.0095x - 0.0000108x^2$ | 0.79 | 206 | 102 | Low | Resistant |

[a] Integration from data points.

**Table 3. Classification for drought reactions of seven common wheat cultivars in terms of productivity, phenology and pedigree.**

| Cultivar | Drought Resistance Class[a] | Phenology | | Pedigree |
|---|---|---|---|---|
| | | Flowering | Plant Height (cm) | |
| Cajeme-71 (Bluebird 4) | S | Late | 95 | Cno "S"/Son-Kl.Rend/8156 |
| BTL | S | Early | 85 | Bluebird/Tordo/ /Lakhish |
| Lakhish | MS | Late | 100 | Yaktana/ /Norin-10/Brevor/ 3/Florence-Aurore |
| H-895 | M | Early | 90 | Sonora/TZPP/ /Nainari/3 Florence-Aurore |
| Barkaee | MR | Early | 80 | Miriam/ /M-66/Florence-Aurore |
| 676 | R | Early | 95 | H574-1-2-6/Lakhish-212 |
| Miriam | R | Early | 105 | Chapingo 53/ /Norin-10/ Brevor/3/Yaqui-54 |

[a]S = susceptible, MS = medium-susceptible, M = medium, MR = medium-resistant, and R = resistant based on Tables 1 and 2.

partly accounted for by their relatively longer growth duration. Drought escape, as promoted by short growth duration, is an important factor in the dryland Mediterranean environment. The range of earliness in the present cultivars does not necessarily involve a negative effect on yield potential as can be seen in 'Barkaee' (Table 1).

Inspection of cultivar parentage reveals some common backgrounds possibly related to cultivar drought reaction. The effect discussed previously of 'H574-1-2-6' parental stock towards drought resistance can be noted again for '676.' The susceptible 'BTL' is noted for having 'Bluebird' (sister selection of Cajeme-71) in its parentage. The 'Tordo' parental line of this cultivar is well reputed to ascribe wide, easily-desiccated leaves to its progeny. The medium-resistant 'Barkaee' has the resistant 'Miriam' in its parentage.

This simple analysis demonstrates that, in terms of productivity, varieties differ in their drought adaptation. Yield potential and drought resistance indeed appear to be independent of each other. The genetic background of a genotype was apparently associated with a given drought response. It is therefore implied that genetic recombination for improved drought resistance through utilization of resistant genetic materials should be feasible, and that drought resistance can be incorporated into an agronomic, high yield potential background.

Further comparative studies with such cultivar sets, with regard to their physiological drought responses, will provide important insights into the physiological nature of drought resistance relative to crop productivity. Such studies also may reveal major physiological selection criteria that are associated with drought resistance and crop productivity under conditions of water stress.

## NOTES

A. Blum, Division of Field Crops, The Volcani Center, ARO, P.O. Box 6, Bet Dagan, Israel.

## LITERATURE CITED

Atsmon, D. 1979. Drought resistance in barley, wheat and related wild species: developmental physiological and metabolic parameters as possible criteria for breeding. Proc. Israeli-Italian Joint Meeting on Genetics and Breeding of Crop Plants, Mong. Genet. Agrar. IV: 189-202.

Blum, A. 1974. Gentoypic responses in sorghum to drought stress. I. Response to soil moisture stress. Crop Sci. 14:361-364.

Blum, A. 1979. Genetic improvement of drought resistance in crop plants: A case for sorghum. p. 430-445. In H. Mussell and R. C. Staples (eds.) Stress physiology in crop plants. John Wiley & Sons, New York.

Blum, A., and A. Ebercon. 1981. Cell membrane stability as a measure of drought and heat tolerance in wheat. Crop Sci. 21:43-47.

Blum, A., K. F. Schertz, R. W. Toler, R. I. Welch, D. T. Rosenow, J. W. Johnson, and L. E. Clark. 1978. Selection for drought avoidance in sorghum using aerial infra-red photography. Agron. J. 70:472-477.

Blum, A., B. Sinmena, and O. Ziv. 1980. An evaluation of seed and seedling drought tolerance screening tests in wheat. Euphytica 29:727-736.

Bradshaw, A. D. 1965. Evolutionary significance of phenotypic plasticity in plants. Adv. Genetics 13:115-155.

Daday, H., F. E. Biner, A. Grassia, and J. W. Peak. 1973. The effect of environment on heritability and predicted selection response in *Medicago sativa.* Heredity 31: 293-308.

Eberhart, S. A., and W. A. Russell. 1966. Stability parameters for comparing varieties. Crop Sci. 6:36-40.

Finlay, K. W., and G. N. Wilkinson. 1963. The analysis of adaptation in plant breeding programme. Aust. J. Agric. Res. 14:742-754.

Fischer, R. A., and R. Maurer. 1978. Drought resistance in spring wheat cultivars. I. Grain yield responses. Aust. J. Agric. Res. 29:897-912.

Frey, K. J. 1964. Adaptation reaction of oat strains selected under stress and non-stress environmental conditions. Crop Sci. 4:55-58.

Hurd, E. A. 1969. A method of breeding for yield of wheat in a semi-arid climate. Euphytica 18:217-226.

Johnson, G. R. 1977. Analysis of genotypic similarity in terms of mean yield and stability of environmental response in a set of maize hybrids. Crop Sci. 27:837-842.

Johnson, G. R., and K. J. Frey. 1967. Heritabilities of quantitative attributes of oats (*Avena* sp.) at varying levels of environmental stresses. Crop Sci. 7:43-46.

Knight, R. 1973. The relations between hybrid vigor and genotype-environment interaction. Theoretic. and Appl. Genet. 43:311-318.

Lomas, J., and Y. Shashoua. 1973. The effect of rainfall on wheat yields in an arid region. p. 531-537. In R. O. Slatyer (ed.) Plant responses to climatic factors. UNESCO, Paris.

Mederski, H. J., and D. L. Jeffers. 1973. Yield response of soybean varieties grown at two soil moisture stress levels. Agron. J. 65:410-412.

O'Toole, J. C., and T. T. Chang. 1979. Drought resistance in cereals—rice: A case study. p. 374-405. In H. Mussell and R. C. Staples (eds.) Stress physiology in crop plants. John Wiley & Sons, New York.

Reitz, L. P. 1974. Breeding for more efficient water use—is it real or a mirage. Agric. Meterorol. 14:3-11.

Richards, R. A. 1978. Genetic analysis of drought stress response in rapeseed *(Brassica campestris* and *B. napus*). I. Assessment of environments for maximum selection response in grain yield. Euphytica 27:609-615.

Roark, B., and J. E. Quisenberry. 1977. Environmental and genetic components of stomatal behavior in two genotypes of upland cotton. Plant Physiol. 59:354-356.

Roy, N. N., and B. R. Murty. 1970. A selection procedure in wheat for stress environment. Euphytica 19:509-521.

Sullivan, C. Y., and J. D. Eastin. 1974. Plant physiological responses to water stress. Agric. Meteorol. 14:113-127.

Sullivan, C. Y., and W. M. Ross. 1979. Selecting for drought and heat resistance in sorghum. p. 264-281. In H. Mussell and R. C. Staples (eds.) Stress physiology in crop plants. John Wiley & Sons, New York.

Townley-Smith, T. F., and E. A. Hurd. 1979. Testing and selecting for drought resistance in wheat. p. 448-464. In H. Mussell and R. C. Staples (eds.) Stress physiology in crop plants. John Wiley & Sons, New York.

Williams, T. V., R. S. Snell, and C. E. Cross. 1969. Inheritance of drought tolerance in sweet corn. Crop Sci. 9:19-23.

17

# BREEDING PROGRAMS FOR STRESS TOLERANCE IN CORN

**R. M. Castleberry**

Selection for tolerance to environmental stresses undoubtedly has taken place since the first progenitors of modern maize *(Zea mays* L.) deliberately were harvested for replanting. The evolution of the crop under the wide array of environments which occur throughout the center of origin for maize (Galinat, 1977), coupled with the plastic and responsive genetic structure of open-pollinated varieties, would lead to continuous, although slow, improvement in stress tolerance. Various forms of this "mass" selection were continued into the modern era as the primary means of improving stress tolerance in maize (Sprague and Eberhart, 1977). The introduction and general use of maize hybrids, combined with the increased understanding of genetics in the early twentieth century, have led to substantial modification of breeding methodology (Sprague and Eberhart, 1977), but a common underlying theme has been maintained. This theme is the evaluation and selection of maize genotypes based on the exposure of an array of germplasm to an array of environments.

## CURRENT BREEDING PROGRAMS

Most current breeding programs are comprehensive in nature. That is, they seek to develop germplasm which has acceptable performance for a large number of traits that are important to the ultimate user of the germplasm. These traits include maturity, stalk quality, insect and disease resistance, and yield as part of a very extensive list. To be effective, selection for stress tolerance must be an integrated part of this comprehensive program, since improved stress tolerance is useful to the final germplasm user only if it comes in a "package" with acceptable levels of performance for all other important traits.

Most, if not all, current maize breeding programs are designed with some awareness of genotype by environment interactions (Robinson and Moll, 1959; Scott, 1973; Sprague and Eberhart, 1977; Sprague and Federer, 1951). Because of the failure of genotypes to perform at relatively equivalent levels under different environments, effective breeding programs must compensate for specific environmental effects in their evaluation and advancement of germplasm. This compensation most often takes the form of a germplasm evaluation in multiple environments (multiple years and/or locations) in an attempt to find germplasm with high levels of general performance (Sprague and Eberhart, 1977). Since environmental stress frequently is a limiting factor in many of these environments, these breeding programs carry out a continuous, if often unstated, empirical selection for acceptable levels of stress tolerance.

Heat, cold, and drought are three environmental stresses commonly considered important in maize breeding programs. Selection for cold tolerance primarily is concerned with seed germination and seedling emergence in cold soils, seedling vigor, and seedling tolerance to chilling temperatures (0–10 C). These traits have been the focus of considerable breeding work (Grogan, 1970; Jugenheimer, 1976) and some are included as criteria for evaluation and selection in most breeding programs. Evaluation of these traits most typically is done by visual ratings following early seeding, but may include counts of germination, seedling mortality, and measures of plant vigor such as height or weight in response to artificially imposed cold treatments (Mock and Bakri, 1976; Jugenheimer, 1976). Because these procedures are well established and described by a number of authors, they will not be described in detail, although a search for useful germplasm sources continues (Mock and Eberhart, 1972; Eagles and Hardacre, 1979; Mock and Skrdla, 1978).

Heat stress frequently is associated with drought stress. Although heat stress has a number of detrimental effects on corn in the absence of drought (Duncan and Hesketh, 1968; Sullivan et al., 1976; Hall et al., 1979; Peters et al., 1971), much of the breeding work for tolerance to heat and drought stress is done without attempting to separate the two. Heat stress, therefore, will be considered to be a part of drought stress in the following discussion.

## BREEDING PROGRAMS FOR DROUGHT TOLERANCE

### Examples of Breeding Programs

Breeding programs for drought tolerance can be described by a number of characteristics. These include the genetic goal of the program, the sources of germplasm, the stage of germplasm development at which selection occurs, the type of stress environment, the selection criteria, and the type of secondary evaluation. Descriptions of four breeding programs utilizing these characteristics are shown in Table 1. These programs are selected as being illustrative examples among the many breeding programs which include drought tolerance as a selection criteria.

The Iowa Stiff Stalk Synthetic project is part of the cooperative corn breeding program of Iowa State University and USDA and is located at Ames, Iowa[1] (Penny and Eberhart, 1971). Iowa Stiff Stalk Synthetic, BSSS (R), is a composite population of elite cornbelt lines which has undergone a number of cycles of reciprocal recurrent selection with the goal of evaluating reciprocal recurrent selection as a breeding procedure and improving population performance[1] (Penny and Eberhart, 1971). Parental lines that have been developed from BSSS (R) are used widely in commercial maize hybrids. Selection of families for recombination in each cycle is based on performance of half-sib testcrosses at nine to ten Iowa locations. Performance is evaluated primarily on mean yield, although an effective index of agronomic traits, including quality of stalk and roots and dropped ears, is included.[2]

The BSSS (R) is a good illustration of a breeding program that, while selection criteria are based empirically on yield and agronomic traits, has an important indirect selection component for drought. Iowa typically has at least some areas, and frequently very substantial areas, of the state where yield is limited by plant available water. Performance of the testcrosses in the multiple location testing program at the sites scattered across Iowa must be influenced strongly by their ability to perform adequately under drought conditions. Selection for yield under these conditions provides a selection index which strongly integrates the many facets of drought tolerance with other desirable yield attributes. Secondary evaluation of exceptional families comes after they are transferred to the parental development program where additional evaluation, multiple location testing, and selection is done prior to release of inbred lines.[2]

The CIMMYT (Centro Internacional de Mejoramiento de Maiz y Trigo) breeding program for drought is carried out at the Tlaltizapan station in Mexico (Drought stress. p. 37-39. CIMMYT Review, 1979). The goal of the program is to find techniques that will allow identification of genotypes with

---

[1] Hallauer, A. R. 1978. Cooperative state-federal corn breeding project, Ames, Iowa. Fourteenth Annual Illinois Corn Breeders School, University of Illinois, Urbana, Illinois.

[2] Russell, W. A. 1978. Some aspects of the applied maize breeding program and related research at the Iowa Agriculture Experiment Station. p. 13-26. Fourteenth Annual Illinois Corn Breeders School, University of Illinois, Urbana, Illinois.

**Table 1. Some characteristics of selected maize breeding programs for drought tolerance.**

| | Breeding Program | | | |
|---|---|---|---|---|
| **Characteristic** | **BSSS (R)** | **CIMMYT** | **Jensen/Welch** | **Castleberry/LeRette** |
| Genetic goal | Improved population | Improved variety | Commercial hybrids | Commercial hybrids |
| Germplasm source | Elite line composites | 'Tuxpeno-1' | Line x line, line x composite, composites | 'Michoacan 21,' line x line, composites |
| Stage of selection | Half-sib testcrosses | Full-sib family | Inbreeding | Inbreeding |
| Stress environment | Multiple environment | Limited irrigation | Limited irrigation | Limited irrigation |
| Selection criteria | Yield, agronomic index | Growth, firing, synchrony, yield | Firing, synchrony, growth, ear | Firing, synchrony, ear, root |
| Secondary evaluation | Line development | Multiple environment | Multiple environment | Multiple environment |

superior drought tolerance and allow recurrent selection for varietal improvement. Full-sib families from a population (Tuxpeño-1) of the widely grown tropical variety 'Tuxpeño' are evaluated for drought tolerance. Drought stress is imposed by withholding irrigation for variable lengths of time following plant emergence. Selection is based on a series of growth measurements. These include "relative extension index," a relative measure of stem and leaf growth under stress; leaf tissue death score, a rating of leaf firing at maturity; and reproductive synchrony, coincidence of pollen shed and silk emergence.

The CIMMYT program has several interesting characteristics. Selection is based on a well defined drought environment imposed by withholding irrigation. This technique potentially can be integrated easily with the already extensive CIMMYT varietal improvement program based on full-sib yield performance under multiple location testing. In addition, selection criteria includes a number of indices based on relatively simple growth characteristics under drought stress. These indices have a number of advantages. They are reasonably rapid and applicable to a large number of genotypes. Some of them can be applied prior to pollen shed. They can be used to categorize both homogeneous and heterogeneous genetic materials. They appear to be effective physiological integrators of a number of potentially important drought-tolerance traits.

The breeding program listed as Jensen/Welch is a composite of two commercial corn breeding programs. These are the programs of Dr. S. D. Jensen of Pioneer Hi-Bred at York, Nebraska, and V. A. Welch of DeKalb Ag-Research at Fremont, Nebraska. These programs have been active in breeding for drought tolerance for a number of years with the goal of developing commercial hybrids with exceptional drought tolerance, particularly for the western cornbelt. These programs utilize line by line crosses, line by composite crosses, and composites as sources of germplasm. Drought selection is done during parental development (inbreeding) under limited irrigation. Selection of families, ear rows, or individual plants is made on the basis of plant growth and development, leaf firing, tassel blasting, reproductive synchrony, and ear development, as well as agronomic traits. Secondary evaluations under fully irrigated and stress conditions are conducted on finished lines. Hybrids and testcrosses are evaluated under multiple environment testing which specifically includes fully irrigated and normally droughty environments.

These are comprehensive programs with the drought-selection work integrated into the overall program. Selection for drought tolerance during inbreeding allows for the selection of a given set of segregating families under a sequence of stress environments. The timing and severity of stress generally is quite variable from year to year and may limit the apparent progress within an inbreeding cycle. However, as inbreeding is completed and lines are evaluated over years and locations, both as lines and in hybrid combinations, truly superior drought tolerant lines with good yield potential and agronomic traits are identified and recycled back into the program. This secondary evaluation is extremely precise in identifying useful lines since many evaluations occur over a wide range of environments. The time between cycles is long, but there is a very high degree of assurance of the usefulness of the germplasm with which the next cycle of selection is begun.

The program listed as Castleberry/LeRette is the current version of the program described recently (Castleberry and LeRette, 1979). Drought selection is carried out at Yuma, Colorado, with the goal of developing parents for commercial hybrids. Germplasm sources are composites and lines developed from a series of U.S. cornbelt line and composite crosses to 'Michoacan 21'. These materials are crossed to elite U.S. lines and composites and selection is carried out during inbreeding (currently, the $S_1$ and $S_3$ cycles). The selection environment is a short, severe drought stress induced by withholding irrigation water, beginning at floral initiation. In the initial cycles of selection, irrigation was resumed prior to flowering and continued until maturity. Currently, irrigation is being withheld until pollination is completed. Selection is based primarily on successful self-pollination. Additional selection is made on the basis of leaf firing, ear development, root type, and agronomic traits. Secondary evaluation for drought tolerance will come from testcrosses of $S_3$ families or lines which will be evaluated in a multiple location format that includes irrigated and stress environments.

This program is unusual in that an exotic source of a particular drought-tolerance trait was utilized to develop the basic germplasm for the program. 'Michoacan 21' is a source of the *latente* trait described previously in Mexico.[3] Selections from this program display extremely slow vegetative growth under preflowering drought stress and rapid growth and synchrony of reproductive development after resumption of irrigation. These attributes appear to be characteristic of *latente* germplasm.[3] Utilizing the ability of a given plant to self-pollinate as the primary selection criterion is useful from both a physiological and a breeding viewpoint. It insures adequate development of both tassels and ear shoots, adequate synchrony between pollen shed and silking, and provides an unequivocal decision point for selection.

### Strengths and Weaknesses of Empirical Selection Programs

Although the maize breeding programs presented as examples are diverse in their conception and breeding approaches, they are indicative of the

---

[3]Muñoz, O.A. 1975. Relaciones aqua-planta bajo sequia, en varios sinteticos de maiz resistentes a sequia y heladas. M. S. Thesis. Post-graduate College, National School of Agriculture, Chapingo, Mexico.

strengths and weaknesses of empirical selection programs for drought tolerance. The strengths of these programs are notable. Each successfully has integrated breeding for drought tolerance into a comprehensive program that produces well defined and directly useful germplasm. They show the diverse sources and types of germplasm that can be utilized, as well as the diverse germplasm levels at which selection can be carried out. Selection is carried out under environments in which the germplasm will be expected to perform and, in either primary or secondary evaluation, in the form in which it will be utilized. This allows evaluation of the material as a performance package. The selection criteria are diverse, but all fit into a standard maize breeding nursery situation. In addition, they are physiologically integrative and indicate that a number of physiological attributes are functioning at levels adequate for effective plant development under drought stress. Finally, all these programs provide means for recycling materials so that desirable drought-tolerance attributes can be accumulated and recombined effectively.

These programs also are indicative of the weaknesses of empirical selection programs. Although multiple location and/or year evaluations provide a powerful tool for identifying desirable genotypes, the sampling of environments is limited. Thus, important environments are sampled infrequently or not at all during the evaluation of a particular cycle of germplasm. The environments under which selection is made occur at random and are poorly defined so that the physiological attributes selected in one environment may be detrimental in another. This leads to the inconsistent ranking of genotypes so frequently found by maize breeders (Jensen, 1971) and the associated difficulties in identifying germplasm with drought-tolerance characteristics which provide a degree of general tolerance to a wide range of drought environments.

The use of multiple environments as an evaluation tool also implies that while the materials developed in these programs will perform well in the environments utilized, acceptable performance may be limited to similar environments. As an example, selection of BSSS (R) under multiple location and year environments in Iowa undoubtedly lead to materials which perform well under the conditions typical for Iowa. The ability of corn to withstand lengthy periods of post-flowering drought stress by utilizing soil moisture stored in the deep soil profiles is undoubtedly important in BSSS (R). Yet, these physiological and morphological adaptations may not be useful in the Southeast which characteristically has soils with low moisture holding capacities and growing seasons with stress situations of shorter duration. Conversely, selection for tolerance to short-term pre-flowering stress in the Castleberry/LeRette program may not lead to materials which are generally useful in Iowa.

Finally, these empirical programs can be quite slow in the identification and selection of desirable genotypes. Undoubtedly, selection of BSSS (R) for many years under the present system has led to substantial improvement in its tolerance to the drought stresses which occur in Iowa. Also, continued development, evaluation, and finally selection of desirable inbreds for recycling in programs like the Jensen/Welch programs will provide continuing improvement. However, both these types of programs may have effective generation times of decades. In addition, as improvements are made utilizing these techniques, the amount of effort needed to make additional improvements goes up rapidly, perhaps exponentially. Obviously, any improvement in techniques which allows for more effective or rapid improvements in drought tolerance will be very useful.

## CONCLUSIONS

A number of factors are needed to improve selection programs for drought tolerance in maize. The type and frequency of drought stresses which occur in definable maize growing areas need to be determined. Definition of the most important environmental attributes that affect performance is needed to provide breeders with well defined target environments. Then, specific physiological characteristics need to be defined which would provide generally improved maize performance in the target environments. These traits do not need to provide improved performance in every stress situation, but should provide improved performance in some reasonably high frequency of stress situations in the target environments. The knowledge that a particular set of drought-tolerance traits will be useful a substantial proportion of the time will allow breeders to continue development of those characteristics over the "background noise" of unimproved performance in some unfavorable environments.

Root size, stomatal response, growth response, metabolic stability, osmotic adjustment, and turgor maintenance are potentially important characteristics in adaptation of maize to drought environments (Shaw, 1977; Fereres et al., 1978; Turner, 1979; Gotoh and Chang, 1979). Target environments may require specific responses, combinations, sequences, or timing for any or all of these. It very well may be that the important root-growth attribute in a specific environment is the timing of root growth or the level of soil water deficit at which root growth is accelerated. These may be more important than actual total root mass. A genotype that maintains strong stomatal control and conserves soil water prior to critical stages of growth and then undergoes strong osmotic adjustment during these critical periods may be advantageous in some environments. Thus, it may be necessary to include in the list of desired physiological traits factors such as the levels of stress necessary to induce specific physiological responses and the ability to respond differently at different growth stages to the same environmental circumstances. Can these types of traits be defined and effective selection techniques be developed?

Once target physiological characteristics are defined clearly, selection environments and selection criteria for development of the desired genotypes can be chosen with greater certainty. This will require an improved ability to control the selection environment. Special field facilities may need to be developed by controlling rainfall or soil moisture holding capacity either in currently nonirrigated production areas or in irrigated environments with very limited rainfall. The latter will be easier operationally, but may have some environmental factors which are very different than the target production area. For example, relative humidity may be much lower. However, the use of a line source irrigation system in a low rainfall western environment with sandy soil (Hanks et al., 1976) allows the imposition of drought stress at several growth stages in a limited experimental area (Stewart et al., 1975), and may allow selection of germplasm targeted for the short, intermittent droughts common in the Southeast.

The selection criteria utilized must be designed to identify the targeted physiological traits. These criteria may be integrative or mechanistic. Criteria such as leaf firing probably provide an integrative empirical estimate of the ability of a genotype to maintain leaf water potential. They may be more applicable and, therefore, more useful in some nursery situations than the more mechanistic traits, such as leaf water potential or diffusive resistance. The converse also may be true, and specific mechanisms of resistance may need to be defined by specific measurements for selection to be effective.

The most difficult and most important factor in developing improved drought selection programs may be developing an effective system. The system should provide the combination of a particular selection environment and selection criteria specifically designed to generate targeted physiological adaptations for drought tolerance.

## NOTES

R. M. Castleberry, Corn Research Center, DeKalb Ag Research, Inc., Sycamore Road, DeKalb, Illinois 60115.

## LITERATURE CITED

Castleberry, R. M., and R. J. LeRette. 1979. Latente, a new type of drought tolerance? p. 46-56. In Proceedings of the thirty-fourth annual corn and sorghum research conference. American Seed Trade Association, Washington, D.C.

Duncan, W. G., and J. D. Kesketh. 1968. Net photosynthetic rates, relative leaf growth rates, and leaf numbers of 22 races of maize grown at eight temperatures. Crop Sci. 8:670-674.

Eagles, H. A., and A. K. Hardacre. 1979. Genetic variation in maize *(Zea mays* L.) for germination and emergence at 10 C. Euphytica 28:287-295.

Fereres, E., E. Acevedo, D. Henderson, and T. C. Hsiao. 1978. Seasonal changes in water potential and turgor maintenance in sorghum and maize under water stress. Physiol. Plant. 44:261-267.

Galinat, W. C. 1977. The origin of corn. p. 1-47. In G. F. Sprague (ed.) Corn and corn improvement. American Society of Agronomy, Madison, Wisconsin.

Gotah, K., and T. T. Chang. 1979. Crop adaptation. p. 234-261. In J. Sneep and A. J. T. Hendriksen (eds.) Plant breeding perspectives. Pudoc, Wageningen, the Netherlands.

Grogan, C. O. 1970. Genetic variability in maize *(Zea mays* L.) for germination and seedling vigor at low temperatures. p. 90-98. In Proceedings of the twenty-fifth annual corn and sorghum research conference. American Seed Trade Association, Washington, D.C.

Hall, A. E., K. W. Foster, and J. G. Waines. 1979. Crop adaptation to semi-arid environments. p. 148-179. In A. E. Hall, G. H. Cannell, and H. W. Lawton (eds.) Agriculture in semi-arid environments. Springer-Verlag, Berlin.

Hanks, R. J., J. Keeler, V. P. Rasmussen, and G. D. Wilson. 1976. Line source sprinkler for continous variable irrigation-crop production studies. Soil Sci. Soc. Amer. J. 40:426-429.

Jensen, S. D. 1971. Breeding for drought and heat tolerance in corn. p. 198-208. In Proceedings of the twenty-sixth annual corn and sorghum research conference. American Seed Trade Association, Washington, D.C.

Jugenheimer, R. W. 1976. Cold tolerance. p. 193-199. In R. W. Jugenheimer (ed.) Corn improvement, seed production and uses. John Wiley & Sons, New York.

Mock, J. J., and A. A. Bakri. 1976. Recurrent selection for cold tolerance in maize. Crop Sci. 16:230-233.

Mock, J. J., and S. A. Eberhart. 1972. Cold tolerance in adapted maize populations. Crop Sci. 12:466-469.

Mock, J. J., and W. H. Skrdla. 1978. Evaluation of maize plant introductions for cold tolerance. Euphytica 27:27-32.

Penny, L. H., and S. A. Eberhart. 1971. Twenty years of reciprocal recurrent selection with two synthetic varieties of maize *(Zea mays* L.). Agron. J. 63:900-903.

Peters, D. B., J. W. Pendleton, R. H. Hageman, and C. M. Brown. 1971. Effect of night air temperature on grain yield of corn, wheat, and soybeans. Agron. J. 63:809.

Robinson, H. F., and R. H. Moll. 1959. The implications of environmental effects on genotypes in relation to breeding. p. 24-31. In Proceedings of the fourteenth annual hybrid corn-industry research conference. American Seed Trade Association, Washington, D.C.

Scott, G. E. 1967. Selecting for stability of yield in maize. Crop Sci. 7:549-551.

Shaw, R. H. 1977. Water use and requirements of maize—a review. p. 119-134. In Proceedings of the symposium on the agrometeorology of the maize (corn) crop. World Meteorological Organization, Geneva, Switzerland.

Sprague, G. F., and S. A. Eberhart. 1977. Corn breeding, p. 305-362. In G. F. Sprague (ed.) Corn and corn improvement. American Society Agronomy, Madison, Wisconsin.

Sprague, G. F., and W. T. Federer. 1951. A comparison of variance components in corn yield trials: II. Error, year x variety, location x variety, and variety components. Agron. J. 43:535-541.

Stewart, J. I., R. D. Misra, W. O. Pruitt, and R. M. Hagan. 1975. Irrigating corn and grain sorghum with a deficient water supply. Amer. Soc. Agric. Eng. Trans. 18:270-283.

Sullivan, C. Y., N. V. Norcio, and J. D. Eastin. 1976. Plant responses to high temperatures. p. 301-317. In A. Muhammed, R. Aksel, and R. C. Von Borstel (eds.) Genetic diversity in plants. Plenum Press, New York.

Turner, N. C. 1979. Drought resistance and adaptation to water deficits in crop plants. p. 343-372. In H. Mussell and R. C. Stables (eds.) Stress physiology in crop plants. J. Wiley & Sons, New York.

# 18

# BREEDING PROGRAMS FOR STRESS TOLERANCE IN FORAGE AND PASTURE CROPS

**G. W. Burton**

Breeding forage and pasture crops for stress tolerance is not new. From the time since man recognized the value of forages and began to introduce them to new environments, he has been concerned with increasing their tolerance to stress. His objective has been to grow and use them as he previously had. If man's new home was colder, hotter, dryer, or wetter than his former home, his forage had to be changed to tolerate the stress of the new environment. It had to be acclimatized or adapted. A brief history of the development of alfalfa as the best forage in much of the world will describe in a general way the early improvement programs for many other forage species.

The word "alfalfa" means "best forage" in Arabic. It generally is agreed that it originated in southwestern Asia and probably was planted in that region ages before history was written. The earliest records indicate that the superior feeding value and soil-building properties of alfalfa already were well understood (Wing, 1912). Pliny records that in 490 B.C. when the Persians invaded Greece, they introduced alfalfa to feed their chariot horses, camels, and domestic animals. From Greece, alfalfa was spread westward to Italy, other European countries, and Spain with the conquering armies as "fuel" for their "war machines."

At the beginning of the sixteenth century, Spanish explorers brought alfalfa to the Americas. Soon it became distributed over Chile and Peru. In the 1850's, gold prospectors going to California by the all-water route around Cape Horn collected alfalfa seed in Chile and planted it in California. From there it moved eastward to Kansas, Nebraska, Iowa, Illinois, and Ohio. But these alfalfas were winterkilled when grown in the northern states.

In 1857, Wendelin Grimm brought his family from the Grand Duchy of Baden, Germany, to Carver County, Minnesota. The next spring he planted about 15 pounds of alfalfa seed that he had brought from his native land. The winters of Minnesota, more severe than winters in Baden, killed most of his alfalfa plants. Seed from the surviving plants gave rise to a population that suffered less from winterkilling. Gradually, by planting seed of the surviving plants again and again, Grimm developed an alfalfa that could become relatively dormant in the fall and resist the cold of Minnesota. Nearly 50 years elapsed before experiment stations proved that Grimm's alfalfa, now bearing his name, could be grown dependably where it was too cold for other alfalfas.

## POPULATION IMPROVEMENT AND SELECTIVE BREEDING METHODS

Development of 'Grimm' alfalfa provides an excellent example of the power of the breeding method that I shall call "population improvement." The necessary ingredients for success are: (1) The population must possess genetic variability. Alfalfa qualified as do many other forages. (2) There must be an effective stress screen that can eliminate the inferior plants. Minnesota's winters supplied the cold screen. (3) Surviving plants must be intermated. Bees, both tame and wild species, intermated Grimm's surviving alfalfa plants. (4) Seed from intermated survivors of the screen must be planted again and again, each planting completing another cycle of what many breeders call "recurrent selection." Grimm practiced such recurrent selection. If the population is variable, the stress screen effective, and the intermating of the survivors complete, each cycle should increase the frequency of those genes that provide greater stress tolerance.

Most of the grasses and legumes planted for pasture and forage in the U.S. are naturalized foreigners. Planted in a new environment by the immigrants who brought them, these plants frequently experienced stresses of cold or drought great enough to kill them. If some plants survived, seed taken from them to plant a new pasture or hay meadow constituted the beginning of a natural recurrent selection program. Years of harvesting and planting seed of a forage plant within a region have improved the stress tolerance of many forages without the aid of controlled plant breeding.

Breeding forages for increased stress tolerance began in a very modest way in the beginning of this century. From 1903 to 1915, alfalfa varieties bearing the names 'Baltic,' 'Cossack,' and 'Ladak' and carrying greater winterhardiness, were selected from Experiment Station introduction nurseries (Tysdal and Westover, 1937). Most of the alfalfas planted in these nurseries had been collected by botanists or agronomists in Asia where winters were colder than those in most of the U.S. It was not surprising, therefore, that these nurseries, located in the northern states of the U.S., should reveal the more winterhardy plants and permit the production of winterhardy varieties.

From 1914 to 1930, agronomists at MacDonald College, Quebec, Canada, developed inbred lines of orchardgrass that survived the winters and then combined the best of them to make a variety called 'Avon' (Vinall and Hein, 1937). They reported that 'Avon' orchardgrass was decidedly more winterhardy and therefore longer-lived than commercial orchardgrass.

## HYBRIDIZATION BREEDING METHODS

Modern forage breeders wishing to increase the water or temperature stress tolerance of a species will search for germplasm in the driest, hottest, or coldest regions where the species or its compatible relatives grow. Occasionally, such germplasm may be good enough to become a new variety. More frequently, however, its genes for stress tolerance are transferred to a superior variety by hybridization and selection. The development of 'Tifton 44' bermudagrass is a good example.

Coastal bermudagrass is a vegetatively propagated $F_1$ hybrid between a South African introduction and 'Tift' bermuda, a local selection (Burton, 1954). It has excellent heat and drought tolerance but lacks enough cold tolerance to grow dependably in the northern third of the bermudagrass belt. Bermudagrass originated in Africa where temperatures are too mild to permit evolution of winterhardy types. Because it is widely distributed across southern Europe, we reasoned that winterhardy plants might be found there at high altitudes or in northern latitudes. In 1966, I collected bermudagrass plants in such sites in Europe and, with the help of agronomists in the U.S., determined their relative winterhardiness. It took a planting made by Milo Tesar at Lake City, Michigan, to reveal a collection from Berlin, Germany, as the most winterhardy of the lot. Prof. V. H. Sukopp had found it growing along a railroad track in Berlin and for 15 years had been taking his ecology class there to see it.

Because the Berlin bermudagrass lacked vigor, we hybridized it with Coastal and several other superior bermudas in order to combine their desirable traits. Hundreds of these $F_1$ hybrids were space-planted at the Mountain Experiment Station in North Georgia in 1970 and 1971 to be screened for cold tolerance. Of the 64 better survivors tested for 3 years in a 9 x 9 lattice square test at Tifton, Georgia, nine were chosen for testing by 30 agronomists in 14 states. Reports from stations as far north as Stillwater, Oklahoma, and Paris, Kentucky, revealed the best of these hybrids to be 'Tifton 44' (Burton and Monson, 1978). Because it could be propagated vegetatively, it was ready for farm distribution and was released in the spring of 1978.

## MUTATION BREEDING METHODS

With a good screen, mutation breeding may be used to improve stress tolerance. Mutation breeding may be the best tool for improving sterile varieties that cannot be hybridized or for adding stress tolerance to a superior genotype without altering its special traits. Our development of 'Coastcross 1-M3' bermudagrass (Burton et al., 1980) is a good example of the methodology involved and some of the problems encountered.

'Coastcross-1' bermudagrass is a sterile $F_1$ hybrid between *Cynodon dactylon* (Coastal bermuda) and *C. nlemfuensis* (a highly digestible cold-intolerant bermudagrass from Kenya) (Burton, 1972). When compared with Coastal bermudagrass, it yields as much dry matter, is 12% more digestible, and therefore gives 30 to 40% better daily per-acre gains by steers. It is propagated easily and is an outstanding forage grass in the tropics. Its lack of cold tolerance restricts its use in the U.S. to Florida and south Texas. The sterility of 'Coastcross-1' prevents improvement by hybridization and selection methods. Further, the plant breeder would like to preserve the rare combination of characters in 'Coastcross-1' as he increases its cold tolerance. If 'Coastcross-1' had the rhizones of its Coastal parent, it would probably survive frosts that kill above-ground growth but do not freeze the soil. Apparently, dominant genes from the Kenya parent mask the genes for rhizome development in 'Coastcross-1.' Thus, irradiation to remove the dominant genes masking rhizome development or cold tolerance was indicated.

On June 21, 1971, we cut and baled about 500,000 green stems capable of developing plants, exposed them to 7000 Roentgens in the Comparative Animal Research Laboratory at Oak Ridge, Tennessee, on June 22, and planted them at the Mountain Experiment Station in North Georgia. Excellent growing conditions enabled a large percentage of them to become well-established before frost. A severe winter killed all but four of the plants. These were taken to Tifton where they were increased and tested. No differences between the survivors and the untreated check were noted until a moderate winter following three mild ones proved that one of the survivors was significantly more winterhardy (Burton et al., 1980). In the more severe winter that followed, 'Coastcross 1-M3' lost stand as badly as the untreated check.

Since 1971, we have planted up to one million treated stems of 'Coastcross-1' in June of six years at Blairsville, Georgia. These have become well-established and have survived the winters in numbers too great to be winterhardy mutants. In only one year out of seven has the weather at Blairsville provided an adequate screen for improving the winterhardiness of 'Coastcross-1' bermudagrass.

## SCREENING METHODS FOR TOLERANCE

In 1976, a plant physiologist, who had spent years studying cold stress in alfalfa, said he had found no substitute for the direct field-planting method of testing varieties for winterhardiness. With so many interacting physical and chemical factors determining cold tolerance, screens using only one or two of them have little chance for success.

Drought is unquestionably the most important environmental factor influencing growth of plants in the semiarid regions of the world. Nevertheless, comparatively little specific breeding for drought resistance has been carried on, probably because of the complex and poorly understood nature of plant reactions to severe moisture stress. After reviewing more than a hundred papers dealing with techniques of breeding for drought resistance, Ashton (1948) concluded: "In general, physical characters such as water requirements and transpiration rate, and anatomical and morphological characters have not been found to provide a simple and practical index of drought resistance in selection work. In the case of physico-chemical characters, there is less general agreement as to their significance in breeding investigations." Thus, most investigators have used the direct method of testing for drought resistance in field or pot experiments and in drought chambers. Wilting tests and techniques that permit rating varieties on their resistance to artificial drought and heat have been promising. Of particular interest to breeders, who must screen large populations, have been methods that involve the testing of seedlings or germinating seeds. A lack of agreement concerning the value of these techniques, however, is proof that a standard method of screening all plant populations for drought resistance has not been established. The complexity of the problem and Ashton's observation, "In all the crops investigated, the capacity to endure drought varies according to the stage of growth," indicate that there may not be one standardized method to test for drought resistance in all plants. I believe, however, that original research in this area can lead to the development of methods that will facilitate greatly the breeding of drought-resistant varieties.

Our discovery of the *tr* gene in pearl millet, which removes all trichomes and makes a cuticle with few cracks and will reduce water loss from pearl millet leaves by as much as 35%, seems to be significant. We have observed that lines and hybrids homozygous for the *tr* gene wilt less and make more growth than their near-isogenic normal counterparts during periods of water stress. The absence of barbs on the edges of the leaves and the balling-up of water on the leaves, as on a freshly waxed auto, make it easy to screen a population for the *tr tr* character. If the *tr* gene will reduce water loss and increase water stress tolerance in all pearl millet genotypes, it can be easily incorporated into outstanding varieties or hybrids in a 2-year backcrossing program. It should provide an easy procedure to increase drought tolerance. If sought, similar genes might be found or induced in other species.

Although there has been comparactively little specific breeding for drought tolerance, much progress has resulted as breeders have improved the adaptation of species to arid environments. Coastal bermudagrass is one of the most drought-tolerant grasses for the southern U.S. It is the best of some 5000 $F_1$ hybrids between a south African bermuda and a very good local selection called 'Tift' (Burton, 1948). Its parents were undoubtedly drought tolerant. It was selected from a space-planting on a deep sandy soil and was tested in replicated plots on similar soil. Its ability to stay green throughout the season and out-yield other bermudas were the main bases for its selection and release. During its evaluation at Tifton (which has an annual rainfall of 120 cm), some pressure for drought tolerance undoubtedly was applied by short periods of drought.

Released in 1943, Coastal bermuda's drought tolerance was not fully appreciated until 1954 when a growing-season rainfall of 34 cm was less than half the amount usually received. Then, Coastal bermuda growing on sand-dune soil remained green and produced six times as much dry matter as the browned common bermuda check. During the perious year of 1953, in which the growing-season rainfall was three times as much, Coastal yielded only twice as much as common in the same test (Burton et al., 1957). A study of root systems showed that while Coastal had less total kg/ha of roots, it had twice as much root mass from the 0.3 to 2.5 m depth (Burton et al., 1954). Coastal roots also penetrated the soil faster and accumulated nearly three times as much $^{32}P$ as common bermuda. Apparently, the heterosis observed in above-ground parts of Coastal bermudagrass extended into the root system below ground. These studies suggested that bermudagrass seedlings could be screened for root-system development and drought tolerance by space-planting them 3 x 3 m apart on a uniform sand-dune soil.

In April, 1960, we planted 3 x 3 m apart on sand-dune soil 530 $F_1$ hybrids between Coastal bermudagrass and two promising introductions of *C. nlemfuensis* from Kenya and Ethiopia. By fall, one-third of the plants had died, and by the following spring, less than half of them remained. Fourteen of the better of these were increased and planted in a replicated plot test with management variables. The best of these fourteen became 'Coastcross-1' bermudagrass described earlier. Observations during droughts in Tifton and reports from Texas indicate that 'Coastcross-1' is equal, if not superior, to Coastal bermudagrass in drought tolerance.

## SUMMARY

Forage breeders developing improved forage varieties for semiarid regions have been plagued by the variability in the climate from year to year. To creen several thousand mature plants for a number of characters including drought and temperature tolerance, they have needed land areas too large for environmental control except for irrigation. In an effort to find natural stresses, they have planted their nurseries in more than one environment. This has added greatly to the cost of their operation.

Effective natural water and temperature stresses occur less than a third of the time on the average in the Great Plains of the U.S. Forage breeders attempting to breed for water and temperature stress in the Great Plains will waste more than half of their time because of the lack of an effective climatic screen. Varieties released after three years of testing in a cool, wet cycle may be inferior to the old standbys when the dry years come.

Forages are grown to feed livestock. When they fail, the livestock farmer must either pay exorbitant prices for replacement feed or sacrifice his livestock for too little return. Consequently, dependability that includes stress tolerance must be one of the most important objectives in any forage breeding program. The need is for more effective screens for stress tolerance. I would like to see a stress screen developed on a deep sand in the desert where summer temperatures are always high and water is almost always short. Here, with the use of some irrigation, I believe a water-temperature stress screen could be created that could be duplicated with little variation annually. With such a screen, I believe forage breeders could increase significantly the stress tolerance of their forages regardless of the species or the breeding methods used. Without such a screen, increasing the stress tolerance of forages by breeding will be slow and very expensive.

## NOTES

Glenn W. Burton, USDA/SEA/AR, Coastal Plain Station, Tifton, Georgia 31793.

## LITERATURE CITED

Ashton, T. 1948. Technique of breeding for drought resistance in crops. Commonwealth Bur. Plant Genet. Tech. Commun. No. 14.

Burton, G. W., E. H. DeVane, and R. L. Carter. 1954. Root penetration, distribution and activity in southern grasses measured by yields, drought symptoms and P32 intake. Agron. J. 46:229-233.

Burton, G. W. 1954. Coastal bermudagrass. Georgia Agricultural Exp. Bul. NS2.

Burton, G. W., G. M. Prine, and J. E. Jackson. 1957. Studies of drought tolerance and water use of several southern grasses. Agron. J. 49:498-503.

Burton, G. W. 1972. Registration of Coastcross-1 bermudagrass. Crop Sci. 12:125.

Burton, G. W., and W. G. Monson. 1978. Registration of Tifton 44 bermudagrass. Crop Sci. 18:911.

Burton, G. W., M. J. Constantin, J. W. Dobson, Jr., W. W. Hanna, and J. B. Powell. 1980. An induced mutant of Coastcross-1 bermudagrass with improved winterhardiness. Environ. Exp. Bot. 20:115-117.

Tysdal, H. M., and H. L. Westover. 1937. Alfalfa improvement. p. 1122-1153. In Yearbook of agriculture. U. S. Government Printing Office, Washington, DC.

Vinall, H. N., and M. A. Hein. 1937. Breeding miscellaneous grasses. p. 1032-1102. In Yearbook of agriculture. U. S. Government Printing Office, Washington, DC.
Wing, J. 1912. Alfalfa farming in America. The Breeder's Gazette, Chicago, Illinois.

# SECTION VI

# STRESS RESEARCH IN CONTROLLED ENVIRONMENTS

*To begin this section, D. Patterson discusses the differences in peak and total irradiance and atmospheric humidity to which plants are exposed in growth chambers and out-of-doors and the effects on leaf structure and photosynthesis. He also describes experiments designed to compare the effects of various temperature regimes on certain dangerous weeds as compared with their host plants. The results of one such experiment indicate that Rottboellia is likely to be a serious pest only in the warmest part of the United States. Another experiment shows that a few cool nights inhibit the growth of cotton more than that of certain competing weeds because leaf expansion is reduced more in cotton than in the weeds. He also reports that growth analysis is useful in this type of study because it indicates what plant organs are being inhibited most by unfavorable temperature.*

*To point out the usefulness of controlled environments for evaluating effects of stress on crop growth and yield, Gold and Raper emphasize that stress conditions in the field are, by definition, uncommon events. Environmental control is necessary because the experimental conditions available under natural field conditions are not likely to be those that give the information needed in the study of stress physiology. Such controlled-environment facilities as phytotrons and rhizotrons provide simple and reproducible observations of plant responses to specific intervals of stress conditions. However, extrapolation of the results to predict behavior at the level of field communities is not straightforward. Mathematical system modeling is a tool to help partition the plant-environment system appropriately into subsystems, to help organize information within the individual subsystems, and to help describe the interface between subsystems within levels and between levels. The information gained at the single plant level in controlled environments thus can be extrapolated to predict crop behavior at the field and management level through crop simulation models such as SOYMOD/OARDC, as discussed by Curry. In the discussion, it was pointed out that mathematical models are essential for experimentally testing the complex hypothesis required for crop reactions to stress. They are valuable in sharpening our understanding of the time when temperature and moisture stress affect yield and point to management and genetic improvements needed to reduce yield losses.*

19

# PHENOTYPIC AND PHYSIOLOGICAL COMPARISONS OF FIELD AND PHYTOTRON GROWN PLANTS

D. T. Patterson

The increasing use of controlled environment facilities in plant research has led to questions concerning the comparability of data collected in such facilities with data collected under field conditions. Knowledge of the differences between plants grown in controlled environments and in the field is necessary before results of controlled environment experiments can be successfully extrapolated to field situations. Controlled environment research is essential in determining the responses of plants to water and temperature stress (Kramer, 1978). Such research can lead to predictions of the stress responses of plants under field conditions.

Various studies have considered aspects of the comparative growth and physiological responses of plants grown in controlled and field environments. For example, photosynthetic rates of plants growing in growth chambers and greenhouses are often less than those of comparable plants in the field (Bazzaz, 1973; Elmore et al., 1967; El-Sharkawy et al., 1965; Hesketh, 1968). However, Ludlow and Ng (1976) reported similar photosynthesis-irradiance response curves for *Panicum maximum* Jacq. var. *trichoglume* grown in growth chamber, greenhouse, and field. Gausman et al. (1971) reported that leaves of cotton, *(Gossypium hirsutum* L.) from a growth chamber had higher water contents and lower light reflectances than leaves from the greenhouse or field. Raper and Downs (1976) summarized results of a comprehensive research program directed at reproducing the field phenotype of tobacco *(Nicotiana tabacum* L.) in controlled environment chambers. Relevant discussions of the "philosophy of phytotronics" may be found in Went (1957) and Evans (1963).

A cooperative study of the comparability of plants grown in the field and phytotron was conducted at the Duke University and North Carolina State University units of the Southeastern Plant Environment Laboratories (Kramer et al., 1970) between 1973 and 1976. Many of the results are already available in the published literature (Raper and Downs, 1976; Patterson et al., 1977; Patterson, Peet, and Bunce, 1977; Davies, 1977; Van Volkenburgh and Davies, 1977). In the following sections, I will summarize and review the major results and conclusions as they relate to the physiological and phenotypic comparability of field and phytotron plants.

## GROWTH AND PHENOLOGY

### Experimental Conditions

Cotton and soybean [*Glycine max* (L.) Merr.] were grown in four environments:

1. Artificially illuminated plant growth room with 26 C day temperature; 20 C night temperature, 12 hour full fluorescent and incandescent lighting (0700 to 1900 hours, 650 $\mu$Einsteins $m^{-2}s^{-1}$ photosynthetic photon flux density, PPFD) with 1 hour extensions with incandescent lighting only (0600 to 0700 and 1900 to 2000) providing 50 $\mu$Einsteins $m^{-2}s^{-1}$ PPFD; daytime relative humidity of 70%, $CO_2$ concentration 300 to 350 ppm.

2. Air-conditioned greenhouse with 26/20 C day/night temperature with 12 hour thermoperiod; natural photoperiod; maximum PPFD of 2000 $\mu$Einsteins $m^{-2}s^{-1}$; relative humidity 70%; $CO_2$ concentration 300 to 350 ppm.

3. Field plot adjacent to Duke University Phytotron ("Duke field").

4. Experiment Station field plot at Clayton, North Carolina ("Clayton field").

In the first three environments, the plants were grown in 25-cm diameter plastic pots in a 1:1 (v:v) mixture of gravel and vermiculite. The pots were watered to excess three times daily with half-strength Hoagland's solution. At the Clayton field site the plants were grown in Wagram loamy sand (Typic Paleudult) and were not irrigated.

A day/night regime of 26/20 C for the growth chamber and greenhouse was chosen to provide 310 degree hours (above 0 C) per day and 240 degree hours per night with the 12 hour thermoperiod. These degree hour totals were equivalent to average field conditions at Durham, North Carolina, during the months of June, July, and August. Application of Went's (1957) method for calculating effective day and night temperatures based on average maximum and minimum temperatures gave similar results. Average daily maximum and minimum temperatures were 28/17 at Duke field and 29/19 at Clayton field. Average daily total photosynthetic photon flux was 37.7

Einsteins $m^{-2}day^{-1}$ at Duke field, 28.3 Einsteins $m^{-2}day^{-1}$ in the greenhouse, and 24.8 Einsteins $m^{-2}day^{-1}$ in the growth chamber. The average maximum vapor pressure deficit for 1974 to 1976 was 15.6 mm Hg at Duke field. In the growth chamber and greenhouse, the vapor pressure deficit at 26 C and 70% relative humidity was 7.6 mm Hg.

**Growth Analysis**

Five plants were harvested from each environment at weekly intervals from about 10 days after emergence until maturity. Leaf areas and weights of plant parts were determined. The integral method of mathematical growth analysis (Ondok and Kvet, 1971) was used to calculate net assimilation rate (NAR = rate of dry matter production per unit leaf area) and leaf area duration (LAD = total amount of leaf area present) for the interval from emergence to Day 120 in soybean and emergence to Day 160 in cotton. The total dry matter production over the specified interval is the product of the NAR and LAD. Phenological events were observed at regular intervals during early growth to determine differences in times of flowering and fruit development in the different environments.

In cotton, differences in growth pattern appeared at about the beginning of boll development at 65 to 70 days after planting (Figure 1). From Day 60 to Day 100, the rate of dry matter production in the Duke field

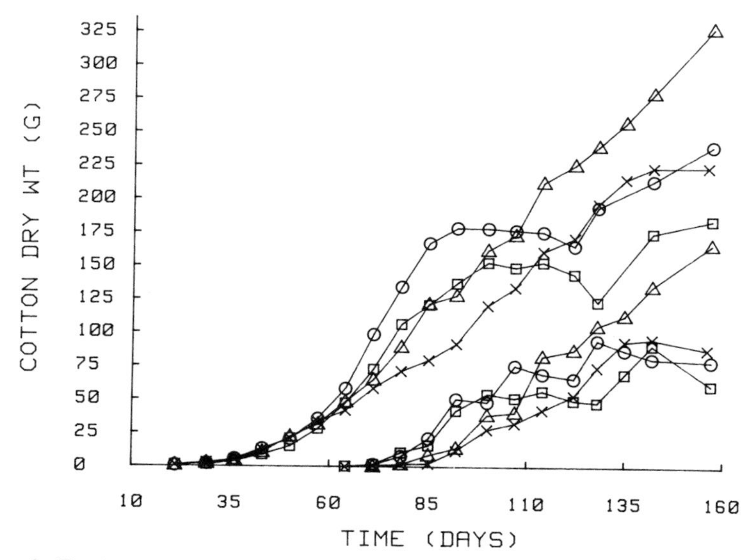


**Figure 1. Total plant dry matter accumulation and boll weight (lower right) of cotton grown in growth chamber (+), greenhouse (△), in pots in Duke field (○), and in soil in Clayton field (□).**

plants exceeded that in the other environments. Dry weight gain in the two field environments ceased from about Day 100 to Day 128 during the period of rapid boll weight gain. This occurred because of a concurrent loss of leaf biomass which had begun earlier at about Day 85 (Figure 2). In the growth chamber and greenhouse, dry weight increased at a fairly constant rate from Day 60 to harvest at Day 157. There was no net loss of leaves although leaf area ceased to increase rapidly after Day 85. Boll weight continued to increase through the entire growth period in the greenhouse, but leveled off after Day 135 in the other three environments.

The timing of reproductive events was similar in all the environments. The first squares appeared at 34 to 36 days after planting, and flowers appeared at 57 to 58 days. Boll opening began about one week earlier in the two field sites (Day 122) than in the growth chamber (Day 128) or greenhouse (Day 131).

Total dry matter production from planting until harvest was significantly greater in the greenhouse than in the other environments (Table 1). Leaf area duration was similar in the greenhouse and growth chamber, and LAD of plants in both these environments exceeded that in the field environments. Net assimilation rate was least in the growth chamber, probably because of the lower irradiance received by the chamber plants. Total boll weight was greatest in the greenhouse and did not differ significantly among the other

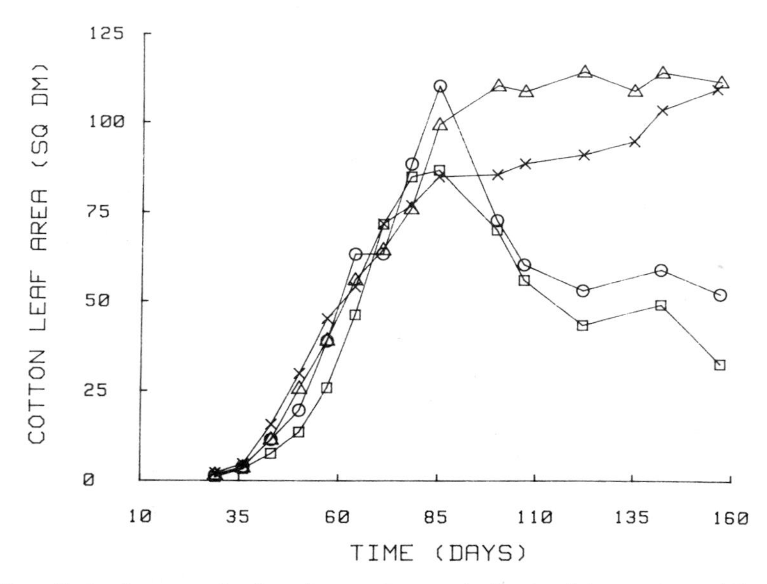


**Figure 2. Leaf area production of cotton in growth chamber (+), greenhouse (△), in pots in Duke field (○), and in soil in Clayton field (□).**

**Table 1. Mathematical analysis of the growth of cotton and soybean in controlled and field environments.**

| Environment | Cotton | | | Soybean | | |
|---|---|---|---|---|---|---|
| | $\Delta W$[a] | NAR[b] | LAD[c] | $\Delta W$ | NAR | LAD |
| | *(g)* | *(g dm$^{-2}$ day$^{-1}$)* | *(dm$^2$day)* | *(g)* | *(g dm$^{-2}$ day$^{-1}$)* | *(dm$^2$day)* |
| Chamber | 227 | 0.024 | 9532 | 157 | 0.021 | 7390 |
| Greenhouse | 342 | 0.032 | 10618 | 288 | 0.023 | 12699 |
| Duke field | 248 | 0.035 | 7084 | 309 | 0.024 | 12952 |
| Clayton field | 192 | 0.033 | 5928 | 138 | 0.018 | 7440 |

[a] $\Delta W$ = dry matter production.
[b] NAR = net assimilation rate.
[c] LAD = leaf area duration.

three environments (Figure 1).

In soybean, the chamber plants began to flower much earlier (Day 37) than the plants in the other three environments (Day 60, 62, and 63 for greenhouse, Clayton field, and Duke field, respectively). Because of this earlier flowering, podding began in the chamber plants about 30 days earlier (Figure 3). Vegetative growth leveled off in the chamber plants as pod-filling

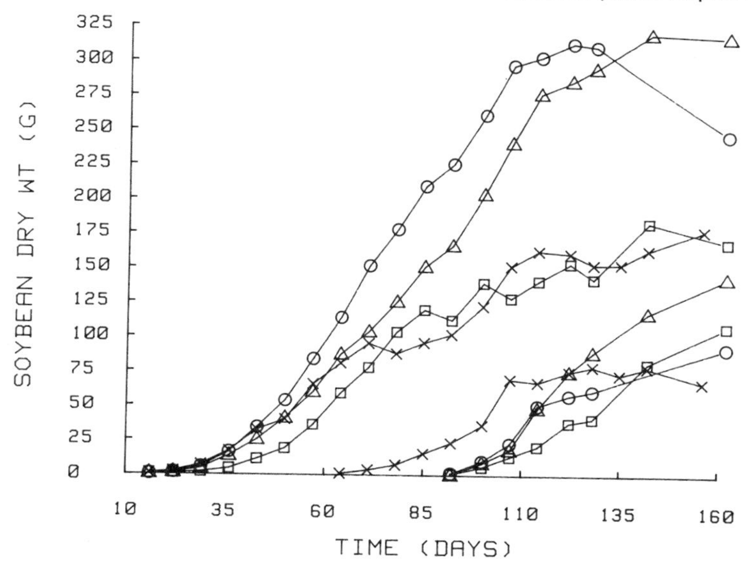


**Figure 3. Total dry matter accumulation and pod weight (lower right) of soybean in growth chamber (+), greenhouse (△), in pots in Duke field (○), and in soil in Clayton field (□).**

occurred (beginning on Day 62) and leaf area declined after about Day 70. Pod-filling began at 88 to 90 days in the greenhouse, Duke field, and Clayton field. Vegetative growth and leaf area production declined during pod-fill (Figure 4).

Total dry matter production through Day 120 was least in the Clayton field plants and greatest in the Duke field plants (Table 1). The chamber and Clayton field plants had similar LAD's, which were only about 58% of the LAD's of the Duke field and greenhouse plants. The NAR's ranged from 0.018 g $dm^{-2}day^{-1}$ in the Clayton field plants to 0.024 g $dm^{-2}day^{-1}$ in the Duke field plants. Thus, most of the difference in total dry matter production was associated with the difference in LAD.

As a result of their earlier flowering and cessation of leaf production, the chamber plants had less leaf area during pod-fill than did the plants in the other three environments. This lower leaf area may have been responsible for the lower final pod yield of the chamber plants in comparison to plants in the other environments. The greatest pod yield occurred in the greenhouse plants, and the plants in the two field environments producted intermediate pod yields.

### Limits to Yield of Chamber-Grown Soybean

The influence of leaf area on pod yield in growth chamber plants was

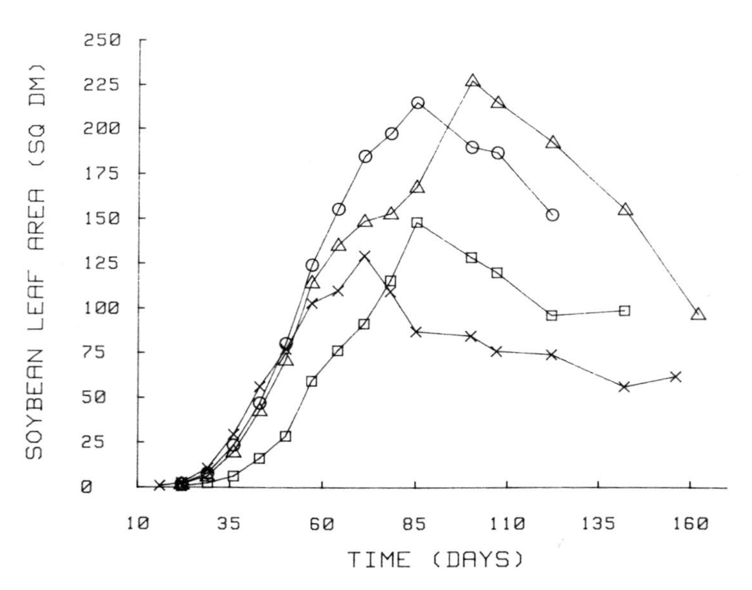


**Figure 4. Leaf area production of soybean in growth chamber (+), greenhouse (△), in pots in Duke Field (○), and in soil in Clayton field (□).**

subsequently investigated by varying the time of photoperiodic floral induction (Patterson, Peet, and Bunce, 1977). The plants were grown in chambers programmed for 14 hours of full fluorescent lighting (treatment L) or for 12 hours of full lighting with 1-hour extensions of incandescent lighting only at the beginning and end of the 12-hour period (treatment S). An additional 1-hour night interruption with incandescent lamps only was provided in each chamber to further insure that floral induction would not occur. The thermoperiod was 12 hours in both chambers, with day/night temperatures of 26/20 C.

To provide plants of three size classes at the time of flowering, a 37-day period of photoinductive 8-hour days was provided, beginning at 22, 30, and 38 days after planting. The plants had three, six, and nine mainstem trifoliate leaves, respectively, at these ages. At the end of the 37-day photoinductive treatment, plants in treatment S were returned to their original photoperiod and plants in treatment L were transferred to a chamber programmed for 16 hours of full fluorescent and incandescent lighting. The S and L treatments allowed an evaluation of the effects of total photosynthetic photon flux received per day.

As expected, increasing plant age at the time of exposure to the 37-day photoinductive treatment greatly increased the size of the plants at the initiation of flowering and podding (Table 2). Delaying photoinduction for

**Table 2. Leaf area at flowering and podding and final yield data for soybean grown in two photoperiods (treatments L and S), given photoinductive 8-hour days starting at 22 (I), 30 (II), or 38 (III) days after planting. (From Patterson et al. 1977b).**

| Plant Characteristic[a] | Treatment L[b] | | | Treatment S[c] | | |
|---|---|---|---|---|---|---|
| | I | II | III | I | II | III |
| Leaf area at flowering ($dm^2$) | 33 b | 63 b | 107 c | 33 a | 62 b | 101 c |
| Leaf area at podding ($dm^2$) | 46 a | 76 b | 131 c | 50 a | 87 b | 129 c |
| Number of pods | 152 a | 233 b | 321 c | 150 a | 198 b | 284 c |
| Pod weight (g) | 97 a | 123 b | 174 c | 81 a | 111 b | 152 c |
| Seed weight (g) | 72 a | 92 b | 129 c | 59 a | 83 b | 111 c |

[a] Values for each characteristic sharing the same letter are not significantly different at p = 0.05 level.

[b] Treatment L received 14 h of full fluorescent and incandescent lighting proceeding the photoinduction period and 16 h full lighting after the photoinduction period.

[c] Treatment S received 12 h of full lighting with 1 h extensions of incandescent lighting only, before and after the photoinduction period.

8 days increased leaf area at flowering by 90%, while a 16-day delay increased leaf area by over 200%. Leaf area at podding was similarly increased by 64% to 72% with the 8-day delay in photoinduction and 156% to 186% with the 16-day delay. Concomitant increases in leaf weight, stem weight, node number, and leaf number also occurred.

The differences in vegetative growth among the three induction groups were maintained during pod-filling and maturation, and at maturity the three groups were well segregated into three classes with respect to both plant size and yield of beans (Table 2). Delaying initial photoinduction for 8 days increased seed yield by 27% to 41%, and delaying photoinduction for 16 days increased seed yield by 78% to 88%.

Final seed weight per plant was closely correlated with leaf area at flowering ($r = 0.97$), leaf area at podding ($r = 0.94$), leaf number at flowering ($r = 0.91$), and leaf number at podding ($r = 0.89$). These correlations also reflect the high correlation between potential reproductive sites (node number or leaf number) and final seed weight.

The range of pod yield per plant in this experiment encompassed all our previously observed pod yields in the greenhouse and the two field environments. We also estimated seed yields per unit area, based on a density of 4.43 plants/$m^2$ in the growth chamber. In the L treatment, plants receiving photoinductive short days beginning at 22, 30, and 38 days produced seed yields of 3,200, 4,000, and 5,700 kg/ha, respectively. In the S treatment, the respective seed yields were 2,600, 3,700, and 4,900 kg/ha. Farm yields of soybean are commonly 3,000 kg/ha with occasional yields of 4,000 kg/ha and a record yield of 5,560 kg/ha (Shibles et al., 1975). Thus, the range of our "chamber yields" encompassed average as well as record farm yields of soybean.

## PHYSIOLOGICAL RESPONSES

Further studies compared photosynthesis and water relations of controlled environment-grown and field-grown cotton and soybeans. These studies were undertaken to provide a better understanding of the overall differences in plant growth which we observed in the different environments.

### Photosynthesis

*In situ* measurements with a hand-held pincer cuvette and infrared gas analyzer (IRGA) (Patterson et al., 1977) indicated that field and greenhouse plants had maximum photosynthetic rates per unit leaf area about twice as great as chamber plants (15 to 16 versus 28 to 32 mg $CO_2$ $dm^{-2}hour^{-1}$) (Table 3). However, these higher photosynthetic rates occurred at irradiances two to three times greater than those in the growth chamber (1000 to 1600 versus 470 to 520 $\mu$Einsteins $m^{-2}s^{-1}$ PPFD).

**Table 3.** **Average maximum *in situ* photosynthetic rates and PPFD during measurement, average and minimum leaf water potentials, average leaf turgor pressures, and individual leaf sizes of cotton and soybean grown in growth chamber (Ch), controlled-environment greenhouse (GH), in pots in Duke field (DF), and in soil in Clayton field (CF).**

| Parameter | Cotton | | | | Soybean | | | |
|---|---|---|---|---|---|---|---|---|
| | Ch | GH | DF | CF | Ch | GH | DF | CF |
| Maximum *in situ* photosynthetic rate (mg $CO_2$ $dm^{-2}h^{-1}$) | 16 | 28 | 32 | 31 | 15 | 29 | 33 | 29 |
| Average PPFD during photosynthesis measurements ($\mu E\ m^{-2}s^{-1}$) | 520 | 1250 | 1175 | 1360 | 470 | 1280 | 1016 | 1560 |
| Average leaf water potential (bars) | -6 | -9 | -10 | -12 | -7 | -8 | -9 | -11 |
| Minimum leaf water potential (bars) | -9 | -14 | -19 | -26 | -11 | -11 | -15 | -11 |
| Average leaf turgor pressure (bars) | 4.0 | 4.0 | 3.0 | - | 4.3 | 3.5 | 2.6 | - |
| Average area per leaf ($dm^2$) | 1.30 | 1.27 | 1.01 | 0.99 | 1.69 | 1.63 | 1.50 | 1.53 |

In order to determine to what extent adaptive leaf characteristics influenced the *in situ* photosynthetic rates, we studied in detail the responses of photosynthesis to irradiance in cotton from the growth chamber and Duke field environments. This was done by enclosing individual cotton leaves in a plexiglass chamber which was illuminated with incandescent lamps (Patterson et al., 1977). Stomatal resistance, chloroplast lamellar characteristics, and leaf anatomy were also examined as possible explanations for the observed differences in photosynthetic rates.

When measured under a range of irradiances, the field plants had higher photosynthetic rates per unit leaf area throughout the range, with the greatest differences occurring above 750 μEinsteins $m^{-2}s^{-1}$ (Figure 5). Light saturation occurred at about 750 μEinsteins $m^{-2}s^{-1}$ in the chamber plants. Light-saturated photosynthetic rates were 28 mg $CO_2$ $dm^{-2}hour^{-1}$ in the chamber plants and 43 mg $CO_2$ $dm^{-2}hour^{-1}$ in the field plants. Total stomatal resist-

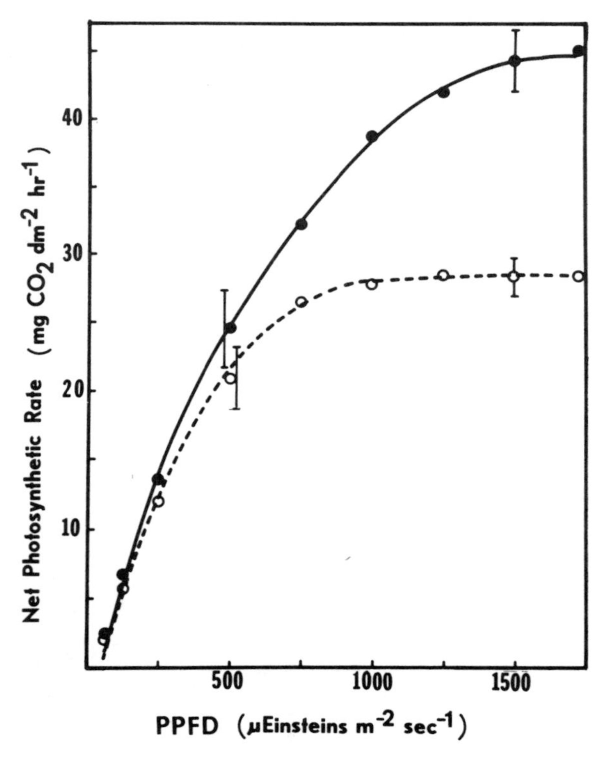


**Figure 5. Response of net photosynthetic rate to irradiance in cotton from growth chamber (open circles, dashed line) and Duke field (solid circles, solid line). Vertical bars indicate $\pm s_x$. (From Patterson et al. 1977).**

tance to $CO_2$ flux at light-saturation was 2.4 s $cm^{-1}$ in both field and chamber plants. However, stomatal frequencies (stomates/$mm^2$) were greater in the field plants. Light-saturated mesophyll resistances to $CO_2$ flux were calculated to be 4.5 s $cm^{-1}$ for chamber plants and 1.8 s $cm^{-1}$ for field plants.

Both the palisade and spongy mesophyll layers were substantially thicker in field leaves, although total epidermal thickness was greater in chamber leaves. The total mesophyll volume per unit leaf area was 2.24 $cm^3$ $dm^{-2}$ leaf surface, compared to 1.70 $cm^3$ $dm^{-2}$ for the chamber leaves. The greater thickness of the field leaves also was reflected in their greater specific leaf weight of 0.534 g $dm^{-2}$ compared with 0.348 g $dm^{-2}$ in chamber leaves.

The field leaves also contained more chlorophyll per unit area than the chamber leaves. Together with a smaller photosynthetic unit (PSU) size, this resulted in almost twice as many PSU's per unit leaf area in the field plants.

These results provide some explanation for the greater photosynthetic rates observed in the field plants *in situ*. The *in situ* rates were in the range that would be predicted from the photosynthesis-irradiance response curves of the field and chamber plants.

Stomatal resistances at light saturation did not differ between the field and chamber plants. Therefore, the greater light-saturated photosynthetic rates of the field plants were due to lower mesophyll resistances. Our estimate of mesophyll resistance includes all nonstomatal resistances to $CO_2$ uptake and does not differentiate between possible limitations associated with respiration, photochemical or biochemical reactions, or diffusive transfer within the leaf. Because the field plants had a considerably greater mesophyll volume per unit area than the chamber plants, photosynthetic rates per unit mesophyll volume were much more similar than were rates per unit leaf area when the two groups of plants were compared. The rates per unit chlorophyll also were much more similar. The amounts of chlorophyll, PSUs, mesophyll surface area, and carboxylation enzyme activity per unit leaf area are presumably all closely correlated with leaf thickness and leaf mesophyll volume per unit leaf area (Nobel et al., 1975; Charles-Edwards and Ludwig, 1975; Patterson, 1980; Patterson et al., 1978; Patterson and Duke, 1979). All of these factors may contribute to the greater photosynthetic rates observed in the field plants both *in situ* and in the laboratory.

Our results, and those of other studies summarized in Patterson (1980), demonstrate the importance of considering quantitative as well as qualitative differences in leaves when photosynthetic rates of plants grown at different irradiances are compared. The choice of the basis for the expression of photosynthetic rates (unit leaf area, unit mesophyll volume, or unit chlorophyll) can facilitate extrapolating phytotron data to field situations. These considerations are particularly important in the development and application of photosynthesis models.

## Water Relations and Leaf Anatomy

Detailed comparisons of the water relations, and stomatal characteristics and leaf anatomy of field-grown and chamber-grown plants have been reported elsewhere (Davies, 1977; Van Volkenburgh and Davies, 1977; Bunce, 1977 and 1978). These studies were conducted during three growing seasons (1974 to 1976), but the same trends generally were apparent each year. Average and minimum leaf water potentials of cotton, measured with thermocouple psychrometers or pressure chambers, were lower in the field environments than in the greenhouse and lower in the greenhouse than in the growth chamber (Table 3). The Clayton field plants experienced a severe drought in 1975, and during this period cotton leaf water potentials fell to -26 bars. However, soybean leaf water potentials remained at -11 bars or greater, even during the drought. Average soybean leaf water potentials ranked the same as cotton (chamber > greenhouse > Duke field > Clayton field), but minimum water potentials were similar in all environments. Leaf turgor pressures in both species were greater in the growth chamber and greenhouse than in the field environments (Table 3). However, relationships between turgor pressure, leaf expansion rates, duration of leaf expansion, and final leaf size were very complex, and no clear-cut differences among the environments were apparent (J. A. Bunce, USDA, Beltsville, MD, unpublished data). Later work by Bunce (1978) demonstrated that leaf expansion was sensitive to atmospheric drought (high vapor pressure deficit) as well as soil drought.

Average individual leaf areas of both species were greater in the growth chamber and greenhouse than in the field (Table 3). These measurements were made at the time of maximum total leaf area in soybean (71, 85, 85, and 100 days after planting for chamber, Duke field, Clayton field, and greenhouse, respectively) and at the time of maximum area or initial plateauing of leaf area in cotton (Day 85). Average leaf areas calculated at later harvests were complicated by the leaf abscission and regrowth that occurred in both species, but particularly in cotton.

Leaf thickness was greater in the field environments than in the growth chamber or greenhouse. This was true for both species and for both years that observations were made. However, when the growth chamber was programmed for a day/night temperature of 28/17 C instead of 26/20 or 30/26 C, leaf thickness was more similar to that of the field plants (Van Volkenburgh and Davies, 1977).

The size and frequency of stomata also varied with the growth environment (Van Volkenburgh and Davies, 1977). In soybean, plants grown in both the 30/26 and 28/17 C day/night chambers had smaller stomata than those grown in the greenhouse at 30/26 or in the Duke field. This was true of stomata on both leaf surfaces. In cotton, the smallest stomata occurred on the upper surface of leaves from plants in the 28/17 C chamber. Stomatal sizes on the lower leaf surface were similar in all environments.

In both species and all environments, stomatal frequency was much greater on the lower than the upper leaf surface. The ratio of lower surface to upper surface stomatal frequency ranged from 2.1 in Duke field cotton to 8.0 in soybean in the 30/26 C chamber.

*In situ* stomatal resistances measured between 1000 and 1500 hours were greatest for plants grown in growth chambers for both species. Resistances were greater on the upper surfaces, but not to the extent that the differences in stomatal frequency between upper and lower surfaces might suggest. Rates of water loss through stomata of well-illuminated, detached leaves were greatest in the 30/26 C chamber plants of both species, even though these plants had the highest *in situ* stomatal resistances. Cuticular water losses were greatest in soybean from the 30/26 C chamber and cotton from the Duke field.

The responses of the stomata to water stress also varied with species and growth environment (Davies, 1977). Stomatal closure occurred at lower leaf water potentials and lower relative water contents (RWC) in soybean than in cotton. In both species, closure occurred at lower leaf water potential and RWC in field plants than in greenhouse plants and in greenhouse plants than in chamber plants. These stomatal responses may be of adaptive significance in view of the greater water stress experienced by the field plants *in situ.*

## IMPLICATIONS FOR FUTURE RESEARCH

The studies summarized here have shown that plants grown in controlled environments programmed to simulate average growing season temperatures can differ significantly from plants grown in the field. Generalizations about the comparability of growth and physiological responses of plants growing in controlled environments and in the field require definitions of typical conditions in field and controlled environments. Such definitions are difficult to make because controlled-environment facilities can provide endless combinations of thermoperiod, photoperiod, temperature, irradiance, light quality, $CO_2$ concentration, vapor pressure deficit, potting media, and watering and nutrient application schedules. Likewise, the "typical" field environment is impossible to define because it varies on a time scale of seconds to years and a space scale of cm to km (Evans, 1963; Kramer, 1978).

This dilemma emphasizes the need for objectives to be clearly defined in controlled-environment research. If perfect simulation of a field environment is desired, why waste controlled-environment space? The experiment should be conducted in the field. If, on the other hand, the researcher desires to determine the effects of well-defined environmental factors acting singly or in combination, carefully planned controlled-environment studies are essential.

Perhaps the most productive research programs have used a combined approach: defining problems and questions in the field, conducting controlled environment experiments to answer specific questions, and verifying results with further field work. This approach is best exemplified by the work of C. D. Raper, Jr., at North Carolina State University (Raper and Downs, 1976) and J. D. Hesketh of the U.S. Department of Agriculture (Hesketh et al., 1975). Both have successfully utilized data collected in controlled-environment facilities to develop models for predicting plant growth and development under field conditions. The modeling approach is an effective means of bridging the gap between phytotron and field research. Hesketh et al. (1975) summarized the role of controlled-environment research in the development of predictive plant growth models.

Another approach has proved useful in the author's research on weed-crop ecophysiology. In this approach, crop plants and their associated weeds are grown under identical controlled-environment conditions, and their responses to selected environmental factors are compared. Mathematical growth analysis techniques (Kvet et al., 1971; Patterson et al., 1979) are used to evaluate the results.

These studies have helped to explain the increased competitiveness of certain malvaceous weeds with cotton following periods of abnormally cool weather early in the growing season (Patterson and Flint, 1979a). A three-day chilling treatment simulating naturally occurring chilling events in May in the Yazzo-Mississippi Delta reduced subsequent dry matter production and leaf expansion of cotton more than that of the weeds.

In research to evaluate the potential impact of the exotic noxious weed itchgrass *(Rottboellia exaltata* L.f.) in the United States, Patterson and Flint (1979b) found that cool temperatures and simulated natural chilling events typical of the first five weeks of the growing season for corn and soybean at Madison, Wisconsin, for soybean at Carbondale, Illinois, and for corn at Waycross, Georgia, reduced the growth of itchgrass more than that of the adapted crop varieties. They concluded that itchgrass is unlikely to be a serious early season pest of corn or of soybean outside the South, even though earlier work (Patterson et al., 1979) had shown that it could grow vigorously and produce seed during the warmest three months of the growing season as far north as southern Wisconsin. Thus, phytotron research can help determine the conditions required for maximum growth and competiveness of weeds (Patterson et al., 1979; Patterson et al., 1980) as well as for the growth and yield of crops (Evans, 1963).

Future work should concentrate on developing simulations of plant responses that have general application in both field and controlled-environment situations. As pointed out by Raper and Downs (1976), these simulations can help circumvent the problems of extrapolating from the phytotron to the field.

## NOTES

David T. Patterson, USDA-SEA-AR, Botany Department, Duke University, Durham, North Carolina 27706.

## LITERATURE CITED

Bazzaz, F. A. 1973. Photosynthesis of *Ambrosia artemisiifolia* L. plants grown in greenhouse and in the field. Am. Midl. Natur. 90:186-190.

Bunce, J. A. 1977. Leaf elongation in relation to leaf water potential in soybean. J. Exp. Bot. 28:156-161.

Bunce, J. A. 1978. Effects of water stress on leaf expansion, net photosynthesis, and vegetative growth of soybeans and cotton. Can. J. Bot. 56:1492-1498.

Charles-Edwards, D. A., and L. J. Ludwig. 1975. The basis of expression of leaf photosynthetic rate. p. 37-44. In R. Marcelle (ed.) Environmental and biological control of photosynthesis. W. Junk, The Hague.

Davies, W. J. 1977. Stomatal responses to water stress and light in plants grown in controlled environments and in the field. Crop Sci. 17:735-740.

Elmore, C. D., J. D. Hesketh, and H. Muramoto. 1967. A survey of rates of leaf growth, leaf aging, and leaf photosynthetic rates among and within species. J. Arizona Acad. Sci. 4:215-219.

El-Sharkawy, M., J. D. Hesketh, and H. Muramoto. 1965. Leaf photosynthetic rates and other growth characteristics among 26 species of *Gossypium.* Crop Sci. 5:173-175.

Evans, L. T. (ed.). 1963. Environmental control of plant growth. Academic Press, New York.

Gausman, H. W., W. A. Allen, D. E. Escobar, R. R. Rodriguez, and R. Cardenas. 1971. Age effects of cotton leaves on light reflectance, transmittance, and absorptance and on water content and thickness. Agron. J. 63:465-469.

Hesketh, J. D. 1968. Effects of light and temperature during growth on subsequent leaf $CO_2$ assimilation rates under standard conditions. Aust. J. Biol. Sci. 21:235-241.

Hesketh, J. D., H. C. Lane, J. W. Jones, J. M. Mckinion, D. N. Baker, A. C. Thompson, and R. F. Colwick. 1975. The role of phytotrons in constructing plant growth models. p. 117-129. In P. Chouard and N. De Bilderling (eds.) Phytotrons in agricultural and horticultural research. Gauthiers-Villars. Paris.

Kramer, P. J. 1978. The use of controlled environments in research. HortScience 13: 447-451.

Kramer, P. J., H. Hellmers, and R. J. Downs. 1970. SEPEL: New phytotrons for environmental research. Bioscience 20:1201-1208.

Kvet, J., J. P. Ondok, J. Necas, and P. G. Jarvis. 1971. Methods of growth analysis. p. 343-391. In Z. Sestak, J. Catsky, and P. G. Jarvis (eds.) Plant photosynthetic production. Manual of methods. W. Junk, The Hague.

Ludlow, M. M., and T. T. Ng. 1976. Photosynthetic light response curves of leaves from controlled environment facilities, glasshouses or outdoors. Photosynthetica 10: 457-462.

Nobel, P. S., L. J. Zaragoza, and W. K. Smith. 1975. Relation between mesophyll surface area, photosynthetic rate, and illumination level during development for leaves of *Plectranthus parviflorus* Henckel. Plant Physiol. 55:1067-1070.

Ondok, J. P. and J. Kvet. 1971. Integral and differential formulae in growth analysis. Photosynthetica 5:358-363.

Patterson, D. T. 1980. Light and temperature adaptation. p. 205-235. In J. D. Hesketh and J. W. Jones (eds.). Predicting photosynthesis for ecosystem models. I. CRC Press, Boca Raton, FL.

Patterson, D. T., J. A. Bunce, R. S. Alberte, and E. Van Volkenburgh. 1977. Photosynthesis in relation to leaf characteristics of cotton from controlled and field environments. Plant Physiol. 59:384-387.

Patterson, D. T., and S. O. Duke. 1979. Effect of growth irradiance on the maximum photosynthetic capacity of water hyacinth [*Eichornia crassipes* (Mart.) Solms]. Plant and Cell Physiol. 20:177-184.

Patterson, D. T., S. O. Duke, and R. E. Hoagland. 1978. Effects of irradiance during growth on the adaptive photosynthetic characteristics of velvetleaf and cotton. Plant Physiol. 61:402-405.

Patterson, D. T., and E. P. Flint. 1979a. Effects of chilling on cotton *(Gossypium hirsutum)*, velvetleaf *(Abutilon theophrasti)*, and spurred anoda *(Anoda cristata)*. Weed Sci. 27:473-479.

Patterson, D. T., and E. P. Flint. 1979b. Effects of simulated field temperatures and chilling on itchgrass (*Rottboellia exaltata)*, corn *(Zea mays)*, and Soybean *(Glycine max)*. Weed Sci. 27:645-650.

Patterson, D. T., E. P. Flint, and R. Dickens. 1980. Effects of temperature, photoperiod, and population source on the growth of cogongrass *(Imperata cylindrica)*. Weed Sci. 28:505-509.

Patterson, D. T., C. R. Meyer, E. P. Flint, and P. C. Quimby, Jr. 1979. Temperature responses and potential distribution of itchgrass (*Rottboellia exaltata)* in the United States. Weed Sci. 27:77-82.

Patterson, D. T., M. M. Peet, and J. A. Bunce. 1977b. Effect of photoperiod and size at flowering on vegetative growth and seed yield of soybean. Agron. J. 69:631-635.

Raper, C. D., Jr., and R. J. Downs. 1976. Field phenotype in phytotron culture—a case study for tobacco. Bot. Rev. 42:317-343.

Shibles, R., I. C. Anderson, and A. H. Gibson. 1975. Soybean. p. 151-189. In L. T. Evans (ed.) Crop physiology—some case histories. Cambridge University Press, London.

Van Volkenburgh, E., and W. J. Davies. 1977. Leaf anatomy and water relations of plants grown in controlled environments and in the field. Crop Sci. 17:353-358.

Went, F. W. 1957. The experimental control of plant growth. Chron. Bot. 17:1-343.

20

# SYSTEMS ANALYSIS AND MODELING IN EXTRAPOLATION OF CONTROLLED ENVIRONMENT STUDIES TO FIELD CONDITIONS

**Harvey J. Gold and C. David Raper, Jr.**

There are two questions that must be considered in discussions of crop responses to stress. The first is how to evaluate the effects of stress conditions on crop growth, particularly upon yield. The second is what to do about it. Our ability to answer the second question depends upon our ability to evaluate the probable outcomes of possible management strategies such as presented by Palmer, Barfield, and Haan (1982). In this paper, we focus on the role of mathematical system modeling in helping to address these questions, and in particular, in helping to interpret the results of controlled environment studies. Our emphasis will be at the broad conceptual level rather than at the level of specific case studies.

## NEED FOR EXTRAPOLATION

The system that we need to understand is one of a community of plants growing under conditions selected jointly by nature and by the farm manager—that is, under field conditions. Given the need for understanding this system, a question that comes to mind is, "Why not take observations under field conditions and by-pass the need for extrapolation?" Indeed, this is the way in which much of agricultural research has been done. The problem is that having made an observation on a particular crop, grown under a specific set of field conditions, we no longer are interested scientifically in that crop or in that particular set of conditions, except to verify the original set of data. As has been discussed by Kramer (1978), our interest lies instead in making inferences to crops that will be grown in the future, under conditions which are known only to the extent that we might place betting odds, i.e., probabilities, upon them. We are rarely interested in data for what they tell us about an experiment that has already been run. Rather, we care about what the data tell us concerning the class of all experiments that might be done. Statistical experimental design consists first of characterizing that class, and then selecting representative members of that class for actual implementation and observation. However, under field conditions, it is nature that presents the overall conditions for the experiment, leaving us to intervene to modify these conditions at a rather gross level. Stress conditions are almost by definition uncommon events, although Patterson (1982) has pointed out that certain kinds of stress might not be as uncommon as we think. When taken in combination with the rest of the environmental setting and the phenology of plant development, we must acknowledge that the conditions of interest are selected only rarely and that the experimental conditions occurring in nature are not likely to be those that give us the information we need.

The follow-up question might then be, "If extrapolation of some sort is needed, what is the need to involve system analysis and mathematics?" The answer is that, as Patterson (1982) has pointed out, the extrapolation is not straight-forward. Presumably the controlled environment systems obey the same physical and biological laws as do field systems, yet there somehow is a difference of behavior. Our problem is how to make use of what is similar (the underlying laws of behavior) to bridge across what is different.

Part of the difficulty in addressing this problem is inherent in the complexity of the system. Biological systems are almost incredibly complex. This complexity is manifest in the large number of interacting variables needed to describe the inner workings of the system, in the large number of possible behavioral responses, and in the large number of environmental regimes to which the system might be subject. Moreover, while the system itself can be highly complex, the experiment must be only as complex as will allow interpretation. In this light, the problem is how to use a finite and relatively small number of observations, made under relatively simple conditions. to make inferences to a nearly infinite set of possible conditions for a vastly more complex system.

## THE ABSTRACTION HIERARCHY

In addressing these problems it will be useful to draw on the concept of the abstraction hierarchy (Mesarovic, 1968; Mesarovic, Macko, and Takahara, 1970). The levels of such a hierarchy refer, in manner of speaking, to how coarse or fine "ground" the system is taken to be. The following levels might be identified: (a) agro-ecosystem level, (b) local (farm) management unit level, (c) plant community level, (d) single plant level, (e) organ structures and elementary functional processes (photosynthesis, translocation, respiration, etc.),

(f) elementary structures (cells, membranes, local structured processes), (g) molecular physics and chemistry, and (h) atomic and molecular physics.

Each of the terms in the hierarchical listing is meant to indicate a level of breakdown or decomposition; it is meant to stand for a collection of subsystems or components that co-exist and possible interact at that level. At the local management unit level, for example, we may conceive of a mosaic of relatively homogeneous local neighborhoods, responding to a common weather front, and interacting via mobile animal species, wind-borne matter, and drainage profiles. Each of these neighborhoods might be a separate plant community with its own management regime, including no management regime. At this same level, one might also consider technological inputs to the system through human decision processes. At the plant community level, we must include at minimum the crop itself, the soil component, the weather system, and such auxiliary biological species as insects, pathogens, soil organisms, competing plant species, etc.

Keeping in mind that each level includes a collection of subsystems, the next higher level may be viewed as a level of aggregation, or more accurately as a level of organizational structure of these components. Clearly, we could proceed to lower (finer) levels as well as to higher (coarser) ones. Just as the plant itself may be more finely decomposed into organ structures, elementary processes, and finally to the molecular level, the other biological and non-biological components can be similarly divided.

If we proceed from bottom to top, that is, finer to coarser levels, it becomes increasingly hard to control the experimental environment. At the atomic level on up through the organ level, experiments and observations are usually made under fairly closely controlled conditions. The breakpoint comes at about the single plant level, for which it is possible, but difficult, to control the environment. This is precisely why devices which control environment at the single plant level, such as the SPAR system, phytotrons, and rhyzotrons, play such an important role in research on plant agricultural systems.

The class of systems to which we wish to draw inferences is at least at the plant community level and possibly higher. The observational level at which we can with accuracy control conditions is limited to that of single plants or very small isolated communities. The processes which give rise to these observations, and upon which extrapolation must be based, are at a finer level yet. Ability to make inferences about the class of systems at the complex level of organization is, therefore, dependent upon ability to draw relations across levels in the hierarchy.

To be sure, the hierarchical structure is an artifact, constructed for human convenience, to facilitate scientific description of the system. The usefulness of such a description hinges on the ability to describe each of the levels and each of the subsystems at a given level independently of the rest. This relative degree of independence results in the development at each level of a separate conceptual framework, and a separate language and methodology.

These differences of language and of concept extend not only to the description of the internal workings of the system components, but also to the way in which we must express the external, environmental interactions. In Tables 1, 2, and 3, for example, the several different levels are compared with respect to energy and material amounts. At the chemical process level (Table 1), all terms are expressed as quantities at a reaction site. At the physiological process level (Table 2), quantities are included that do not pertain to the reaction sites or even to the plant. They pertain to the chamber in which the study is being done. Moreover, even when special instrumentation permits direct measurement, there is still the question of energy and concentration gradients, which are of importance in describing physiological processes, but which have no analogs at the molecular level. At the community level (Table 3), we may measure ambient temperature and incident light. It is the gradients, however, that are important in considering interactions between plants and between species. and in turn which influence the intra-plant gradients.

In principle, the temperature and its gradients in a crop could be completely described at an instant of time by a three-dimensional array showing

**Table 1. Inputs at chemical process level.**

| | |
|---|---|
| Heat energy | Temperature at reaction site |
| Light energy | Radiation intensity at reaction site |
| Nutrients & metabolites (reactants), including water | Concentration or chemical potential at reaction site |

**Table 2. Inputs at plant physiological process level.**

| | |
|---|---|
| Heat energy | Temperature of growth chamber may be taken as the temperature at all reaction sites, but temperature gradients (i.e., point to point difference) may be of importance in plant behavior. |
| Light energy | Incident light may be measured. Light gradients are input variables. |
| Metabolites (including water) | Nutrients supplied in growth chamber. Gradients are of clear importance in driving plant processes. |
| Wind, humídity | Important to include in input space because of effect on temperature and metabolic gradients. No analog at molecular level. |

**Table 3. Inputs at community level.**

| | |
|---|---|
| Heat energy, light energy | Ambient temperature and incident light. However, horizontal and vertical distributions affect relationship between plants. The structure of the community then feeds back to affect input to individual plants. |
| Metabolites (including water) | Average concentration; however, distribution within soil component feeds back to affect input to individual plants. |
| Wind, humidity | Wind and humidity above canopy may be uniform, but structure of the community affects input to individual plants. |

the temperature at every point, with points taken, perhaps, 1 nm apart. To attempt to do so, however, is both a theoretical and practical absurdity. The concept of temperature at the crop level must be of a different sort from the concept of temperature at the molecular level. The biochemist, the physiologist, and the agronomist all understand something different by the word, "temperature." Drawing the relation between these concepts is one of the functions of a mathematical model. Such a model can also serve as a communication tool, since it allows the differences between the concepts and their relation to be made explicit.

## RELATION BETWEEN THE LEVELS

As we proceed up the hierarchy, it is essential to bear in mid that the detail which describes the lower level is specifically not wanted for the description of the higher level. The point may be illustrated by looking at the simple and well-characterized example from elementary thermodynamics of the relation between a volume of gas and the molecules it comprises. The molecules themselves are described by their positions and their energy levels. Since any reasonable volume of gas will contain on the order of $10^{23}$ molecules, the complete molecular description at any instant is more than we want to deal with. The total collection of molecules, however, is described simply by three numbers to define volume, temperature, and pressure. The individual molecules are free to change position and exchange energy, while the volume, temperature and pressure remain constant. The theory of statistical mechanics deals with the question of how to abstract from the molecular detail just that information needed to describe the aggregate level. The variables of volume, temperature, and pressure that describe the higher level are functions of the collection of variables which describe the lower, but these functions have no inverse and they cannot be run backward. That is, if we are given the molecular description, we could deduce the aggregate description; but, given only the aggregate description, we cannot recapture the detail of the molecular level. One may say that information is lost as we go up the hierarchy. It is the role of mathematical modeling to help determine and to help express the relevant information functions and, thus, to help determine what information needs to be retained.

To carry this thermodynamic example just one step further, it is clear that the structure of the whole collection imposes certain boundary conditions or constraints on the behavior of the individual molecules. In the classical thermodynamic case, these constraints are that the total volume occupied remains constant as delimited by the walls of the container, and that the total energy stays the same. Any change at the aggregate level, such as a change in the relative position of the walls, or a change in energy, such as by some heat input, will change these constraints.

While the behavior of the system at the higher level is a function of the behavior at the lower level, the lower level operates under constraints imposed by the organizational structure of the higher level. In a biological system, there is an additional complication. Part of the behavior of the system at any level is to alter its own structure to cause a feedback relation between constraints and processes. This relation is illustrated in a formal way in Figure 1 which is patterned after the treatment in Mesarovic, Macko, and Takahara (1970). The figure shows a system at level *i*, designated $S_i$. It has input and output appropriate to its level. The relation between the input and output depends on four things: (a) the organization structure designated $O_i$, (b) the instantaneous state of the system, designated $\sigma_i$, (c) the processes occurring at the next lower level; the relevant information concerning these processes is symbolized by the function $L_{i-1}$, and (d) the constraints imposed by the next higher level, represented as $H_{i+1}$. If we focus on the development of the organizational structure $O_i$, any changes from a time $t_1$ to a later time

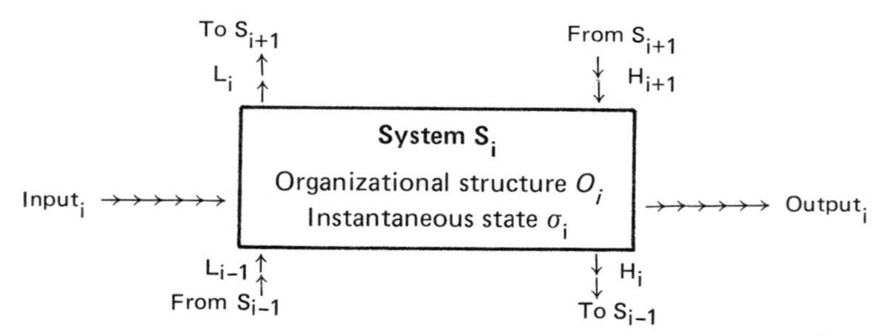


**Figure 1. Schematic representation of interlevel dependencies. $L_i$ or ($L_{i-1}$) = function of process at level *i* (or i–1) which influences dynamics at level i+1 (or at level *i*). $H_i$ (or $H_{i+1}$) = function of structure at level *i* (or i+1) which provides boundary conditions for level i–1 (or for level *i*).**

$t_2$ depends upon: (a) the input during the time interval; (b) the lower level process, represented by $L_{i-1}$, during the time interval; (c) the higher level constraints $H_{i+1}$, as they operate, and possibly change, during the interval; and (d) the initial organizational structure and state at time $t_1$.

An example of an attempt to distill the relevant information at the lower process level, so as to obtain that which is relevant at the crop level, is the definition of the energy-crop growth variable by Coelho and Dale (1980);

$$ECG = \frac{SR}{600}(1 - e^{-.79LAI})\frac{ET}{PT}FT$$

where SR = daily solar radiation, LAI = leaf area index, ET/PT = ratio of actual to potential evaporation, and FT = a daily dimensionless temperature scaling factor. This variable uses a summary of the light energy input (SR), a relevant characteristic of the leaf structure (LAI), a variable which summarizes an enormous amount of detail governing water movement (ET/PT), and a variable which summarizes heat input (FT). It has the dimensions of energy per unit time and is one of the *'L'* functions of Figure 1. The information it carries is the amount of energy per day available for crop growth. Presumably, this information can be entered into a description of the growth dynamics of the plant to arrive at a description of the evolution of the plant structure, which provides the *'H'* functions that feed back to govern the elementary processes.

Several important conclusions may be drawn from these general arguments. First is that interactions between subsystems at the same level tend to proceed through the interaction of lower level component processes. This is illustrated in a schematic fashion in Figure 2. A specific example is the interaction between a plant and the soil system, which occurs via the molecular processes of water and solute movement within both components and between them (Figure 3). This movement is governed by the constraints of

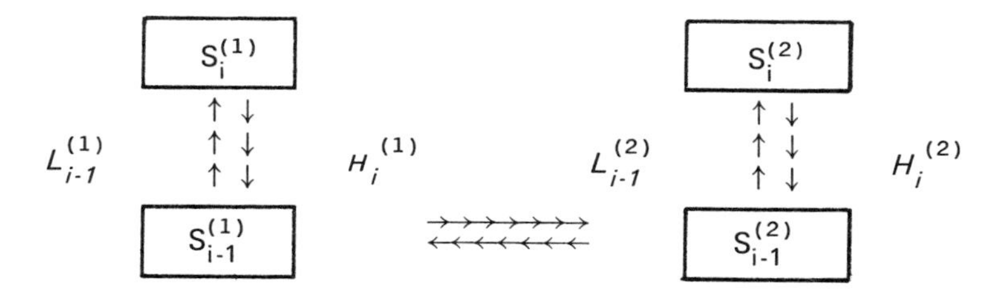


**Figure 2. Representation of interaction between $S_i^{(1)}$ and $S_i^{(2)}$, proceeding through interaction at level i-1. The interaction arises because $L_{i-1}^{(1)}$ and $L_{i-1}^{(2)}$ are functions of the interaction between $S_{i-1}^{(1)}$ and $S_{i-1}^{(2)}$.**

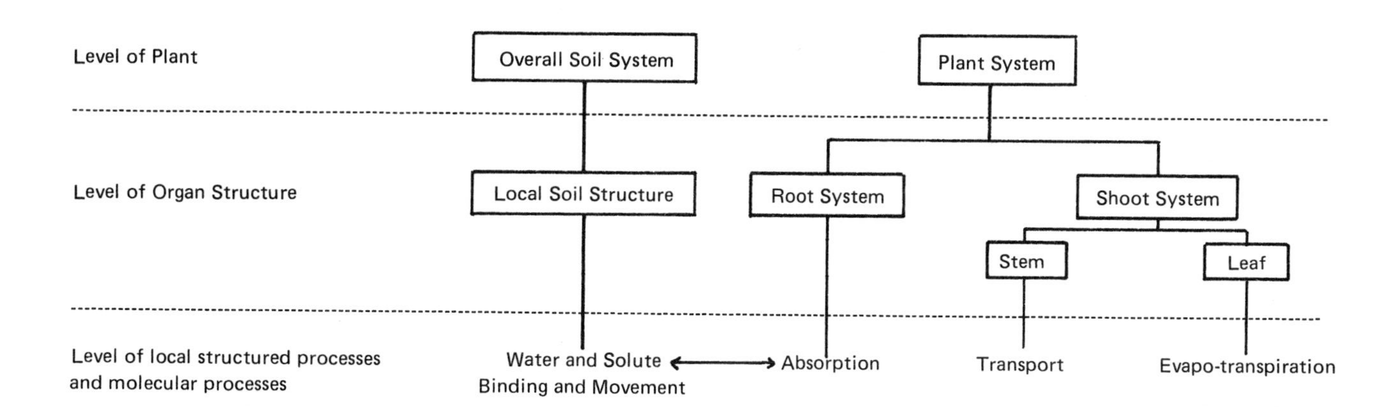


**Figure 3. Example: the absorption through the root system depends upon its overall state, which is influence by aggregate of processes in leaf and stem. Direct interaction between root and soil is governed by local molecular processes.**

the internal structures of the two components, i.e., the soil and the plant structure, and the structural relation between the two, which is a constraint imposed by yet a higher level of organization.

A second conclusion is that subsystems at different levels tend to operate on different time scales. This should be intuitively reasonable, since changes at one level are functions of processes at the lower level. This time scale separation of processes at different levels is one of the factors that makes it possible to study different levels in a quasi-independent fashion. It is also important in considering the role of the system history in the development of the structural constraints. Note that structural or organizational changes may take place on a variety of time scales. At the one extreme, there are structural features, such as the position of a particular leaf, which once built into the growth of the plant remain constant. At the other end of the time spectrum, there are structural characteristics, such as metabolite gradients and degree of stomatal opening, which are reversible on a relatively short time scale. Other features are altered on intermediate time scales, and it is these that may be reffered to adaptive processes. It is unfortunate that computer simulation of systems involving processes on different time scales is difficult and often involves numerical instabilities.

A third conclusion follows from the nature of the observation process. An observation intrinsically involves the interaction, at some level, between the system itself and the observer, i.e., the measuring device. The previous argument indicates that these interactions involve lower level processes. It follows that observations made at any given level of the system must depend upon processes at the lower level.

A final conclusion is that because of the feedback loop involving the $L$ and $H$ functions of Figures 1 and 2, and because we can never with reason hope to know everything at any given level, it is hubris to consider extrapolation from one level to any other. The interlevel relationship, as expressed by $L$ and $H$ must be viewed as being derived from interpolation between levels, rahter than on the basis of extrapolation from one level to the other. This interpolation is based upon our understanding of the rules (that is, the physical laws) which govern the behavior of the system at the different levels and upon the interlevel interactions.

Most crop models are based on models for single plants, sometimes as modified by the presence of other plants. These models express growth and metabolic activity as a function of environmental variables, using mathematical forms suggested by our knowledge of basic biochemistry and plant physiology. Verifying the applicability of these mathematical descriptions and determining the values of the parameters to use generally requires replicated experimentation under as carefully controlled conditions as possible. The experimental conditions are chosen not with an eye to direct extrapolation to the field, but for the purpose of verifying the mathematical formulation of the underlying plant physiology, for the purpose of testing the formulation of how the levels relate to each other, and for the purpose of learning the values of the appropriate parameters.

As an example, we take the vetetative plant growth model of Wann and Raper (1978, 1979). The fundamental processes that were considered are: (a) interception of photosynthetically active radiation (PAR) by leaves, (b) absorption of PAR, (c) photosynthesis, (d) flow from carbohydrate pools for respiration, (e) flow from carbohydrate pools for growth, and (f) translocation. These processes occur at the level of organ structure and of elementary function process. The inputs, that is, the driving variables, are incident PAR and ambient temperature. These driving variables pertain to the whole plant level. The output of the model is accumulated dry matter partitioned among leaves, stems, and roots. The elementary processes clearly operate under the constraints of the plant structure, which is itself the output of the model. Moreover, the mathematical description of the elementary processes depends upon parameters, such as rate constants, that cannot be directly measured as they operate in the intact plant. Their values must be inferred from estimation procedures based on observing the system at the whole-plant level, under controlled and repeatable experimental conditions. The role of the mathematical model is to relate the individual processes to the whole-plant behavior.

It is important now to point out that the availability of methods for handling complex models, taken together with the availability of methods for controlled experimentation at the whole-plant level, serve as the link between plant physiology and agronomy. These methods, in a short span of time, have made these sciences relevant to each other in new and essential ways. The result bids to cause a revolutionary change in agronomic science. As never before, the plant physiologist is being called upon to produce research results that relate to agronomic decision making. Some physiologists will see this as irksome and distracting; others will see it as a challenging and rewarding opportunity. Taking advantages of this opportunity, however, will require a combination of skills that are new to both agonomist and physiologist.

## CONCLUDING REMARKS

Our starting point was that the results of an experiment are usually of interest only insofar as they are representative of the results of a class of possible experiments. Any scientists in planning and interpreting experiments does so in the light of a model which characterizes the class of systems and/or conditions which the individual experiment is intended to represent. At the simplest level, that model need not be made explicit. It is just there, and it forms the cornerstone of the processes of reasoning and of inference. However, as we move toward more complex levels, the process of reasoning becomes more complex, and there is greater need for explicit model formulations.

We conclude with the warning that we must be careful not to expect too much too fast from our models. An explicit model contains at best only that knowledge of the system that we are able to make explicit. It does not contain our knowledge about the system that we have not yet learned to make explicit, and it surely does not contain what we don't know. In that sense, at least, the model can always be counted upon to be wrong. In using a model, whether as a tool for extrapolation or for decision making, it is therefore a serious error to substitute the model for biological intuition. Properly used, explicit models are valuable, and sometimes indispensible, aids to intuition but almost never a valid substitute.

## NOTES

Harvey J. Gold, Biomathematics Program, Department of Statistics; and C. David Raper, Jr., Department of Soil Science, North Carolina State University, Raleigh, North Carolina 27650.

We are pleased to acknowledge the helpfulness of critical remarks by Dr. Mein Wann, who reviewed an earlier draft of this paper.

## LITERATURE CITED

Coelho, D. T., and R. F. Dale. 1980. An energy-crop growth variable and temeprature function for predicting corn growth and development: planting to silking. Agron. J. 72: 503-510.

Gold, H. J. 1977. Mathematical modeling of biological systems: An introductory guidebook. John Wiley & Sons, New York.

Kramer, P. J. 1978. The use of controlled environments in research. HortScience 13: 447-451.

Mesarovic, M. D. (ed.). 1968. System theory and biology. Springer-Verlag, New York.

Mesarovic, M. D., D. Macko, and Y. Takahara. 1970. Theory of hierarchical multilevel systems. Academic Press, New York.

Palmer, W. L., B. J. Barfield, and C. T. Haan. 1982. Simulating the economics of supplemental irrigation for corn. p. 151-166. In C. D. Raper, Jr., and P. J. Kramer (eds.) Crop reactions to water and temperature stresses in humid, temperate climates. Westview Press, Boulder, Colorado.

Patterson, D. T. 1982. Phenotypic and physiological comparisons of field and phytotron-grown plants. p. 299-314. In C. D. Raper, Jr., and P. J. Kramer (eds.) Crop reactions to water and temperature stresses in humid, temperate climates. Westview Press, Boulder, Colorado.

Wann, M., and C. D. Raper, Jr. 1979. A dynamic model for plant growth: Adaptation for vegetative growth of soybeans. Crop Sci. 19: 461-467.

Wann, M., C. D. Raper, Jr., and H. L. Lucas, Jr. 1978. A dynamic model for plant growth: A simulation of dry matter accumulation for tobacco. Photosynthetica 12: 121-136.

## SECTION VII
# RESEARCH NEEDS AND PRIORITIES

*This interdisciplinary workshop was successful in identifying some of the problems that must be solved in order to increase crop yields in temperate climates. It also indicated the usefulness of bringing together people from various disciplines to discuss and find solutions for these complex problems.*

*It is clear that research to lessen the crop losses caused by environmental stresses requires an interdisciplinary approach. Meteorologists and soil scientists must define the nature and severity of the environmental stresses and plant physiologists then must determine the plant processes that are being inhibited by these stresses. The crop production specialists must determine whether the damage can be alleviated by cultural practices while plant breeders must attempt to produce new cultivars with greater stress tolerance. Organization and direction of such interdisciplinary research teams requires broad vision and better than average managerial ability, but the results promise to be very rewarding.*

*In this section, research needs and priorities that emerged from the workshop are presented. The discussion leaders during the several sessions of the workshop, Joe T. Ritchie, J. Vieira da Silva, John S. Boyer, and Oliver E. Nelson, have contributed papers expressing their responses and summaries. The other participants of the workshop collectively also have contributed suggestions of research needs and priorities. Several remained after the workshop to discuss recommendations. Others have expressed their opinions by correspondence. As editors, we have compiled the principal conclusions into an epilogue for the workshop. In assuming responsibility for omissions in the content, we gratefully acknowledge the contributions of all participants.*

21

# REDUCTION OF STRESSES RELATED TO SOIL WATER DEFICIT

Joe T. Ritchie

The reduction or elimination of relatively short periods of drought in humid and sub-humid climates is extremely important when thinking realistically about increasing crop productivity in the United States and throughout much of the world. The use of irrigation water is approaching its maximum availability in arid and semi-arid regions, and in some instances the availability is diminishing; thus, the more humid regions where water is more abundant almost certainly will have to provide a major part of increased demand for food, feed, and fiber production in the foreseeable future.

World population increases coupled with changes in food habits are expected to cause world food requirements to approximately double in 20 to 25 years. Because of the apparent potential for increased production in humid regions of the United States, no one can look with indifference upon the problems of eliminating production barriers caused by water and temperature stresses. Evidence that this increased production is possible comes from comparing actual yields with record yields. Grain yield in more developed countries averages about 2,700 kg/ha, and is about half that level in less developed countries. Yet, record yields for grain crops vary from about 10,000 kg/ha for rice to about 20,000 kg/ha for maize. To obtain near-record yields, elimination of water stress and other yield reducing factors is necessary.

## SOIL WATER BALANCE AND CROP GROWTH

In the papers of Section I, Decker and Dale point out that two important factors, drought and temperature, must be considered in evaluation of crop production potentials. Decker pointed out that drought was not a rare occurrence in humid and semi-humid regions. Of the three approaches he discussed to define drought, two—the climatic expectations (mainly rainfall probability) and the impact of dry weather on crop yields—have been used extensively, but their value for planning and managing water resources is limited. The third definition regards the shortage of water in agricultural soils. In recent years, great strides have been made in the evaluation of the soil water balance using known information about the weather, crops, and soils. Decker pointed out the excellent early work of Shaw (1962) and van Bavel (1959) on soil water balance evaluations for humid regions. Work of the type they did has not been used as extensively as it could have been for water management and planning, but present trends indicate an upsurge of interest in water balance evaluations when coupled with crop growth models.

The paper by Barfield is an example of what can be done when the technology for soil water balance is coupled with crop growth models to provide answers to critical questions regarding the feasibility and economics of water management in humid regions. Decker, Shaw and Cassel pointed out that a difficulty in evaluating soil water balance is the lack of ability to define the capacity of the soil water reservoir. This factor is highly important because, when combined with the evapotranspiration rate, it basically determines the frequency with which the soil must be replenished with water to prevent stress.

The dynamics of the soil water balance and crop growth is demonstrated in Figure 1 with four possible scenarios representing combinations of two soil storage capacities and two evaporation rates. The crop represented has an adequate supply of nutrient, maximum rooting depth, and full canopy cover, and its dry matter accumulation is influenced only by soil water deficits. Initially the soil reservoir is at field capacity and receives three inputs of water of equal amounts. The water holding capacities of the soils (50 mm for a shallow soil and 300 mm for a deep soil) represent a realistic range for agricultural soils. The assumed maximum evaporation rates ($E_{max}$) of 10 mm/day for high demand and 6 mm/day for low demand also represent a range of values that can be obtained within the same radiation flux density of a clear day and an average temperature but a considerably different combination of vapor pressure deficit and wind.

Approximate lower limits of water available to plants are indicated in Figure 1 as specific fractions of the total storage capacity. The lower limits include thresholds for the beginning of stress for expansion growth (extension threshold), the beginning of stress for the stomatal functions of photosynthesis and transpiration (stomatal threshold), and the lower limit of plant survival. Scientists studying plant response to water deficits have demonstrat-

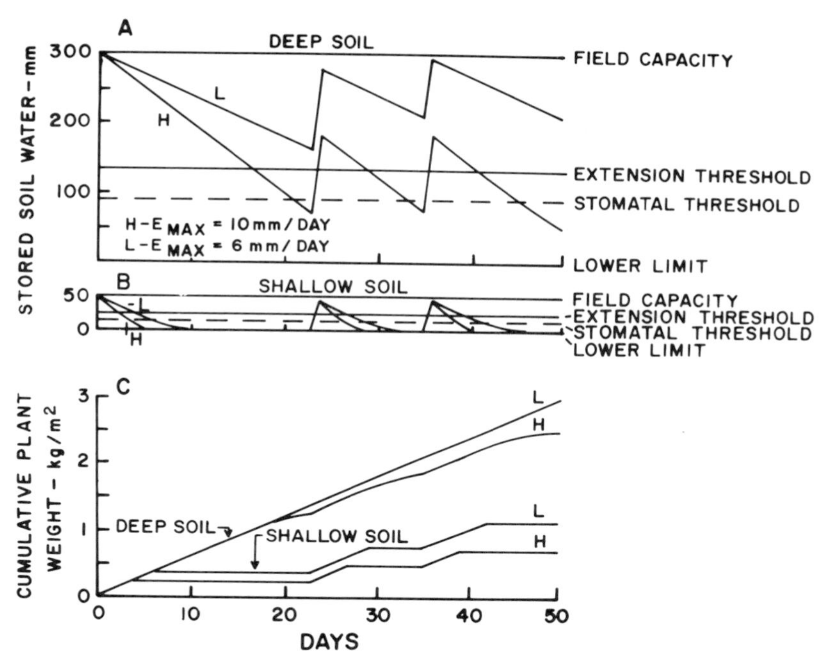


**Figure 1.** **Four scenarios of seasonal patterns of stored water for plants growing in deep soil (1-A) and shallow soil (1-B) with high (H) and low (L) maximum exaporation ($E_{max}$) rates. Expected cumulative plant growth (1-C) is shown for each soil depth and $E_{max}$. The extension and stomatal threshold for amount of stored soil water represent limits where leaf extension growth and stomatal diffusion begin to be reduced by soil water deficit.**

ed the distinction between these limits (Hsiao, 1973). The precise fraction of the total storage at which these treshold limits occur may differ somewhat among soil types and evaporative demand. Plants can survive because these different threshold limits allow adaptation to periods of water deficiencies. During periods of water deficit, photosynthesis and transpiration continue, but at limited levels, cell expansion is reduced, and storage of carbohydrates increases in already developed plant tissue. These adaptations often are overlooked when studying plant response in relatively small containers and the differences in these processes occur within such short time intervals that they are not observed.

Note from Figure 1-A that the deep soil with low $E_{max}$ has adequate storage capacity in the root zone to maintain the water content above the threshold for expansion growth; thus, a high dry matter accumulation is obtained with time (Figure 1-C). At the high $E_{max}$, however, there are several days during which water deficits in the soil fell below the threshold for expansion growth. During a few of these days, water deficits also fell below the stomatal threshold and affected rates of transpiration and photosynthesis. Thus, for the deep soil with high $E_{max}$ plants accumulated less dry matter than with low $E_{max}$ (Figure 1-C). For the shallow soil, water deficits remained below threshold values for both expansion growth and stomatal function during extended periods (Figure 1-B); thus dry matter production was reduced with either high or low $E_{max}$, but the reduction was greater with high $E_{max}$ (Figure 1-C). These example scenarios illustrate the value, when evaluating crop production systems, of considering both a water balance for the soil and the threshold levels for various plant processes. The example scenarios also provide insights into some of the possibilities for avoiding water stress in crops, such as timing of irrigation and matching cycles of crop growth with the period of water availability.

## DEFINING THE LIMITS OF SOIL WATER AVAILABILITY

The most accurate determination of water capacity limits for drought analysis in humid regions is needed in the shallow, sandy soils because the probability of a water deficit for such soils is the greatest. However, determining the soil reservoir available to plants is not as simple as determining the capacity of a container. First, some water may percolate, under the influence of gravity or other forces, below the root zone to become unavailable to plants. Part of the water that is retained within the root-zone reservoir, although available to plants, cannot be taken up as rapidly as needed because it is held too tightly by soil particles.

The traditional approach of evaluating the limits of soil water availability has been to use pressure chambers to measure the water contents at "wilting point" and "field capacity" of soil samples removed from the field. Usually water content at a pressure potential of –15 bars has been considered as the wilting point and water content at –0.33 or –0.10 bar has been considered as field capacity. To determine plant available water within the root zone, the difference between these two limits are summed for soil samples taken at various depths down to the measured or estimated maximum depth of rooting.

Several criticisms can be made of these rather static traditional definitions of the available soil water limits. The criticisms concern: (1) the amount of water used above field capacity while drainage is occurring, (2) plant growth being retarded before the soil has dried to the lower limit, (3) extraction of water below the –15 bar range, (4) incomplete extraction because of low root density in lower parts of the root zone, and (5) the possibility of upward flow into the root zone. I discussed several aspects of these problems in a recent review (Ritchie, 1981). From my experience, defining the limits of soil water availability to use in field water balance evaluation is not so much related to relevant soil water potential at the upper and lower limits for plant use as it is to obtaining accurate estimates of soil water potential on disturbed samples removed from field sites. In cases where field and laboratory measurements of the limits of water availability are made, the two types of estimates often are not in good agreement. Thus, rather serious errors are possible when laboratory pressure extraction equipment is used to imply plant available water limits. This possible error, when combined with the problems of incomplete extraction at the lower part of the root zone and the possibility of drainage rates being quite variable between soils, has led me to conclude that accurate water balance models need to have field measured limits of water availability.

Field measurements of the lower limit of water availability often reveal the approximate rooting depth and how much water uptake at the lower depths is decreased by root density. The influence on rooting patterns by tractor compaction, as discussed by Cassel, almost certainly would impact on the amounts of water that could be extracted by plant roots from the soil. In such cases, the dynamics of root growth play a dominant role in determining the amount of soil water removable by the root system. It is obvious from the regional maps prepared by Cassel that a majority of the soils in the humid regions of the U.S. outside the Corn Belt are affected by some physical or chemical barrier that restricts rooting depth. These root-restricting barriers often are difficult to evaluate quantitatively with physical or chemical tests. A measure of the lower limit of water content in the field thus can be valuable in interpreting the limitations of rooting depth in the soil and in providing the lower limits needed for water balance evaluations.

Lack of aeration in poorly drained soils can affect rooting depth. According to Shaw, a frequent contribution to summer drought in maize production in parts of the Corn Belt is the lack of deep root penetration caused by poor aeration during wet periods when root growth is active. Later in the year, when rain decreases and aeration in the root zone improves, plants are in a later stage of development and there is practically no new root growth. Under such circumstances, water cannot be extracted from the lower depths when needed although it is there in abundant supply.

Evaluating the upper limit of soil water holding capacity, even in the field, is more difficult because of drainage. There are useful and relatively simple functions than can be fit from field water content measurements to describe drainage. With such functions, the amount of water available for root absorption above the upper limit can be calculated.

Hopefully, accumulated experience in determining the extractable water limit and field drainage functions will provide reasonable extrapolation to situations where field data are not available. Until then, field measurements are necessary for accurate water balance evaluations.

## TEMPERATURE EFFECTS ON DURATION OF GROWTH

Dale pointed out the need to separate temperature effects on plant growth and plant development. Plant growth is concerned with the accumulation of biomass and is primarily related to net photosynthesis. Temperature influence on photosynthesis has a plateau over a fairly wide range of temperatures. Therefore, temperature is important in plant growth only in the context of how unusually high and low temperatures affect biomass accumulation. Plant development, however, is related to temperature in a more direct way. The development of vegetative and reproductive parts and the phasic development of plants is highly responsive to temperature. Crop development processes, such as the rate of leaf appearance, are related linearly to temperature over practically all the temperature ranges plants would be expected to experience.

In phasic development, the degree day concept, mentioned by Dale, often is used to estimate when plants change developmental stages. Within this context, development is related directly to temperature as long as it is within the thresholds that are established for those developmental processes. However, genetic differences in the maturity type and photoperiod sensitivity cause variations in the actual number of degree days for different phases of development. Photoperiod sensitive plants like corn and grain sorghum have a proportionally longer developmental phase in the vegetative stage during long days than they do during short days. Likewise, short maturity-type plants have fewer growing degree days during the vegetative phase than plants with longer maturity genotypes. These genetic-climatic interactions are probably important reasons why a constant number of degree days for a particular variety and stage of growth are not common to different climatic regions of the country.

## WATER USE EFFICIENCY AT HIGH YIELD LEVELS

In contrast to the general regional approaches to analysis of climate and yield estimation as presented in papers of Dale, Decker, and Shaw, the paper presented by Barfield is a site-specific analysis of how climate affects water requirements of plants and the economics of irrigation. In my opinion, the type of analysis reported by Barfield will be essential for evaluating the probable impact of supplemental irrigation on yields and profits for farmers in humid regions. Developing systems for handling the types of analyses done by Barfield and his co-workers for other soils, crops, and management strategies should receive high priority for future evaluation of cost effectiveness of irrigation in humid and sub-humid areas.

One management factor used in the model by Barfield was a plant population of 74,071 plants/hectare. When there is an adequate water supply for corn, the yield potential of corn at this population likely is lower than possible for higher plant populations. A recent report by Wright, Rhodes, and Stanley (1978) indicated the plant population for maximum yields of fertilized and irrigated corn in Florida was about 90,000 plants/hectare. When lower plant populations were used, even with high fertilizer rates, yields were lower than in the high population plots. If the results of the Florida test were applied to the Barfield analysis, projected yields might have been even higher. Such yield increases would improve the cost effectiveness of using irrigation, and would make it an even more desirable management option.

There is little question that supplemental irrigation is required in order to ensure high yields in humid regions where there may be short periods of water deficits. Generally, a crop whose yield is reduced by a low level of management will use about the same amount of water as one that is well managed. Therefore, when water is kept in adequate supply by irrigation, other management factors should be optimized to increase cost effectiveness. Risk analyses of the type done by Barfield and his colleagues easily could evaluate other management schemes to optimize management strategies of farmers. For highest yields, the work reported from Florida by Wright et al. (1978) has indicated that high levels of nitrogen application can be effective only if applied at several times and applied through the irrigation water. Other management factors that must be considered for high yields include pest control, adjustment for proper pH values, and the application of adequate amounts of secondary and micro-nutrients. In general, it can be stated that any management factor that promotes better yields will increase water use efficiency.

When humid regions produce as much economic yield as the more arid regions, the water use efficiency is usually higher in the humid regions because the higher humidity and lower advective energy in the humid regions reduce the evapotranspiration rates. Thus, even for irrigated agriculture, the greatest water use efficiency can be expected in humid regions if production can be maintained at a high level.

## IRRIGATION SYSTEMS FOR HUMID REGIONS

Sneed pointed out that supplemental irrigation produces significant increases in yield and that increasing costs of land, equipment, and fertilizer make supplemental irrigation even more profitable for the humid Southeast. Sneed also stated that varieties with higher yield potentials can be used with irrigation because there is no concern about water stress. However, there are special problems with irrigation in humid areas. One is the possibility of crop damage from saturated soil if irrigation is followed by heavy rainfall. This emphasizes the need to irrigate with amounts less than required to fill the soil profile. Barfield's analysis for the climate and soil at Lexington,

Kentucky, showed that, if the soil reservoir is filled to 67% of the water deficit rather than 100%, the cost effectiveness of the irrigation is greater.

There are two types of irrigation systems not mentioned by other speakers in this symposium that seem particularly adaptable to the types of farming systems used in much of the humid region. Many fields in the humid region are relatively small and have more irregular topography and shape than the large open fields in the western United States that can accommodate large center-pivot irrigation systems or surface flow systems. These smaller, irregular fields are adaptable to use of lateral and reversible drainage irrigation systems.

Lateral irrigation systems have an advantage over center pivot systems because they can apply water at uniform rates and are more adaptable to the usual rectangular shape field. As reported in *Irrigation Age* (Anon., 1980), a new refinement in the art of lateral irrigation systems was developed by an Idaho farmer named Allen Nobel. This lateral system should be a promising possibility for use in humid areas, even when land is not level enough for straight, open ditches. The system operates from a buried pressurized pipe. The system automatically couples and uncouples to the buried main line without interrupting water pressure or machine movement. A computer controls the operation of the system. Two large arms that "bend at the elbows" each have a two-wheel tractor at the terminal end and a water supply device that "finds risers on the main line and automatically couples with them to maintain a continuous water supply." While one arm is supplying water to the system, the other is moving into position to capture the next riser. Electronic sensors guide the tractor units along a steel cable stretched parallel to the main line. A small generator on the central unit supplies the electrical energy to operate the computer and to move the tower and terminals. This system uses nozzles, rather than conventional sprinkler guns, enabling the water to be broken into fine droplets and to be directed downward to the crop at low pressure. Thus, the energy requirement is less than for a high pressure sprinkler irrigation system. The electronic control system allows the irrigator to apply water at any rate desired between 0.2 to 5.0 cm/hour in a single pass. Although this has been used and tested in the more traditional irrigated areas of the western United States, it should be adaptable for irrigation in humid regions.

A second type of irrigation system that has promise for use in humid regions is a drainage system used in reverse (Skaggs, 1975; Doty and Parsons, 1979; Doty, 1980). Controlled and Reversible Drainage (CaRD) is a new name for an old system whereby soils that need drainage for maximum production can be managed in such a way that water can be applied back through the drainage system as sub-irrigation. This system is limited to certain types of soils and to areas where land is relatively flat. Such systems have been used for many years in Europe and are very efficient systems. Doty, Currin and McLin (1975) reported that the CaRD system is applicable to about 650,000 ha. in both North and South Carolina and can be utilized in Mid-Atlantic and Southeastern Coastal Plain states from Delaware to Texas where the water table should be controlled between 60 to 100 cm deep to provide increased yields and adequate recharging of the water during dry periods. The CaRD system provides a possible means to minimize irrigation water requirements and is gaining farmer acceptance in several areas of the Southeast. One reason is that the soil is an economical storage reservoir for water. However, when the soil reservoir remains full too long, crops suffer from oxygen stress. Drainage is necessary under those conditions and the CaRD system provides the means of sub-irrigation through the same tile system when needed. This has proven to be beneficial and leads to increased crop production.

Further research will be needed to evaluate the feasibility of using the CaRD system for large farming operations. On some farms it should be possible to drain water from crops where aeration might be a problem and transport it to fields that are fallow or where crops would not be sensitive to a high water table. Later, the water can be transported in the reversible mode back to the field for use during short periods of water deficits. Water from wells or reservoirs could also be used in the reversible sub-irrigation system. Research will also be necessary to determine how the CaRD system affects water quality, ground water recharge, and crop yields within the context of the entire farm operation system. At present, the CaRD system seems to be a favorable way to eliminate short periods of water stress at low cost where the soil and topography are favorable.

## CONFLICTS BETWEEN HIGHER TECHNOLOGY FARMING AND ENVIRONMENTAL QUALITY IN THE EXPANSIONS OF IRRIGATION IN HUMID REGIONS

There are institutional problems which inhibit the use of irrigation in humid areas. There is a short history of agricultural water use; thus, the priority of water use in agriculture is generally low. From past experience, we know that rapid increases in irrigation often have followed legislation providing increased access of irrigation water to farmers. In humid regions, however, most development of water resources is left to the individual rather than implementation of a large water resource scheme. Expenditures on clean-up of municipal waste usually have been related to problems of environmental quality rather than for a possible use as irrigation water in the humid regions. There is no question that the role of irrigation in these areas needs improved institutional and economic evaluation. Better evaluations will demand better technological understanding of the problems associated with irrigation and water quality in these regions.

Many principles of irrigation in humid regions are different than those that have been well established in drier regions. Technology will be needed to minimize the leaching of nitrogen into the ground water while maintaining an adequate supply of nitrogen and water in the rooting zone of the soil for crop use. This is especially important in shallow soils where root systems are restricted or in sandy soils. Irrigation practices on sloping land must be researched more intensely to ensure that erosion is not increased. Erosion hazards are large in humid areas where unpredicted precipitation can follow irrigation. In addition, irrigation in humid areas can lead to problems in plant disease, insect, and weed control that are not present in more arid areas. In many soils with trafficability problems for field operations under normal rainfall, the problems are intensified when crops are irrigated. Electrical power use for pumping water is seasonal and highly variable in humid regions. Thus, management of power generation and distribution for peak loads will require attention. Improved forecasting of rainfall could help decrease the vulnerability of soils and the environment to problems associated with irrigation in humid regions, but chances for this are probably slim until the reliability of forecasts is greatly advanced.

Even though there are institutional and technical problems associated with irrigation in humid regions, there has been a considerable expansion of irrigation in these areas. Irrigated land in the southern humid states has increased by about 5-fold since 1950, and the increase has been about 9.5-fold in northern humid states. In Georgia alone, irrigation has increased from about 120,000 ha. in 1975 to about 360,000 acres in 1980.

## CONCLUSIONS

If we are to increase food production on a national level, the greatest increase should be possible where high potential productivity is limited by management rather than by water resource. Therefore, increasing of expenditures seems essential on research for improved water management in humid regions where water is available. The southeastern part of the United States has a tremendous potential for expansion of agricultural production in the future. This expansion is possible because of the availability of land resources, the long growing season, the favorable climatic conditions, an excellent location for ease of commerce, and the availability of rechargeable and manageable water systems. This growth, however, cannot occur without creating many problems in water resource management, environmental concerns, and land resource utilization. Natural resource management systems are in early stages of development in the humid and sub-humid areas, and research associated with them must be intensified to provide guides for future generations.

It is evident from the papers in the symposium dealing with the impact of weather, climate, soils, management, and plant characteristics on crop yield that an integrated approach is needed to understand stress problems in humid regions. Evaluation of crop systems through computerized yield models provides a means of integrating the behavior of crop plants in response to weather, soils, and management.

A difficulty in applying today's modeling technology to present or previous problems is that some necessary climate, plant, or soil information is not available. Thus, it is important for agricultural researchers to recognize the need to make the necessary, extra measurements to evaluate results from experiments within the context of weather, management, and genetics. It often is argued that the systems approach can work only when we know everything about an entire system. In my opinion, we shall never know everything about a system. Fortunately, for any level of organization it is neither essential nor practical to model at a level of detail greater than required for useful predictions. Simple models can have power predictive value when one or two major factors dominate the performance system. Acceptance of the systems approach will be hastened by convincing demonstrations of its use and relevance to real world problems.

Development, testing, and application of a crop production model requires a coordinated effort by a team of people representing an array of disciplines and including some with practical experience with the system under study. Development of a crop production model can take advantage of what is already known, whether published or unpublished, and often makes what is known more explicit. For such models, the systems approach can be used in attempts to quantify existing knowledge of crop production, the major components and processes involved, and feedback mechanisms. The models of crop systems than can aid in identifying significant problems to crop performance and in devising management strategies for relief or modification of such problems. Simulation of a large number of possible alternative strategies needed in risk analysis is possible with a model when coupled with historical weather records. Any predicted management strategies that result must be tested. Experimentation must continue, but balanced crop modeling can provide a basis for more objective field and laboratory research to fill gaps in present knowledge.

Evaluation of crop systems generates a demand for data that directly relate to the functioning of the system under study. Conventional experiments rarely provide all the necessary data, but a relatively small additional effort in recording a specified minimum set of crop, weather, and management data can greatly increase the value of experimental results with the benefit of strengthening existing agricultural research by improved performance at experimental sites. Often, recording the additional data requires neither major change in experimental plans nor much additional investment in capital or labor.

The problems of eliminating stress and maintaining environmental quality in humid regions is complex. Almost certainly, the systems approach will be required for evaluating the most appropriate cropping and management strategies and in guiding researchers toward priority needs.

## NOTES

Joe T. Ritchie, USDA-ARS, Grassland, Soil and Water Research Laboratory, Temple, Texas 76503.

## LITERATURE CITED

Anon. 1980. New lateral operates from buried line. Irrigation Age. P. 20.

Doty, C. W. 1980. Crop water supplied by controlled and reversible drainage. Trans. ASAE 23: 1122-1130.

Doty, C. W. and J. E. Parsons. 1979. Water requirements and water table variations for a controlled and reversible drainage system. Trans. ASAE 22: 532-536.

Doty, C. W., S. T. Currin and R. E. McLin. 1975. Controlled subsurface drainage for southern coastal plains soil. Soil Water Conserv. 30: 82-84.

Hsiao, T. C. 1973. Plant responses to water stress. Annu. Rev. Plant Physiol. 24: 519-570.

Ritchie, J. T. 1981. Soil water availability. Plant Soil 58: 327-338.

Shaw, R. H. 1963. Estimation of soil moisture under corn. Iowa Agr. and Home Econ. Exp. Sta. Res. Bul. 520.

Skaggs, R. W. 1973. Water table movement during subirrigation. Trans. ASAE. 16: 988-993.

van Bavel, C. H. M. 1959. Water deficits and irrigation requirements in the southern United States. J. Geophys. Res. 64: 1597-1604.

Wright, D. L., F. M. Rhodes, and R. L. Stanley. 1978. Corn in the Southeast. Crops Soils 31: 10-11.

## 22

# ENVIRONMENTAL CONDITIONS AND CROP CHARACTERISTICS IN RELATION TO STRESS TOLERANCE

**J. Vieira da Silva**

This workshop confirmed the fundamental idea that the humid, temperate regions have much higher productive possibilities than are usually obtained from crop plants. These potential yields are several-fold the usual mean crop yield, and Boyer stressed the point that environmental stresses are responsible for most of the decreases below the potential. It is urgent, therefore, that cooperative research efforts be made in order to obtain information on how to avoid such reductions in yield. I have reviewed some of the research needs and priorities linked with environmental conditions responsible for lower than potential yields.

## TEMPERATURE

Most of the crops reviewed during this workshop, corn, soybeans, sorghum, etc., are of tropical and subtropical origin and are, therefore, very sensitive to temperatures below 10 to 5 C. Temperatures within this range are relatively frequent both at planting time and by the end of the crop cycle. In some European regions such low temperatures can even occur in mid-season.

McWilliam and Wilson discussed the physiological basis of chilling injury and tolerance. They have shown that part of the injury can be attributed to water stress resulting from an inability to absorb water and from lack of stomatal closure. For some tropical plants, such as *Espicia,* Wilson indicated that injury results directly from chilling and cannot be averted by relieving the water stress. In the crop plants emphasized during this workshop, there is also a direct chilling injury effect involving changes in metabolism linked with alterations in membrane composition and enzyme conformation.

McWilliam indicated that many weeds of tropical origin have accompanied crop plants to the temperate regions and that natural selection has been successful in creating strains able to stand lower temperatures. It seems, therefore, that a good prospect exists for selection for chilling resistance, and some species of cotton (Muramoto et al., 1971; Smith et al., 1971) and strains of corn from high altitudes in the tropics (Duncan, 1975), among many other species, can provide germplasm useful for understanding and improving chilling resistance. Moreover, Duncan (1975) states that whole ears of corn buried in the soil at the onset of cold weather will have almost complete germination in the spring, but single seeds under the same circumstances will not germinate. It would be interesting to know more about the mechanisms involved here.

High temperature stress often occurs in mid-summer and is usually associated with water stresses. Even if very high temperature stress is not as frequent in temperate climates as it is in arid regions, Mederski and Eastin have shown that high night temperatures can increase respiration rate and decrease yield in plants such as soybean and sorghum. The decreased yield probably is associated with protein turnover and it is certainly possible to select for slower turnover rates.

## WATER

In the humid, temperate regions, precipitation usually provides adequate water for the growth of high yielding crops, and according to Shaw, plants are damaged more often by excess water in the spring than by drought in summer. Excess water decreases soil aeration, and even though the physiological basis for tolerance of anoxia seems to be known (Crawford and Tyler, 1969), selection for such traits is rare in breeding programs Tolerance of anoxia is, however, a very important factor for obtaining high yields in humid regions.

Lack of water often results from the inability to exploit water stored in the soil, and Boyer shows that the more productive, modern strains of soybeans have better root systems and avoid the low mid-day water potentials of older strains. The variation in rooting ability is high among most crop strains, and it would seem that there is a good possibility for direct selection for improved rooting characteristics.

Our work with cotton (unpublished) has shown that root elongation during the first five days following germination varies from 4 to 25 cm within the same strain. This elongation is not correlated with dry weight of either the seed itself or of the shoot.

Mechanical resistance to root growth can sometimes be corrected by subsoiling, as shown by Cassel, or by using a species with strong roots that can penetrate the hardpan, as demonstrated by Kriedemann with safflower. The decaying root channels can be used afterwards by roots of other crop species. I wonder if, in the temperate regions, winter oil-seed rape could be used for this purpose.

Aluminum and manganese was shown by Cassel to be present frequently in toxic concentrations in the subsoils of the humid regions and to contribute to limited and shallow root penetration and to an inability to use water stored in subsoil horizons. Selection for aluminum and manganese tolerance is possible (Foy et al., 1978) and would increase greatly the water use by crops under these conditions.

Shallow or sandy soils cannot store enough water for a prolonged dry period. The only ways to cope with this situation are irrigation, use of drought-tolerant strains, or reduction of water loss by the plants. However, postponement of water loss by stomatal closure can result in lower yields if closure occurs at the time of sink filling when remobilization of stored assimilates is not possible. Reyniers (private communication) observed for rice in the Ivory Coast that carbon mobilization from the culms during grain filling increases with water stress. This situation seems to be generally applicable for many plant species. Short-stemmed varieties often suffer during stress periods because they possess less storage space in their culms. Blum pointed this out for wheat. It therefore is necessary to select for remobilizable reserves, even if these reserves are not used during years without water stress and sometimes seem wasteful.

It is probable that the reduction of transpiration that results from stomatal closure would reduce nitrate uptake and the level of nitrate reductase activity in the plant. This could have an effect on nitrogen metabolism and yield.

Drought delays silking in corn more than it delays tasseling, and this is one of the reasons that Shaw presented for the sensitivity of corn to drought at flowering time. Genetic modification of this response would confer to corn a much higher adaptability.

Eastin discussed how the components of yield in sorghum can be changed by environmental conditions during development of the crop. Modification of sink size and rate of sink filling by water stress make determinate crops particularly sensitive to periods of environmental stress.

The effect of even small amounts of water stress on the photosynthetic apparatus is known (Vieira da Silva, 1976). Boyer considers that the improved yield of modern soybean varieties is at least partially due to a better mid-day water status and better photosynthesis. A direct connection between photosynthesis and yield of sorghum is considered by Eastin to be possible. This indicates a potential for increasing the grain filling period that often is curtailed in sorghum by formation of the black layer association with cessation of translocation into the grain.

It seems, nevertheless, that as record yields are approached, photosynthesis becomes limiting. Hunter (1980) suggests that in early varieties of corn reductions in leaf area prevent adequate photosynthesis for high yields. In the soybean leaf, more over, photosynthesis cannot cope with both pod filling and nitrogen fixation simultaneously so that any reduction in photosynthetic rate resulting from stress becomes critical. Also, in cotton (McArthur et al., 1975) the remobilization of nitrogenous compounds from the leaf to the nearest boll can be associated with the onset of senescence of that leaf and with a reduction in its ability to sustain the high rates of photosynthesis required to supply the boll. The necessary carbohydrates must be supplied by another leaf, and so more than one leaf is used to feed a single boll. Interruption of the already limited photosynthate supply by a stress creates additional pressure on the leaves to maintain yield. Perhaps the ability to recover from stress without long-lasting after effects is one of the best characteristics to incorporate in crop plants for a humid environment where water stress is only occasional.

## NITROGEN

Record yields can be attained in cereals only with very high levels of nitrogen application (Ritchie, 1980). Perhaps much of the environmentally related suppression of yield is the result of modifications in nitrogen metabolism.

Nitrogen fixation in legumes is very sensitive to drought because, as pointed out by Valentine, *Rhizobium* has a weak capacity for osmoregulation and, as discussed by Mederski, photosynthate available for nitrogen fixation is limited during drought. Breeding *Rhizobia* that are better adapted for osmoregulation combined with breeding for host plants that are photosynthetically more tolerant of water stress could alleviate limits to yield caused by nitrogen deficiency.

## CONCLUSIONS

Even small effects of each of several environmental factors can combine to give substantial yield reductions. Although this situation can be corrected by breeding and by farming practices, there is still a considerable amount of work that only can be accomplished by cooperative research for each particular crop within each environment. The use of controlled environment facilities and simulation models seems essential for the success of such a colossal enterprise.

It seems to me, however, that some of the difficulties encountered in obtaining record yields in temperate, humid regions are inherent to the type of crop cultivated. In cereals and seed legumes, flowering and seed filling are generally the periods most sensitive to stress during the life of the plant; thus, it seems much easier to obtain reliable, high yields of vegetative parts than of reproductive parts. Harvesting the stems, tubers, roots, or leaves of a vegetative crop would give yields of starch, sugar, or protein that would be both higher and less affected by environmental factors than the yields obtainable from harvest of seed crops. From the biochemical point of view, it is very inefficient to go through a complicated series of phenomena such as flowering, fertilization, and embryo development to accumulate such a simple molecule as starch during the short period of effective photosynthesis for grain filling. Protein accumulation in seeds also is a wasteful process. As suitable technology is developed for extraction of protein, it seems much more efficent to obtain protein directly from leaves which already contain the high levels of protein necessary for high photosynthetic rates. We should not forget that alfalfa, for instance, without added nitrogen fertilizer produces yields of dry matter as high as a record crop of corn and has a high protein content per hectare.

Research on production and development of vegetative crops for humid, temperatre regions must be initiated now if we are to have the highest crop yields for the third millenium. Substitution of vegetative for seed crops is certainly the direction for enhancing productivity of the tropical regions.

**NOTES**

J. Vieira da Silva, Laboratoire d'Écologie Générale et Appliquée, Université Paris VII, 2 Place Jussieu, 75221 Paris Cédex 05, France.

**LITERATURE CITED**

Crawford, R. M. M., and P. D. Tyler. 1969. Organic acid megabolism in relation to flooding tolerance in roots. J. Ecol. 57: 237-246.

Duncan, W. G. 1975. Maize. p. 23-50. In L. T. Evans (ed.) Crop physiology. Cambridge University Press, London.

Foy, C. D., R. L. Chaney, and M. C. White. 1978. The physiology of metal toxicity in plants. Ann. Rev. Plant Physiol. 29: 511-566.

Hunter, R. B. 1980. Increased leaf area (source) and yield of maize in short-season areas. Crop Sci. 20: 571-574.

McArthur, J. A., J. D. Hesketh, and D. N. Baker. 1975. Cotton. p. 297-325. In L. T. Evans (ed.) Crop physiology. Cambridge University Press, London.

Muramoto, H., J. D. Hesketh, and D. N. Baker. 1971. Cold tolerance in hexaploid cotton. Crop Sci. 11: 589-591.

Ritchie, J. T. 1980. Plant stress research and crop productivity: The challenge ahead. p. 21-29. In N. C. Turner and P. J. Kramer (eds.) Adaptation of plants to water and high temperature stress. John Wiley & Sons, New York.

Smith, E. W., R. C. Fites, and G. R. Noggle. 1971. Effects of chilling temperatures on isocitratase and malate synthetase levels during cotton germination. 1971 Cotton

Defoliation-Physiology Conference, Atlanta, Georgia. p. 47-54.
Vieira da Silva, J. 1976. Water stress, ultrastructure and enzymatic activity. Ecological Studies 19. Springer Verlg, Berlin.

23

# MECHANISMS CONTROLLING PLANT PERFORMANCE

J. S. Boyer

As plants invaded the land over evolutionary time, they met new environmental extremes. As scientists, we need to understand how they cope with these extremes and how plant adaptation can be accelerated. With this knowledge, man can intensify the selection of desired plant attributes. Without it, he must rely on traditional, often slow approaches to plant inprovement and management.

What then are the routes to this goal? Research probably should be of three general types. First, there should be investigation of the mechanisms by which environment alters plant performance. This involves careful analysis of whole plant behavior followed by detailed biophysical and biochemical investigations at the subcellular level. The emphasis should be on identifying the specific factors actually controlling plant performance. Second, those mechanisms providing improved capability for reproduction should be identified. This phase of the research requires the growth of plants under controlled conditions so that the supposed mechanism can be perturbed and measured in the appropriate environmental extreme, and the eventual outcome on plant reproduction can be determined. Third, for agricultural improvement, those characteristics having positive adaptive value should be intensified by genetic screening procedures based on the desired trait.

For agricultural research, such an approach has the benefit that desired performance is fostered genetically and is therefore accessible both to capital-poor and capital-rich agriculture. This approach also requires little energy or environmental disturbance.

There are several areas of research that deserve attention at the mechanistic level. First, water deficiency caused by both soil and atmospheric phenomena are so inhibitory to plant productivity that considerable research should be done in this area. Particularly important questions are: Why is growth inhibited at low water potentials; and Why does reproductive development often fail under moderately dry conditions? Secondly, plant nutrition also deserves attention. Important questions in this area of research include: Can plant utilization of limited nutrients be made more efficient; What is the mechanism of mycorrhyizal enhancement of plant growth; and What are the mechanisms of salinity tolerance in plants? Thirdly, the response of plants to extreme temperatures, both hot and cold, requires further investigation. Questions of concern for this area are: What is the molecular explanation of how plant tissue tolerates freezing temperatures; and Why is reproductive development so sensitive to heat and cold?

As mechanisms controlling various aspects of plant performance are understood, it is important to identify those that contribute to reproductive success. This point cannot be overemphasized. In natural communities, the measure of success is the ability to produce progeny. For much of agriculture, the criteria are similar. Consequently, knowledge of reproductive success is vital to any understanding of how plants cope with environmental extremes. Too many experiments deal with only a short portion of the life cycle and the experiments are terminated before reproductive performance is evaluated. What is needed are measurements that expose plants to a particular environmental extreme, demonstrate a perturbation in physiological performance, and measure the reproductive consequence of the perturbation. Those physiological attributes that increase reproduction are then considered to have adaptive value whereas those that do not are considered to have less adaptive value.

The factors identified in this latter type of research form the basis for selecting improved genotypes. If selection can be made at an early stage of the life cycle, the selection rate can be enhanced manyfold. This increases the rate of plant improvement and is clearly dependent upon the knowledge from the research described above.

Occasionally it has been argued that nature already has selected for all of the environmental possibilities in plants, or that selection under optimum environmental conditions automotically will select for those plants that do well under nonoptimum conditions. These arguments, in my view, have flaws. The weakness in the first is that natural selection is not always intense enough to remove large numbers of undesired genes from the population. In other words, selection pressures are often mild and variable. The second argument has a similar weakness, because selection under optimum environmental conditions provides no screen for those factors which are important only under nonoptimum conditions.

It is becoming increasingly clear that knowledge generated by investigating plant-environment interactions will be of use only if the three steps outlined above are incorporated into our research effort. The areas needing investigation represent major problems in developmental biology and agriculture that ought to be attractive to many scientists. However, regardless of this point, one fact stands out: agricultural resources may change as energy, nutrients, and capital become scarce, but the environment will remain. Over the long term, concerted efforts to understand plant-environment interactions and manipulate them genetically will be the only way for continued agricultural improvement and management of natural ecosystems.

**NOTES**

J. S. Boyer, USDA/SEA/AR, Departments of Botany and Agronomy, University of Illinois, Urbana, Illinois 61801.

24

# GENETICS AND PLANT BREEDING IN RELATION TO STRESS TOLERANCE

Oliver E. Nelson

Being charged with the task of emphasizing the common themes apparent in the papers reporting on plant breeding approaches to alleviating stress and having the opportunity to suggest research priorities in the area of genetics and plant breeding, I find myself in the somewhat anomalous position of asserting that the outstanding needs for plant breeders are better methods of applying selection pressure as well as increased knowledge of the basic mechanisms by which a plant resists stresses. In the reports and discussions of the workshop, I have been impressed by the paucity of basic information concerning mechanisms of stress resistance in plants. This lack of information underlies the emphasis in both Dr. Blum's and Dr. Castleberry's papers on the empirical nature of their approach to breeding for stress tolerance in wheat and corn. No other avenue is realistically open. In the remainder of my discussion I shall be examining this point and the manner in which physiologists and plant breeds can most profitably interact in pursuit of a common goal.

## BREEDING FOR STRESS TOLERANCE

In an interesting presentation, Dr. Boyer drew attention to the marked discrepancy for several crops between record yields and the average yields for those crops. Assuming that the record yields represent reasonable indications of what elite genotypes can produce under the most favorable conditions of soil fertility, climatic circumstance, and relative freedom from diseases and insects, the lower yields in previous or subsequent years on the same farms where record yields were attained in a given year and where differences in soil fertility can be discounted, we are left with less favorable genotype-environment interactions as a reason that potential yields are not attained. Of the possible stresses impinging on a population of plants, Dr. Boyer has indicated that drought is the most important, and subsequent discussion will center on the problem of breeding for drought tolerance. It should be clear, however, that the same general principles apply to selection for tolerance to any environmental stress.

There is little doubt that superior genotypes for drought tolerance exist within the germ plasm of every crop species grown in the temperate zones and that these genotypes could constitute the starting point for systematic efforts to incorporate a meaningful degree of drought tolerance into elite varieties of those species. In instances where such tolerant hybrids or varieties have been identified, their tolerance is inadvertent in the sense that usually no deliberate selective pressure for tolerance was applied during the plant breeding manipulations that led to the production of those varieties. It is also true that we do not understand in most instances why a particular variety possesses a greater degree of tolerance than others that apparently have equivalent yield potentials under favorable conditions.

Since drought-tolerant varieties can be shown to be present among varieties selected without reference to drought tolerance (Hurd, 1976), it is interesting to learn of the strategies employed in plant breeding programs where deliberate selection for drought tolerance is practiced. Both Dr. Blum and Dr. Castleberry have described such programs. While we should keep in mind that one program seeks to improve a self-fertilized small grain (wheat) while the other deals with a cross-fertilized plant (maize) where the breeding practice is to develop elite inbreds that are crossed to produce the hybrids that are grown by the farmers, the two strategies being pursued are not related to these differences. In the wheat breeding program described by Dr. Blum, selection for drought tolerance comes only after several cycles of selection have identified those lines which possess superior yielding ability. Dr. Castleberry's approach to the problem with maize involves the inclusion of drought tolerance as one of the factors on which selection in each generation would be based. In his presentation, however, Dr. Castleberry emphasized the comprehensive nature of the breeding program; i.e., drought tolerance as a criterion for selection must be kept in perspective as being only one of the factors that contributes to a more or less formal selection index.

While Blum and Castleberry discussed strategies for identifying drought tolerance and implementing its use in plant breeding programs, Brim (in a presentation not included in these proceedings) concentrated attention on a crop production strategy in which varietal mixtures of soybeans are planted. This practice takes advantage of differences in flowering dates and other attributes of existing varieties to ensure that an ephemeral stress would not affect all the plants in a population at a crucial stage as regards seed setting and seed filling.

It usually has been assumed that the most effective programs in breeding for drought resistance would be those in which breeders and plant physiologists combined their efforts in a collaborative probem, and such joint efforts should have the greatest probability of success. I am aware that many collaborative programs have not been as fruitful as hoped when they were initiated. There are various reasons why such cooperative programs have often not fulfilled the initial hopes, and it is likely that some blame accrues to participants from both disciplines. Plant breeders often assume that measuring yield at the end of the growing season constitutes a sensitive measure of the action and interaction with each other and with the environment of all genetic factors influencing yield. It follows then that no individual component of yield is worth measuring as a basis for selection. On the other hand, plant physiologists often seem irritated by the request for simple selective screens that can be applied to the very large populations that plant breeders wish to handle and by the repetitive, routine aspects of many plant breeding operations.

## BENEFITS OF COOPERATIVE PROGRAMS TO GENETICISTS

Given these strains in collaborative programs aimed at identifying stress-tolerant plants of a species and utilizing these plants in breeding acceptable varieties, I want to outline the specific instances in which each discipline can contribute materially to the collaborative effort. Let me concentrate first on the useful information which the physiologist can contribute.

In the first place, it would be most useful to understand the physiological (or anatomical) basis of the distinctly stress-tolerant lines that have been identified since it is clear that tolerance exists for various, different reasons. It is not necessary, of course, to understand the physiological basis(es) of a desirable attribute in order to incorporate it into commercially utilizable varieties. Given the fact that plants may be tolerant of drought stress for obviously different reasons, it would be most useful to know if qualitatively different sources of stress resistance exist within a species and if so, how many. It then would be possible to combine these different types of tolerance in populations in which selections were being made. As a corollary of this approach, the breeders then could ascertain whether combining two different physiological bases for tolerance produced progeny in which the effects on drought tolerance were simply additive, synergistic, or even antagonistic as might conceivably be the case.

It is noteworthy that all the plant breeders emphasized the importance to their programs of selective screens and the desirability for simpler, more effective screens. Dr. Burton concentrated on this aspect of breeding for drought tolerance and described the conditions which he found most effective in selecting for tolerance in his Bermuda grass breeding program. It is with respect to the design of innovative selection systems that physiologists can be of most immediate assistance to plant breeding programs.

Dr. Valentine reported on bacterial mutants with altered osmoregulatory properties and raised the question of whether such variants exist in higher plants and might aid in drought tolerance. In bacteria, the phsyiological basis of such enhanced regulation is understood to be the greater synthesis of glutamine to serve as a counterion to the $K^+$ influx. Proline can be accumulated to serve the same purpose if it is present in the medium, and Dr. Valentine is attempting to introduce in plasmids genes that catalyze steps in proline synthesis to ascertain whether osmoregulation is enhanced by extra copies of these genes. We do not know as yet whether substantial and usable differences for osmotic regulation exist within a crop species, and it is clear that for higher plants we presently do not have the flexibility in genetic manipulation that would allow us at will to introduce extra copies of a specific gene via a cloning vehicle. However, greater contents of compounds increasing osmotic regulation could be sought by attempting to identify overproducers of that compound (via the derepression of the gene catalyzing the limiting step in its synthesis). Overproducers of particular amino acids have been selected in plant cell cultures by challenge with amino acid analogs (Widholm, 1974).

The subject of osmoregulation and its genetic control has been the subject of a recent volume, *Genetic Engineering of Osmoregulation,* edited by Rains, Valentine, and Hollaender (1980), which brings the subject up to date. The extent to which it can be utilized in higher plants depends upon the extent to which the resources of the plant are drawn upon to maintain the regulatory properties. As Valentine pointed out in his presentation, there appear to be metabolic costs associated with enhanced osmotic regulation in bacteria. If this were true also in higher plants as is likely, it might not be an acceptable route to drought tolerance for plants grown in regions where stress is of short duration and appears sporadically during the growing season.

The question of possible metabolic costs associated with enhanced osmoregulation is a special case of a more general problem which is the extent to which selection for drought tolerance identifies genotypes that are unable fully to utilize optimal conditions. A limited scan of the literature concerning selection for drought-tolerant plants has indicated both instances in which drought-tolerant lines have been deficient in yield under favorable conditions and other reports of such lines that yielded well (Hurd, 1976). In this workshop, Dr. Blum has indicated that some wheat varieties selected for superior yielding capability have been shown to have above-average drought tolerance so that the two attributes are not mutually incompatible. The reported observations that some drought-tolerant selections do not yield as well as elite varieties under good conditions while others do are not surprising since drought-tolerant plants may be so for any one of a variety or morphological or physiological adaptations or a combination of several adaptations (Parker, 1968). Some of these adaptations may be antithetical to maximum yielding capacity under favorable conditions of moisture supply. Certainly some evolutionary adaptations to xerophytic conditions (CAM metabolism, for example) fall into this category. It would be most useful to plant breeders considering programs in drought resistance to learn what types of adaptations disqualify a plant from yielding at optimal levels for the species under favorable conditions.

Several other areas of physiological research have the potentiality of producing basic information of relevance to plant breeding programs. Superior osmotic regulatory capability has been briefly discussed as a possible contributory factor in producing drought-tolerant plants. For species in which plants can be regenerated from cells in cultures, the cell culture conditions would be the simplest selective system. Then, if the changes to superior osmotic regulation that could be identified proved to confer that level of osmotic regulation on the regenerated plant and to be heritable, the plant breeder would have a source of genetic variability that could be exploited in plant breeding operations. I assume that most other factors affecting drought tolerance will be attributes of organized systems and hence not subject to selection in cell cultures.

It would also be valuable both for plant breeders and plant physiologists to have a more complete understanding of the effects of water stress on plants. It would be particularly useful to understand which effects are the direct effects of low water potential and which are mediated by the action or non-action of plant hormones.

## BENEFITS OF COOPERATIVE PROGRAMS TO PHYSIOLOGISTS

The focus in this discussion to this point has been on the possible contribution of plant physiologists to plant breeding programs. It would be a mistake, however, to assume that cooperative interactions between plant geneticists and plant physiologists benefit only the geneticists. Substantial benefits can accrue to the plant physiologists. As a starting point for investigations into the bases of drought tolerance, plant breeders can furnish, in a number of economically important species, lines or variants that have been identified as having various degrees of drought tolerance. Beyond this contribution, however, geneticists can furnish an entrée to the singularly underutilized resource of the existing mutants of higher plants. Although the fusion of biochemical and genetical investigations of prokaryotic organisms has resulted in substantial progress over the last three decades and spawned the new discipline known as molecular biology, plant physiologists have not, in general, appreciated the experimental advantages of mutant stocks. In part, this reluctance to approach problems via mutant analysis stems from the greater difficulty in eukaryotes than in prokaryotes of devising selective screens to detect mutants affecting a particular biosynthetic pathway. Nevertheless, in several species, mutants already exist which bear on problems of significance, and this is certainly true with respect to drought tolerance.

In several species, there exist simply inherited wilted mutants. While these mutants do not initially extend our ability to breed for drought-tolerant plants, their analysis can be most informative concerning the mechanisms by which a normal plant counters the vicissitudes of its environment. The best analyzed example of this class of mutant is the *flacca* mutant of tomato investigated by Tal and his colleagues in Israel (Tal et al., 1979). The investigations in comparison to its normal counterpart of this mutant, which is deficient in its ability to synthesize normal amounts of quantities of abscisic acid, demonstrate clearly the central role of this compound in mediating the response of the plant to its environment as well as the profound effects on plant growth owing to hormonal imbalance. There are also reports of chronically wilted mutants in other organisms stemming from primary defects other than the inability to synthesize normal quantities of abscisic acid (e.g., in maize, see Postlethwait and Nelson, 1957).

Other interesting mutants exist whose characterization should shed light on various aspects of the ability of plants to adopt to a water-deficient environment. Since I do not intend this to be an extended discussion of such mutants, I shall note the existence of only two particularly interesting mutants. In maize, Jenkins (1930) reported the isolation of a mutant which he designated as *rootless.* Mutant plants produce a primary root but secondary roots are either lacking or few in number. The mutant plants grow to maturity although they must be supported. It is clear that such a mutant would allow investigations to ascertain the effect of a natural restriction in root development on the absorptive capacity of that plant.

In garden peas, plants homozygous for the mutation, *af,* have all leaflets modified to tendrils. It is interesting that such leafless pea plants yield as well as normally leafed plants of the same maturity class (Snoad, 1974). While this mutant has been used to some extent for investigations of carbon assimilation (Harvey, 1972), it has not, to my knowledge, been used for studies of water economy which one would expect to be altered in the mutant plants.

## CONCLUSIONS

In concluding, I want to suggest how cooperative programs between plant breeders and plant physiologists can most effectively be fostered in pursuit of the common goal of drought-tolerant plants. Priority should be given to support of those programs in which both disciplines are represented and where the initial approach on the part of the plant physiologists is to characterize the physiological (and/or morphological) bases of those drought-tolerant selections with which the plant breeders can furnish them. It may then be possible to devise selective screens that take advantage of new knowledge concerning the basis of drought tolerance and that are simpler to implement than field trials. Since, by design, plant breeders would be an integral part of the program, the findings of the physiologists could then be utilized without lag time which often is a factor when investigators must first be apprised of findings at another institution and then decide that these findings are important enough to constitute a basis for their actions.

## NOTES

Oliver E. Nelson, Department of Genetics, University of Wisconsin, Madison, Wisconsin 53706.

Contribution from the College of Agricultural and Life Sciences.

## LITERATURE CITED

Harvey, B. M. 1972. Carbon dioxide photoassimilation in normal-leaved and mutant forms of *Pisum sativum* L. Ann. Bot. 36: 981-991.

Hurd, E. A. 1976. Plant breeding for drought resistance. p. 317-353. In T. T. Kozlowski (ed.) Water deficits and plant growth. Vol. IV. Soil water measurement, plant responses, and breeding for drought resistance. Academic Press, New York.

Jenkins, M. T. 1930. Heritable characters of maize. XXXIV—Rootless. J. Hered. 21: 79-80.

Parker, J. 1968. Drought-resistance mechanisms. P. 195-234. In T. T. Kozlowski (ed.) Water deficits and plant growth. Vol. 1. Development, control, and measurement. Academic Press, New York.

Postlethwait, S. N., and O. E. Nelson. 1957. A chronically wilted mutant of maize. Amer. J. Bot. 44: 628-633.

Rains, D. W., R. C. Valentine, and A. Hollaender. (eds.). 1980. Genetic engineering of osmoregulation: Impact on plant productivity for food, chemicals, and energy. Plenum Press, New York.

Snoad, B. 1973. A preliminary assessment of leafless peas. Euphytica 23: 257-265.

Tal, M., D. Imber, A. Erez, and E. Epstein. 1979. Abnormal stomatal behavior and hormonal inbalance in *flacca,* a wilty mutant of tomato. V. Effect of abscisic acid on indoleacetic acid metabolism and ethylene evolution. Plant Physiol. 63: 1044-1048.

Widholm, J. M. 1974. Cultured carrot cell mutants: 5-methyltryptophan-resistance carried from cell to plant and back. Plant Sci. Lett. 3: 323-330.

# 25

# RESEARCH NEEDS AND PRIORITIES: EPILOGUE

Paul J. Kramer and C. David Raper, Jr.

In the long term, mild, humid regions of the world such as the central and southeastern United States have the highest potential for large crop yields. High energy costs, competing demands for a limited supply of water, and salt accumulation are beginning to limit the area which can be kept in production by irrigation agriculture. The high rainfall and the long growing season of the southeastern United States make it a particularly favorable area for food and wood production.

Even in temperate, humid climates it is well established that crop yields often attain less than their full genetic potential because of various soil and atmospheric constraints, especially drought and short periods of abnormally high or low temperatures. As pointed out in the "Introduction," losses in yield from environmental stresses generally are much greater than losses from diseases and insects.

The reduction in yield caused by environmental constraints can be materially reduced by environmental modification, better crop management practices, and development of cultivars with more tolerance of specific stresses. However, additional research is needed to provide more specific information concerning both the environment and the physiological responses of plants to the environment. To obtain support for this kind of research will depend on persuading administrators to reorder their research priorites, because at present little attention is being given to study of environmental constraints on yield. Some examples of specific needs will be described under several headings.

## METEOROLOGY AND CROP DAMAGE

Apparently, the best criterion for agricultural drought is depletion of soil water. However, monitoring the available water content of the soil requires more knowledge than is generally available of the soil water storage capacity, the depth of rooting, and the rate of water extraction by particular crops. If these data were known for particular soil types, it would be possible to predict the onset of drought more accurately and thus schedule the timing of irrigation and even predict the probable yield in the absence of irrigation.

It also is clear that knowledge of short-term variations in both air and soil temperatures and of both day and night temperatures are necessary in order to predict the effects of temperature on crop yield. It is reported that day and night temperatures have important effects on soybeans, and a few chilly nights affect the growth of beans, cotton, and tobacco, while hot weather at pollination time reduces corn yields. It seems clear that the effects of droughts and temperature perturbations depend on the stage of plant growth at which they occur. Therefore, controlled-environment research is needed to learn more about the effects at various stages of growth of water stress and temperature perturbations on the growth and yield of crops.

## PHYSIOLOGICAL EFFECTS

At present, plant breeding is largely an empirical process guided by yield data. However, it was indicated that if morphological and physiological characteristics could be identified as increasing tolerance of stress, plant breeders would proceed more efficiently toward development of varieties likely to produce higher yields under stress. Unfortunately, too little of the desired information is available.

It is still uncertain how much of the damage caused by water stress is caused simply by reduced growth and stomatal closure resulting from loss of turgor, and how much is caused by disturbance of nitrogen metabolism, translocation, and injury at the cellular level, such as damage to the photosynthetic machinery. Does high temperature injury result in part from excessive use of energy in respiration? How much chilling injury is the direct effect of low temperature on membrane structure and metabolic processes, and how much is caused by water stress injury resulting from decreased permeability of the roots to water? Thus, much more physiological research is needed at all levels, the whole plant, plant organs, and cells, to learn if there are physiological or biochemical characteristics that are sufficiently related to stress tolerance to be useful in plant breeding programs.

## MANAGEMENT PRACTICES

Supplemental irrigation is becoming increasingly useful and profitable in the humid Southeast. However, better weather predictions and better indicators of the need for irrigation are required to increase the efficiency of supplemental irrigation. More information on available water storage capacity of soils would be particularly useful. In some soils the need for irrigation can be decreased by subsoiling, because it makes a larger volume of soil available for occupancy by roots. However, more research is needed in the timing of subsoiling, and the need for it might be decreased if plants with roots capable of penetrating impermeable soil layers could be developed.

There is still uncertainty concerning the best method of applying water; whether by flooding, sprinkler, or trickle irrigation. There also needs to be more study of the feasibility of applying fertilizer, especially nitrogen, in the irrigation water. This may be difficult with supplemental irrigation in humid climates because of the infrequent and irregular scheduling. It is suggested that in the long run simulation models will become very useful as management tools for the timing of irrigation and the prediction of yields. However, more information about plant development and the timing of critical events such as tasseling of corn and pod filling of soybeans is needed to improve the usefulness of such models. Better long-term weather forecasts also would be very useful for crop management.

## PLANT BREEDING

Several speakers stated that considerable capacity for stress tolerance is present in existing cultivars, but more effective screening methods for its detection are needed. Apparently, a combination of yield tests in field plots and exposure to stress in controlled environments is most effective, but much more research of this kind is needed. It was pointed out that cell cultures are being used effectively to test salt tolerance and osmotic regulation in microorganisms and a few seed plants. This method of testing for tolerance of environmental stresses should be energetically investigated to determine if it is widely applicable to seed plants. If so, it would greatly decrease the time and cost of screening populations for tolerance of particular stresses.

Plant breeders are operating primarily on an admittedly empirical basis, with yield as their only criterion. This empirical approach is necessitated by lack of information concerning morphological and physiological characteristics that are the basis of drought tolerance. There is urgent need for methods of screening for stress tolerance based on biochemical or physiological characters which can be measured in the laboratory.

Another problem important to plant breeders is whether selection for high yield under stress also will select for varieties with lower yield potentials under favorable environmental conditions. There is a difference of opinion on this point, and some ecologists argue that plants well adapted to water stress will never yield as well when well watered as those adapted to a moist environment. This would mean that the high yielding varieties under irrigation might not be good varieties to use where droughts can be expected and irrigation is not practiced. This problem of yield stability needs further investigation.

Because the damage caused by drought or abnormal temperatures depends on the stage of plant development at which the stress occurs, timing of processes such as flowering and pod filling in relation to weather may be very important. The timing of planting and time from emergence to the critical stage may affect yield; thus, for example, if droughts are more common late in the season, early planting may be desirable. However, this requires seeds which germinate and grow well in cold soil. Here again, knowledge of the physiological requirements at various stages of growth, combined with knowledge of the meteorological probabilities during the growing season can be of assistance to plant breeders and agronomists.

## SUMMARY

Most discussions and symposia on water and temperature stress have concentrated on the severe conditions found in hot, arid or semiarid regions. However, most of the world's food is produced in mild, humid climates where droughts occur at random throughout the growing season, though most commonly in the latter part. Low temperatures and chilling injury are most common during the early and late parts of the growing season, while damage from high temperature, often combined with drought, is most common in midseason. There is much evidence that relatively short periods of stress cause substantial, although often unrecognized, reductions in growth and yield. In view of these facts it seems important to make more intensive studies of the effects of environmental stresses on crops in humid climates than have thus far been made.

There is considerable discussion of the effects on plant growth and crop yield of possible climatic changes, such as an increase in global temperature caused by the increasing $CO_2$ concentration of the atmosphere. However, the extent and timing of climatic changes remain uncertain while there is no doubt that unfavorable weather frequently reduces yields. It therefore seems more important to study the effects of short-term perturbations in weather on crops than to worry about the effects of uncertain long-term changes in climate. Furthermore, careful study of the effects of short-term changes in environmental conditions will provide much of the information needed to predict the effects of long term climatic changes.

# INDEX

# PARTICIPANTS

DR. B. J. BARFIELD
Department of Agricultural Engineering, College of Agriculture, University of Kentucky, Lexington, KY 40546

DR. A. BLUM
Agricultural Research Organization, Volcani Center, P.O.B. 6, Bet Dagan, Israel

DR. JOHN S. BOYER
USDA-SEA-AR, 289 Morrill Hall, University of Illinois, Urbana, IL 61801

DR. C. A. BRIM
Funk Seeds International, P.O. Box 2911, Bloomington, IL 61701

DR. GEORGE E. BROSSEAU, JR.
National Science Foundation, Washington, DC 20550

DR. JAMES BUNCE
Light and Plant Growth Laboratory, USDA-SEA, Beltsville, MD 20705

DR. G. W. BURTON
USDA-SEA Coastal Plain Station, Tifton, GA 31794

DR. MARY CARTER
USDA-SEA, Beltsville, MD 20705

DR. D. K. CASSELL
Department of Soil Science, North Carolina State University, Raleigh, NC 27650

DR. R. M. CASTLEBERRY
Corn Research Ctr., DeKalb AgResearch, Sycamore Road, DeKalb, IL 60115

DR. MERYL N. CHRISTIANSEN
Plant Physiology Institute, USDA-SEA, Beltsville, MD 20705

DR. R. B. CURRY
Ohio Agricultural Research & Development Center, Wooster, OH 44691

DR. ROBERT F. DALE
Department of Agronomy, Purdue University, West Lafayette, IN 27907

DR. WAYNE L. DECKER
Department of Atmospheric Science, University of Missouri, Columbia, MO 65201

DR. J. D. EASTIN
Department of Agronomy, University of Nebraska, Lincoln, NB 68583

DR. HARVEY GOLD
Biomathematics Program, Department of Statistics, North Carolina State University, Raleigh, NC 27650

DR. MORRIS G. HUCK
Department of Agronomy & Soils, Auburn University, Auburn, AL 36830

DR. K. R. KNOERR
School of Forestry and Environmental Studies, Duke University, Durham, NC 27706

DR. T. T. KOZLOWSKI
Department of Forestry, Russell Laboratories, University of Wisconsin, Madison, WI 53706

DR. P. J. KRAMER
Department of Botany, Duke University, Durham, NC 27706

DR. P. E. KRIEDEMANN
Division of Irrigation Research, Griffith, N.S.W. 2680, Australia

DR. D. T. KRIZEK
Plant Stress Laboratory, USDA-SEA, Beltsville, MD 20705

DR. J. R. McWILLIAM
Department of Agronomy and Soil Science, The University of New England, Armidale, N.S.W. 2351, Australia

DR. ALBERT H. MARKHART, III
Department of Horticulture, University of Minnesota, St. Paul, MN 55108

DR. H. J. MEDERSKI
Ohio Agricultural Research & Development Center, Wooster, OH 44691

DR. OLIVER E. NELSON
Department of Genetics, University of Wisconsin, Madison, WI 53706

DR. JAMES PALLAS
Southern Piedmont Conservation Research Center, Watkinsville, GA 30677

DR. LAWRENCE R. PARSONS
Agricultural Research & Education Center, Lake Alfred, FL 33850

DR. D. T. PATTERSON
USDA, Department of Botany, Duke University, Durham, NC 27706

DR. R. P. PATTERSON
Department of Crop Science, North Carolina State University, Raleigh, NC 27650

DR. BOYD POST
Program Planning Staff, USDA-SEA, Beltsville, MD 20705

DR. C. DAVID RAPER, JR.
Department of Soil Science, North Carolina State University, Raleigh, NC 27650

DR. JOE T. RITCHIE
USDA-SEA-AR, Grassland Soil and Water Laboratory, P.O. Box 748, Temple, TX 76591

DR. ROBERT H. SHAW
Climatology and Meterology, 310 Curtiss Hall, Iowa State University, Ames, IA 50011

DR. NASSER SIONIT
Department of Botany, Duke University, Durham, NC 27706

DR. R. E. SNEED
Department of Biological and Agricultural Engineering, North Carolina State University, Raleigh, NC 27650

DR. BOYD R. STRAIN
Department of Botany, Duke University, Durham, NC 27706

DR. R. C. VALENTINE
Plant Growth Laboratory, University California, Davis, CA 95616

DR. J. VIEIRA DA SILVA
Laboratoire d'Ecologie Generale et Appliquee, Universite Paris VII, 2 Place Jussieu, 75221 Paris Cedex 05, France

DR. M. T. VITTUM
USDA-SEA, Cooperative Research, Washington, DC 20251

DR. J. M. WILSON
School of Plant Biology, University College of North Wales, Bangor, Wales

DR. CHUNG LUN WU
Chinese Academy of Forestry, Beijing, Peoples Republic of China

## Other Titles of Interest from Westview Press

*New Agricultural Crops,* edited by Gary A. Ritchie

**Farming Systems Research and Development: Guidelines for Developing Countries,* W. W. Shaner, P. F. Philipp, and W. R. Schmehl

*Readings in Farming Systems Research and Development,* edited by W. W. Shaner, P. F. Philipp, and W. R. Schmehl

*Wheat in the Third World,* Haldore Hanson, Norman E. Borlaug, and R. Glenn Anderson

*Third International Symposium on Pre-Harvest Sprouting in Cereals,* edited by James E. Kruger and Donald E. Laberge

*Azolla as a Green Manure: Use and Management in Crop Production,* Thomas A. Lumpkin and Donald L. Plucknett

**Science, Agriculture, and the Politics of Research,* Lawrence Busch and William B. Lacy

*The Role of* Centrosema, Desmodium, *and* Stylosanthes *in Improving Tropical Pastures,* edited by Robert L. Burt, Peter P. Rotar, and James L. Walker

*Tomatoes in the Tropics,* Ruben L. Villareal

*XIV International Grassland Conference,* edited by J. Allan Smith and Virgil W. Hays

*World Soybean Research Conference II: Abstracts and Proceedings,* edited by Frederick T. Corbin

*Successful Seed Programs: A Planning and Management Guide,* edited by Johnson E. Douglas

*Rice in the Tropics: A Guide to Development of National Programs,* Robert F. Chandler, Jr.

*Agroclimate Information for Development: Reviving the Green Revolution,* edited by David E. Cusack

*Available in hardcover and paperback.